T0376551

Human Factors and Ergonomics for the Gulf Cooperation Council

Processes, Technologies, and Practices

Human Factors and Ergonomics for the Gulf Cooperation Council

Processes, Technologies, and Practices

Edited by

Shatha N. Samman

CRC Press is an imprint of the
Taylor & Francis Group, an **informa** business

CRC Press
Taylor & Francis Group
6000 Broken Sound Parkway NW, Suite 300
Boca Raton, FL 33487-2742

Printed on acid-free paper

International Standard Book Number-13: 978-1-4987-8189-3//978-1-138-59798-3 (Paperback/Hardback)

Visit the Taylor & Francis Web site at
http://www.taylorandfrancis.com

and the CRC Press Web site at
http://www.crcpress.com

I dedicate this book to every HFE advocate.

Let's make a better world.

Contents

Foreword

I am delighted for the privilege of forwarding the first edition of this very important book, *Human Factors and Ergonomics for the Gulf Cooperation Council: Processes, Technologies, and Practices*. The book is published at an imperative time. Very special and sincere thanks go to the book's editor, Dr. Shatha Samman, for such a thoughtful book. Numerous domains are demonstrated in the chapters written by expert authors from all over the world, and significant case studies from the GCC countries are presented from respected practitioners and organizations. It presents a novel approach to highlight the implementation of the processes, technologies, and practices of human factors/ergonomics (HFE) within the GCC region.

It is of paramount importance to recognize the role of human performance, safety, and well-being in the design and development of today's technology, artifacts, and environment. While this is an obvious credence that the human is at the center of originations, many of today's innovations went oblivious to this fact and focused solely on the technology. Ergonomics (or the study of human factors) is the scientific discipline concerned with the understanding of the interactions among humans and other elements of a system, and the profession that applies theory, principles, data, and other methods to design in order to optimize human well-being and overall system performance.* In other words, it is the science of fitting the environment to the safety, efficiency, and comfort of people.

Among the many priorities of HFE is a healthy work environment. In fact, leading organizations are integrating HFE deeply into their operations. Having a healthy, safe, and strong workplace as a core value for any organization would lead to a better safety culture. This would reduce costs, mitigate risk factors, lower turnover, decrease absenteeism, improve quality and productivity, enhance morale, and promote employee involvement and increase efficiency, all of which lead to better human performance for the organization. Therefore, it is important that HFE becomes more than just a stand-alone safety program, but is included in the overall daily business activities. GCC organizations have to be concerned with the optimization and integration of HFE aspects into their core operating strategies and system performance to effectively drive profitability and to achieve wide success.

HFE is essential and can contribute to effective standards and regulations. International standards aim to ensure that the design of products, processes, and environments results in optimized performance, safety, and well-being of the user. Most standardization organizations have taken this issue well within their working programs. There are several hundred relevant standards in the International Standards Organization (ISO) to help HFE professionals and the organizations they support in different fields, such as the physical environment, healthcare and medical ergonomics, accessibility, occupational health and safety, smart cities, information technology, transportation, energy, children products and environments, etc. Among them are Ergonomics general approach, principles and concepts (ISO 26800);

* International Ergonomics Association.

Ergonomics of human-system interaction (ISO 9241); Ergonomics of the physical environment (ISO 28802); Ergonomics of the thermal environment (ISO 13732); Ergonomics principles related to mental workload (ISO 10075); Ergonomics design of control centers (ISO 11064); Usability of consumer products and ease of operation of everyday products (ISO 20282); and Human-centered organization (ISO 27500).

Within the GCC region, the GCC Standardization Organization (GSO) is a Regional Standardization Organization (RSO) established by the resolution of the GCC Supreme Council in 2001 and assumed its operation in 2004 with membership of the governments of The State of the United Arab Emirates, The Kingdom of Bahrain, The Kingdom of Saudi Arabia, The Sultanate of Oman, The State of Qatar, The State of Kuwait, and the Republic of Yemen. It has more than 80 standards that deal directly with the issue of ergonomics and its influence on the workplace, health, and safety. All are the result of a consensus-based process that draws on the collective wisdom of technical experts in the interaction of people with products, systems or processes, and research evidence on the strengths and abilities of people.

The best way to implement a comprehensive HFE program is to integrate it into all business initiatives. For instance, GCC programs in design, manufacturing, and other similar programs need to be influenced and must be regulated by adding value and new quality in products. Market access and exports of products will benefit from having human-centered design appealing features. Scalable and sustainable workplace improvements can dramatically improve their bottom lines. Therefore, effective HFE requires a proactive, multi-disciplinary, systemic approach that addresses productivity and total quality management.

His Excellency Mr. Nabil Molla
GSO Secretary General
GCC Standardization Organization

Preface

Helping others solve problems is one of my core values. When I was introduced to human factors and ergonomics (HFE) and learned about its foundational power of putting people and their needs at the center of problem solving, I knew that was my calling. HFE and its cross-disciplinary scope resembles my background—both academic (Computer Science, Psychology, Industrial Engineering, and HFE) and professional (teaching, research, industry). With my bi-cultural background (USA, Saudi Arabia), I am constantly inspired by the impact of culture on our attitude, behavior, and cognition, and its importance in human-centered design. I am in awe of the fact that HFE is a unique discipline with a variety of practices and a high impact on our way of life.

Throughout my undergraduate/graduate school and my professional life in the USA, I was surrounded by a scientific community that was aware of the tremendous value of HFE and its significant role in human interaction. This was changed when I came to the GCC region. Every time someone asked me about my background and I attempted to describe HFE, I would either get a confused reaction or a response such as "yes, human resources." I attended a professional meeting a couple of years ago and the speaker was describing how ergonomics is a new science related to office/workspace best-practices. Mainly, my encounter with frequent bad design and frustrating incompatible user-interactions that could have been easily avoided by simple HFE solutions was my strongest motivator to bring this book to the GCC region.

The objective of the book is to introduce HFE to the GCC. The approach taken was a holistic view to emphasize the breadth and depth of HFE. The book is designed for academics, professionals, and practitioners in the GCC. It introduces them to the HFE discipline providing both theory (basic research) and application (applied research) of the field. The book represents diverse perspectives on HFE from multidisciplinary and culturally diverse backgrounds and nationalities of expert academics. Following each chapter, case studies are written by industry professionals highlighting their work from the following GCC countries: Bahrain, Kuwait, Oman, Saudi Arabia, and the United Arab Emirates (UAE).

The book is written at a level that can be easily understood by readers with no prior knowledge or formal education in HFE. To assist the reader with the content of the book, all chapters are consistent in their formats discussing the fundamental principles and concepts of HFE theories, methodologies, applications, future trends, and a list of key terms to enhance the reader's understanding of the chapter's content. The book provides many topics of interest to anyone curious in discovering HFE and learning about the discipline and profession. Each chapter is not comprehensive of the domain science covered; however, the book as a whole gives the reader exposure to the variety of concepts and methodologies within HFE that are transferable and applied to most domains. Furthermore, the domains selected in this book are not exhaustive of the numerous fields of HFE, but only provide a general overview of the discipline and selected domains relevant to the GCC countries.

Readers including students, academics, researchers, practitioners, policy makers, and the general public can learn about HFE. All readers can apply the principles, methods, and applications in this book to their respective field. Readers will have a better understanding of the diversity of HFE theories and practices in various domains and HFE's holistic systematic processes in terms of individual and organizational dimensions.

This book can be used as an undergraduate- or graduate-level textbook to introduce HFE to students from related fields of study. Equally, practitioners interested in introducing HFE to their field of work to respond to challenges in their own practice, and/or enhance the work environment of others, will benefit from reading this book. These may include allied professionals whose jobs are impacted or can be enhanced by HFE principles. Policy- and decision-makers may use this book as a tool translating the principles and practical implications of HFE into standards and regulations. Savvy consumers and consumer advocates will appreciate how employing HFE will result in good design that is compatible with human needs, capabilities, and limitations.

The growth in the GCC is towards a knowledge-based economy that encourages scientific innovation and entrepreneurship through improved access to education with greater efforts in research and development. The information in this book will contribute to fostering a knowledge society that is more active and instrumental in seeking an environment where human-system interactions are evidence-based practices.

Coordinating and editing this book with expert authors from diverse backgrounds spanning the globe was challenging at times, but the vision of having a valuable HFE book for international and GCC readers was worth all the effort. I have learned so much from the expert authors of this book and am very grateful for their contributions. My aspiration is for every reader of this book to become an HFE advocate of the scientific discipline, spreading the benefits it brings to make this world more human-centric.

Acknowledgments

I thank God for his abundant blessings. My thanks and gratitude to those that helped me on this book journey. To the authors in this book, I thank you for your incredible contributions. I hope this work stands as testament to your excellence. My deep gratitude to Dr. Haydee Cuevas, who was my sounding board throughout the process. I appreciate Cindy Carelli's constant guidance and support in bringing this book to fruition. Thank you to my colleagues Dr. Mohammed Al-mumin, Mr. Nawaf Al-Sahhaf, and Ms. Deenah Alhashemi for connecting me with the case study authors.

When you undertake a book project, it takes away so much of your family time. With my deepest gratitude and unconditional love I thank Baba, Mama, Awny, Taghreed, Rowa, Reem, Jude, Rahaf, Rana, Sadeem, Haneen, and Yasmeen. I appreciate your support, encouragement, comfort, and unconditional love that brings happiness in my life and joy to my heart. Your continuous prayers point my compass toward greater good.

Shatha N. Samman

Editor

Shatha N. Samman, PhD, is a human factors engineer and research scientist investigating a broad range of human performance issues in normal and complex operational environments. She is the founder of Global Assessment, a Human Factors and Ergonomics research, design, training, and assessment firm working in the USA and GCC region. Previously, she was the Research Division Head at the King Abdulaziz Center for World Culture, Saudi Aramco and Adjunct Professor at the University of Central Florida (UCF).

She is a founding member of the Human Factors and Ergonomics Society GCC Chapter headquartered in Dubai and is currently serving as the President. She is a member of the International Ergonomics Association (IEA) communication and public relations committee and is a Dubai Association Center (DAC) Ambassador. Shatha received her PhD in Applied Experimental and Human Factors Psychology from UCF. She holds a Master of Science degree in Industrial Engineering and Management Systems, and two Bachelors of Science degrees in Psychology and Computer Science.

Contributors

Cakil Agnew, PhD, is Assistant Professor of Psychology at Heriot Watt University Dubai Campus. She joined Heriot Watt University in 2014. Cakil received her PhD from the University of Aberdeen (UK) in 2010. Following her PhD, she worked as a postdoctoral research fellow in the Industrial Psychology Research Centre at the University of Aberdeen. Her research focus is on the impact of safety culture and leadership on performance at work, and in high-risk industries. Dr. Agnew has published and presented her work in international journals and conferences.

Ali al-Azzawi, PhD, is the City Experience Advisor at Smart Dubai Office, responsible for developing the Happiness Agenda, and ensuring the validity of the science that underpins it. Dr. Ali brings his international, academic, and industrial, hands-on work experience as a technologist in physics, electronics, and computing, along with his psychology PhD and design work on projects such as DubaiNow, Dubai Pulse, and the Happiness Meter, to ensure a cross-disciplinary and pragmatic approach to his work.

Mostafa Aldah, PhD, is Senior Researcher at the Traffic Institute for the General Administration of Traffic at Dubai Police General Headquarters. He has worked in areas related to vehicle and materials engineering, transport safety and policing, quantitative and qualitative research, and joint projects at the European level, and is familiar with the latest advances in 3D printing, 3D laser scanning, and CAD modelling.

Ali M. Al-Hemoud, PhD, CIH, CSP, CMIOSH, is an associate researcher at the Kuwait Institute for Scientific Research in the Environment and Life Sciences Research Center. He is a Certified Industrial Hygienist (CIH No. 9614), a Certified Safety Professional (CSP No. 18699), and a Chartered Member of IOSH CMIOSH (No. 166075). He earned his degrees from the University of Cincinnati (PhD), Cleveland State University (MS), and the University of Pittsburgh (BS), all in industrial engineering. Dr. Ali's research interests are in the field of macroergonomics and human factors engineering. He has published several articles in areas related to physical ergonomics, manual material handling, musculoskeletal disorders and repetitive strain injuries, and behavior-based safety. His recent research interest is in the area of reviewing international health, safety, and environment (HSE) standards and guidelines, and developing a new ergonomics standard for Kuwait.

Fay Abdulla Al Khalifa, PhD, is an Assistant Professor at the Department of Architecture and Interior Design in the College of Engineering at the University of Bahrain. Her research interests focus on the interrelationship between cultural change and urban sustainability, particularly in the Arabian Gulf Context. Her research aspires to contribute to the understanding of sustainable urbanism within the context of transformed cultures, urban islands, and urban archipelagos.

Dr. Al-Khalifa is also interested in the importance of sustainability in real estate, as an asset class, the dynamics of considering sustainability assessment tools in real estate development, and the influence of visualization on the opinions and actions of decision-makers. Dr. Al-Khalifa is now leading a $22,000 study that looks into visual discomfort and architecture.

Waleed A. Alsaleh, B.S., is Manager of the Marketing and Communications Department at the Badir Program. He joined Badir as Team Leader of Content Marketing in 2015. In his current role, Waleed is responsible for the management of the marketing and communications department and his primary focus is developing an integrated communications strategy for the program incorporating PR, marketing, and online and communicating the program message to the media, investors, and the public. Prior to joining the Badir program, Waleed served as Technical Manager in Saudi Arabian parsons LTD. He has also worked at Saudi television and Saudi radio, spending nine years producing and presenting numerous programs. Waleed holds a BS degree in Electrical Engineering from King Saud University.

Reem F. Al-Shammari, PhD, is Information Security Team Leader at the Corporate Information Technology Group in Kuwait Oil Company (KOC). Dr. Reem's extensive experience includes leading and managing projects in various information technology areas such as cyber security, digital oil fields, wireless and telecommunication technologies, corporate systems and applications, real-time and industrial control systems and applications, and web development and design. She is head of the K-Companies cyber security committee, an IT lead for the Kuwait Intelligent Digital Oil Fields (KwIDF) projects in the KOC, a certified Project Management Professional (PMP), a member of the Project Management Institute, and a member of the Chapter Affiliations Arabian Gulf Chapter for Project Management. Dr. Reem received her BS in Computer Engineering and MBA from Kuwait University, and a PhD in Business Studies from Aston University, UK.

Mariam Al Shamsi, BS, is Acting Director of the Space Science Department at the Mohammed Bin Rashid Space Centre (MBRSC). Mariam is also leading the Instrument Science team involved in the Emirates Mars Mission (EMM) project. Mariam graduated from the American University of Sharjah (AUS) with a BS in Computer Engineering. She directly joined MBRSC in July 2013 as an Associate Ground Software Engineer working in the Mission Control Section. In 2014, she joined Advanced Aerial Systems Program (AASP) as an Avionics System Engineer and worked with a team on a collaboration project with Airbus DS to fly a High-Altitude Pseudo Satellite (HAPS), Zephyr UAV. After the announcement of the EMM in 2015, Mariam joined the Space Science section to support promoting science sector in the UAE.

Yousuf M. Al Wardi, MD, is an Aviation Medicine Consultant with the Royal Air Force of Oman. He received Doctorate in Medicine (MD) from the Sultan Qaboos University and a Diploma in Aviation Medicine from King's College-London. He obtained an MS in Human Factors and Safety Assessments in Aeronautics and an MS in Safety and Accident Investigation (Air Transport) from Cranfield University. Dr. Al Wardi's research interests are in human factors, accident investigation, and culture studies.

Erminia Attaianese, Architect, PhD, is a European ergonomist and a human factors expert. She received the National Academic Qualification as Associate Professor in 2014. She was a full-time Assistant Professor of Architectural Technology in 2002 at the Department of Architecture of the University of Naples Federico II. Dr. Attaianese is Head of the LEAS (Laboratory of Applied and Experimental Ergonomics) at the DiARC (Department of Architecture) in 2011 and was an Interdepartmental Research Centre LUPT in 2015, both at the University of Naples Federico II (www.leas.unina.it). She is a member of the IEA Technical Committee on Human Factors for Sustainable Development in 2015, where she was involved in green buildings issues. Dr. Attaianese is also the coordinator of the "Ergonomics of Built Environment" working group of the Italian Society of Ergonomics and Human Factors (SIE).

Catherine M. Burns, PhD, is a Professor in Systems Design Engineering at the University of Waterloo, Canada where she directs the Advanced Interface Design Lab and the Centre for Bioengineering and Biotechnology. Catherine's research is in human factors engineering, where she is well-known for her work in cognitive work analysis, ecological interface design, and in the development of decision support systems. In this area she has contributed over 250 publications and is the co-author of seven books. Dr. Burns is a Fellow of the Human Factors and Ergonomics Society.

Barrett S. Caldwell, PhD, is Professor of Industrial Engineering (and Aeronautics & Astronautics, by courtesy) at Purdue University. He earned BS degrees in Aeronautics and Astronautics, and Humanities, at MIT (1985) and a PhD in Social Psychology from the University of California, Davis (1990). His research team, the Group Performance Environments Research (GROUPER) Laboratory, studies how people get, share, and use information well in settings including aviation, event response, health care, and spaceflight. He is a Fellow and Past Secretary-Treasurer of the Human Factors and Ergonomics Society (HFES). Professor Caldwell is a 2016–2017 Jefferson Science Fellow at the US Department of State, Office of Japanese Affairs.

Nicklas Dahlstrom, PhD, is Human Factors Manager at Emirates. He oversees CRM training and integration of human factors in the organization. He was previously a researcher and instructor at Lund University in Sweden and still holds a part-time position as Assistant Professor there. His research work has focused on aviation, but also on maritime transportation, health care, and the nuclear industry. His research areas have been on mental workload, training, and simulation. Nicklas has written research articles and book chapters, and delivered presentations and training in 20 countries.

Rhona Flin, PhD, FBPsS, FRSE is Professor of Industrial Psychology, Aberdeen Business School, Robert Gordon University and Emeritus Professor of Applied Psychology, University of Aberdeen. Her work examines human performance in high-risk work settings, such as health care, aviation, and the energy industries, with studies focusing on leadership, safety culture, team skills, and cognitive skills.

Current projects include product safety culture, managers' safety leadership, and non-technical skills in safety-critical tasks. Her books include *Safety at the Sharp End: A Guide to Non-Technical Skills* (2008, with O'Connor and Crichton) and *Enhancing Surgical Performance: A Primer on Non-Technical Skills* (2015, with Yule and Youngson).

Alexandra Fernandes, PhD, is a researcher in the Industrial Psychology Department at the Institute for Energy Technology (IFE) in Norway. She has a master's degree in Cognitive Psychology and a PhD in Experimental Psychology and Cognitive Sciences and has been working on HFE topics since 2012. She has co-authored multiple publications, including technical reports for the industry. At IFE she has been working with the nuclear industry on topics such as human performance assessment, control room verification and validation, control room modernization, and human-system interfaces evaluation. She has also collaborated in HFE projects for the petroleum and aviation industries.

Mazin Gadir, PhD, is Senior Specialist at the Executive Office for Organizational Transformation, Dubai Health Authority. He is a management consultant with wide experience in organizational transformation at government authorities and health care provider enterprises. He is an expert in digital health and innovation strategy development/implementation and organizational future transformation. Dr. Gadir was a Senior Manager in the Healthcare Technology and Disruptive Innovation Consulting Practice at PwC Middle East in Dubai for two years (2014–2016), overseeing and leading projects across Europe, USA, and the EMEA. He has over 15 years of digital health innovation, health informatics, e-health frameworks, health information systems, clinical innovation, project and change management experience in health care IT transformation.

Cleotilde Gonzalez, PhD, is Professor of Decision Sciences and the Founding Director of the Dynamic Decision Making Laboratory at Carnegie Mellon University. She is affiliated with the Social and Decision Sciences department and has additional affiliations with the Center for Behavioral Decision Research, Human-Computer Interaction Institute, and others. She is a Fellow of the Human Factors and Ergonomics Society; Associate Editor of *Cognitive Science* and the *Journal of Cognitive Engineering and Decision Making*; and part of the Editorial Board of journals including the *Journal of Behavioral Decision Making*, the *Human Factors Journal*, *Decision*, and the *System Dynamics Review*.

Sue Hignett, PhD, is Professor of Healthcare Ergonomics and Patient Safety at Loughborough University (UK). Over the last 30 years she has experienced the health care industry as a clinician, ergonomist, researcher, and patient. Her research looks at a wide range of HFE issues including design of safer systems, building and vehicle (ambulance) design, emergency and CBRNe response, and staff well-being. Professor Hignett is an Editor for *Ergonomics*; Chair of the Professional Affairs Board at the Chartered Institute of Ergonomics and Human Factors; and past Chair of the International Ergonomics Association Technical Committee on Healthcare Ergonomics.

Samantha J. Horseman, BPhty, CCP (erg), MBA, DBL, leads the Human Energy Management, Talent & Leadership Department at Saudi Aramco. She has been involved in the HSE, innovation, and human performance industry for over 20 years. A well-known international speaker, she is a certified Ergonomics Professional (CCP), has a bachelor's in Physiotherapy (Otago), master's in Business (Hull), and doctorate in Business Leadership (UNISA). She has 27 publications and 22 regional and international awards. She also has an impressive intellectual property portfolio of 15 prototypes in the pipeline and 45 granted patents globally. In 2016, six of her patents were also licensed to a local startup.

Gyuchan Thomas Jun, PhD, is Lecturer in Human Factors and complex systems at Loughborough Design School, Loughborough University, UK. He is a founding member of the Human Factors in Complex Systems Research Group and runs SystemsThinkingLab. His research interest is in integrating systems thinking, design thinking, and resilience thinking into the participatory design of complex socio-technical systems like health care. His particular expertise is in representing/analyzing complex socio-technical systems for participatory design. He led the production of two highly praised story-based short animations on system safety (www.systemsthinkinglab.com).

Jussi Kantola, PhD, is Professor in the Department of Production at University of Vaasa, Finland. Previously, he was Associate Professor in the Department of Knowledge Service Engineering at KAIST, Korea. From 2003 to 2008 he worked in various research roles at Tampere University of Technology and University of Turku, Finland. He has PhD degrees (IE) from University of Louisville, KY, USA, 1998 and Tampere University of Technology, Finland, 2006. From 1999 to 2002 he worked as an IT and business and process consultant in the USA and in Finland. His research interests include organizational resource management, human factors, and product design and development.

Alexandra Lang, PhD, is Healthcare Human Factors Research Fellow at the University of Nottingham. Her research interests include investigating the impact of mHealth interventions on clinical pathways, health care technology design and use in the context of health promotion in specialist populations. Alexandra is a member of the Chartered Institute of Ergonomics and Human Factors (CIEHF) and Chair of the Healthcare Steering Committee. She sits on the Faculty of Engineering Ethics Committee at the University of Nottingham and is a consultant Human Factors Expert for the Centre for Healthcare Equipment and Technology Adoption (CHEATA) at Nottingham Universities Hospitals Trust.

Guofa Li, PhD, is Assistant Professor at the Institute of Human Factors and Ergonomics, Shenzhen University. He received his BS degree from Beijing Institute of Technology, Beijing, China in 2010, and his PhD degree from Tsinghua University, Beijing, China in 2016. From 2012 to 2013, he was a visiting scholar in Driver Interface group in the University of Michigan Transportation Research Institute (UMTRI). His current research interests include driver behavior modeling, autonomous vehicle technologies, and human factors in transportation systems. He has authored and co-authored 24 peer-reviewed journal and conference papers and serves as the reviewer for ten journals.

Carlos A. Lucena, PhD, spent the year of 2016 accomplishing a Postdoctoral Fellowship at the University of Waterloo, Canada, sponsored by the Science Without Borders program (CNPq, Brazil). His research was supervised by Professor Catherine Burns, Director of the Advanced Interface Design Lab and Executive Director of the Centre of Bioengineering and Biotechnology. Lucena received his PhD at the Pontificia Universidade Catolica do Rio de Janeiro, Brazil (PUC-Rio) in 2014. His thesis explored the aspects of online and mobile collaborative learning in the medicine field. Lucena was also one of the co-founders and shareholders of AfferoLab (www.afferolab.com.br), Brazil's a leading corporate education provider.

Sean A. McGlynn, MS, graduated from the University of Connecticut with degrees in Cognitive Science and Psychology and a minor in Neuroscience. He received his master's degree in Engineering Psychology at the Georgia Institute of Technology in the Human Factors and Aging Laboratory under the advisement of Dr. Wendy Rogers and is currently pursuing a PhD in this program. His research interests include design for older adults, presence in virtual reality, human-robot interaction, technology acceptance, and cognitive aging.

His Excellency (H.E.) Mr. Nabil Ameen Molla has been the Secretary General of GCC Standardization Organization (GSO) since April 2012. Previously, he was the Governor of Saudi Standards, Metrology and Quality Organization (SASO) (2005–2012). He led the GCC Accreditation Center (GAC) to be an internationally recognized body in (ILAC), when he was the Director General of the GCC Accreditation Center (GAC) until 20 June 2016. H.E. Mr. Nabil is a member of many national and international organizations, such as the ISO Council (2011–2012), the Council of the Saudi Food and Drug Authority (2005–2012), is Secretary-General of King Abdul-Aziz Quality Award (2005–2012), a member of the Saudi Consumer Protection Association (2008–2012), a member of the American Society for Quality (ASQ) (2008–2011), and a member of the Board of ASTM International (2014–2016). H.E. Mr. Nabil represented the Kingdom of Saudi Arabia in many meetings of regional and international standardization organizations (GSO, AIDMO, ISO, IEC, OIML, BIPM, ILAC, IAF). He received a BS in Chemistry/Biology, from King Saud University, Riyadh Kingdom of Saudi Arabia in 1974 and M.A. in Chemistry, University of Denver, Colorado, USA, 1978.

Beena Prabhu, MS, is the co-Founder of DesignFold. Her journey has taken her across the design, user experience, and engineering disciplines. This has enabled her to bring together systems-level thinking and empathy for users that supports creation of compelling products and services. In the last two decades she has applied her understanding of people and their contexts towards the creation of solutions that have a positive impact on users and their environments. Apart from consulting, Beena has been evangelizing user experience design through various training programs for start-ups, designers, product managers, engineers, and design and management students. She has also been a key player in launching Designfold's HeyDesign program for high school students.

Girish Prabhu, PhD, is the Director of Srishti Labs, Dean at the School of Design, Business and Technology, Srishti Institute of Art, Design and Technology, and Co-Founder of Institute for Futures, Design, and Entrepreneurship. He has held various management positions in new product development both in the USA and India, and was instrumental in implementing human-centered design and innovation in multiple product areas. Girish has 25+ years of experience in design-led innovation, new product and business development, and product localization. His expertise covers various consumer product and service domains such as education, health care, entertainment, and retail. Girish has published referred papers in international journals and holds 16 US patents. Dr. Prabhu is always keen on improving methods, frameworks, and approaches in stakeholder-centric innovation and experience design.

Prashanth Rajivan, PhD, is a Postdoctoral Research Fellow in the Department of Social and Decision Sciences, Carnegie Mellon University, Pittsburgh, USA. He is an interdisciplinary computer security researcher with a research agenda centered on the intersection of security and human behavior. He holds a PhD in Human Systems Engineering from Arizona State University, USA (2014), MS in Computer Science from Arizona State University and B.E. in Information Technology from India. His main areas of interests include cyber security, human factors, decision-making, teamwork, adversarial behavior, usable security, simulation, and modeling. He is the author of several peer-reviewed publications and book chapters.

Kine Reegård, MS, is a researcher at the Institute for Energy Technology, department of Industrial Psychology. She has a master's degree in Work and Organizational Psychology from the University of Oslo. Her main body of work is on how new technologies change ways of working, and in particular in the petroleum industry, focusing on how new ways of working can be implemented by aligning people, technology, and organizations. She has co-authored several publications on this topic, including manuals and technical reports developed for the industry.

Wendy A. Rogers, PhD, is Khan Professor of Applied Health Sciences at the University of Illinois Urbana-Champaign. She is Program Director of CHART: Collaborations in Health, Aging, Research, and Technology, and Director of the Human Factors and Aging Laboratory (www.hfaging.org). Her research interests include design for aging; technology acceptance; human-automation interaction; aging-in-place; human-robot interaction; and aging with disabilities. She is funded by the National Institutes of Health (National Institute on Aging) as part of the Center for Research and Education on Aging and Technology Enhancement (www.create-center.org); and the Administration for Community Living (National Institute on Disability, Independent Living, and Rehabilitation Research; NIDILRR) Rehabilitation Engineering Research Center on Technologies to Support Successful Aging with Disability (www.techsage.gatech.edu).

Atul Saraf, BS is an architect and a user experience design professional with 20 years of collective industry experience. Atul is passionate about design that is strategic, delightful, and meaningful. As a designer, Atul has engaged with multinational corporates, NGOs, and educational institutes to create solutions for complex problems. His area of work includes space design, branding, visual communication, and digital products and service design. His work at the Srishti Institute of Art Design and Technology includes leading UX and Innovation projects as well as teaching the Interaction Design program. As part of DesignFold consulting, he currently designs experiences for digital systems and conducts training for interaction design.

Steven Seay, PhD, is the Head of the Leadership Development Division for Saudi Aramco. He is a product of Silicon Valley, California and an MIT-trained strategist. Steven has also served as Vice President of Human Resources at Best Buy and Jostens. Steven holds a B.S. degree in Organizational Psychology, an MBA in Strategy, and a PhD in Management Psychology and Organization Psychology. Dr. Seay also serves as an Advisory Board member for the Twin City Company—Pinnacle Services, Inc., and as a board member for the Minnesota Mental Health Association. Steven helped develop the first PHR/SPHR exams for the Society of Human Resource Management.

Ari Sivula, PhD, is a University Lecturer of the Industrial Systems Analytics Master's Program in the School of Technology and Innovations at the University of Vaasa. He received his MS in Computer Science from the University of Vaasa, Finland, in 2011, and his PhD in Industrial Management from the University of Vaasa, Finland, in 2016. He is experienced in software engineering, project management, regional and business development, and innovation management and has been active in the public and private sector. His current research interests include innovation management, new product and service development, crowdsourcing, and project management.

Colin M. Sloman is the Talent & Leadership Department Director for Human Resources at Saudi Aramco. Colin has 25 years of experience in human resources, talent, and organization consulting and has worked in the oil and gas industry for most of this time. Previously, Colin was an organization strategy and change consultant at Accenture, leading the Global Human Capital practice. He has worked with board-level executives to design people strategies to deliver business goals and has led global transformation programs across the enterprise including human resources, finance, and procurement. Colin has published in the areas of organizational change, talent, leadership, and culture.

Ying Wang, PhD, is a human factors and traffic safety expert. She holds a doctorate degree from Tsinghua University, Beijing, China (2009). She is currently the Principal of United Human Factors, a consulting service based in Illinois, USA, and used to be an Associate Professor in School of Transportation Science and Engineering at Beihang University, Beijing, China (2013–2016) and a Postdoctoral Research Associate at MIT AgeLab (2009–2012). Her research seeks to build safety and well-being in surface transportation systems from a human factors perspective. She also served the International Traffic Medicine Association as the Secretary from 2011 to 2016.

1 Human Factors and Ergonomics
History, Scope, and Potential

Wendy A. Rogers and Sean A. McGlynn

CONTENTS

INTRODUCTION

Historical Overview

The field of human factors and ergonomics (HFE) revolves around optimizing human performance in systems and reducing errors by designing those systems to accommodate the capabilities and limitations of humans from a perceptual, cognitive, and physical perspective. An early example of this type of thinking goes back to the scholar Hippocrates, who suggested optimal positions for the surgeon, patient, and light source to facilitate operations, and provided guidance about surgical tool shapes and sizes that would be easiest to use (Marmaras, Poulakakis, & Papakostopoulos, 1999). These ideas were precursors to the philosophy underlying HFE of designing for human use.

Human factors really began to take shape in the early 1900s with the work done by Frank and Lillian Gilbreth (reviewed in Sanders & McCormick, 1993), which elaborated on Frederick Winslow Taylor's scientific management method for finding the optimum way of completing a given task (Taylor, 1914). The Gilbreths conducted "motion studies" on skilled performance and equipment/workspace design, with the primary goal of understanding how the "materials element" of any system could be adapted in such a way that the energy expended by the "human element" could be put to optimal use (Gilbreth & Gilbreth, 1919). They did this by filming workers completing tasks to identify areas where motion could be reduced. These films also served as template videos for providing training of best-practices to new workers. One practical application of their work was in their reduction of the number of motions involved in the task of brick-laying, which increased workers' productivity by nearly 200% (Gilbreth, 1911).

The discipline of HFE as it is recognized today became prominent during World War II, wherein advances in aircraft systems introduced added cognitive loads on pilots that could not be overcome solely through personnel selection and training. This resulted in the realization that performance could be enhanced by applying the principles set forth by scientific management and time and motion studies, which emphasized designing the system around the capabilities and limitations of the human user. A major addition to the field during this time period was the expansion of principles of physical ergonomics into the realm of cognitive ergonomics. A classic HFE study conducted by Fitts and Jones (1947) sought to understand human interactions with cockpit controls and subsequently implemented an efficient configuration that minimized the cognitive load placed on the pilot.

In the decades following WWII, individuals such as Alphonse Chapanis used the knowledge gained during the war to push the field of HFE as a profession outside of military operations (Chapanis, 1999). The year 1949 brought about two important milestones in HFE; the formation of the Ergonomics Research Society in Britain (now the Chartered Institute of HFE), and the publication of the first book on human factors, *Applied Experimental Psychology: Human Factors and Engineering Design* (Chapanis, Garner, & Morgan, 1949, discussed in Salvendy, 2012). Military research in HFE continued to be stimulated by circumstances surrounding the Cold War, such as the arms race and space race in the 1950s and 1960s. The increasing awareness of practical HFE applications resulted in the creation and growth of other professional and governmental organizations during this time period, many of which continue to be prominent agencies in the field today (as evidenced by the chapters in this book).

The primary nonmilitary boom in the field came with the digital revolution during the Information Age, beginning around the early 1980s. Smaller and more affordable computers became increasingly prevalent in companies, sparking a need for ergonomically designed interfaces, software, and input devices (Sanders & McCormick, 1993). Not surprisingly, this trend toward user-centered design has continued to grow with the rapid expansion of technologies such as the internet, cell phones, and mobile devices. A widely disseminated book by Don Norman helped to popularize concepts of user-centered design (*The Psychology of Everyday Things*, 1988, later reissued as *The Design of Everyday Things*, 1990). HFE will continue to permeate society alongside the development of increasingly complex technologies such as smart devices and vehicles, connected homes, robots, and virtual/augmented reality.

The establishment of HFE as a scientific domain can be largely attributed to practical implementations of user-centered design principles. Cooke and Durso (2007) provided excellent examples of HFE successes and failures throughout history across a variety of domains such as perception, performance, team coordination, training, and safety. The evolution of HFE and the relevant practices, processes, and technologies that will be discussed throughout this book has been brought about by learning from these and many other stories of success and failure, conducting systematic research studies, and disseminating knowledge via scientific publications and professional meetings.

Professional and Governmental Organizations

One avenue for a discipline to flourish is through a professional association that provides opportunities for scientists and practitioners to interact and share their experiences; for students to network and receive guidance; and for the development of shared goals and objectives. In the United States of America, the primary organization for HFE is the Human Factors and Ergonomics Society (HFES), which hosted its first meeting in 1957, emerging from joint discussions of the Aeromedical Engineering Society of Los Angeles, CA and the Human Engineering Society of San Diego, CA (www.hfes.org//web/AboutHFES/History.html). The Society has continued to expand with local and student chapters in Australia, Canada, Europe, and the Gulf Cooperation Council region. The journals that are managed by HFES (*Human Factors*, *Cognitive Engineering and Decision Making*, *Ergonomics in Design*) along with the proceedings from the annual conferences, serve as a repository for a wealth of information about HFE in a broad range of areas (evidenced by the HFES technical groups—see Table 1.1); several of these areas are described in greater detail in this book.

There are HFE professional organizations throughout the world, affiliated with the International Ergonomics Association (IEA), which is a federation of organizations in human factors and ergonomics. IEA held its first general assembly meeting

TABLE 1.1
Human Factors and Ergonomics Society Technical Groups

Aerospace systems	Environmental design	Perception and performance
Aging	Forensics	Product design
Augmented cognition	Health care	Safety
Children's issues	Human performance modeling	Surface transportation
Cognitive engineering and decision-making	Individual differences in performance	System development
Communications	Internet	Test and evaluation
Computer systems	Macroergonomics	Training
Education	Occupational ergonomics	Virtual environments

Source: www.hfes.org/web/TechnicalGroups/descriptions.html.

TABLE 1.2
International Ergonomics Association Technical Committees[2]

Activity theories for work analysis and design	Ergonomics in advanced imaging	Musculoskeletal disorders
Aerospace HFE	Ergonomics in design	Organizational design and management
Affective design	Ergonomics in design for all	Process control
Aging	Ergonomics in manufacturing	Psychophysiology in ergonomics
Agriculture	Gender and work	Safety and health
Anthropometry	Healthcare ergonomics	Slips, trips, and falls
Auditory ergonomics	Human factors and sustainable development	Transport ergonomics and human factors
Building and construction	Human simulation and virtual environments	Visual ergonomics
Ergonomics for children and educational environments	Mining	Work with computing systems

Source: www.iea.cc/about/technical.php.

in 1961 and today includes 50 Federated Societies (http://www.iea.cc/about/council.html), and spans 27 Technical Committees (Table 1.2), which speaks to the breadth of HFE worldwide.

Another mechanism to ensure application and dissemination of the methods and findings of a discipline is for government agencies to adopt the practices and/or to impose regulations that require it. This process has occurred broadly for HFE, which is perhaps not surprising given its relevance to so many aspects of everyday life. One example is the Federal Aviation Administration (FAA), which has a Human Factors Division with a goal "to support the attainment of high levels of human system performance across all aviation domains" (www.hf.faa.gov/media/RoleOfHF-FAA.pdf). Another is the Food and Drug Administration (FDA), which requires human factors evaluations at both the pre-market and post-market stages for medical devices. There are numerous other examples, worldwide, from a variety of sectors wherein HFE principles, best practices, and standards are being followed to ensure safety and effectiveness. These were reviewed in Rogers' HFES Presidential Address in 2005 (www.hfes.org//Publications/ProductDetail.aspx?ProductID=66) and have continued to grow since then. One simple illustration of this growth is that a Google search on the terms "human factors" and "ergonomics" at the time of Rogers' 2005 address yielded ~3 million and ~4.5 million results, respectively. Those same searches now produce ~7 million and ~19 million results.

Education and Training in Human Factors and Ergonomics

To meet the needs of the increasing requirements for HFE inclusion in the development cycle, the number of educational programs has increased and there are multiple opportunities for short-term training. With respect to formal education, there are 17 undergraduate and 15 graduate programs of study accredited by HFES spanning locations in the USA and Canada. Outside of North America, countries with educational programs in HFE and related topics include Australia, China, Portugal, Saudi Arabia, Sweden, and the United Kingdom (https://ergoweb.com/university-programs-ergonomics-human-factors/). These university programs are housed in a variety of departments such as psychology, engineering, and computing.

FUNDAMENTALS

The basic tenet of the discipline of HFE is "designing for human use." A detailed definition was provided by the IEA as:

> The scientific discipline concerned with the understanding of interactions among humans and other elements of a system, and the profession that applies theory, principles, data and methods to design in order to optimize human well-being and overall system performance. Practitioners of ergonomics and ergonomists contribute to the design and evaluation of tasks, jobs, products, environments and systems in order to make them compatible with the needs, abilities and limitations of people. (http://www.iea.cc/whats/)

The fundamental components of the field may be grouped into: physical (e.g., anthropometry, biomechanics); cognitive (e.g., learning capabilities, memory limitations); and organizational (e.g., policies, culture). However, these broad categories describe the field only generally. HFE is perhaps best considered as an approach—a systematic process to understanding the human in the loop performing a given task, to achieve a goal, in a context. As such, HFE draws from multiple disciplines including engineering, psychology, occupational health, environmental design, medicine, and others. There are well-documented methods (described in the next section) that must be used to guide the design and deployment of products for people in particular situations.

There are standards that can be used to guide initial designs. The International Standards Organization (www.iso.org) and the American National Standards Institute (www.ansi.org/) provide broad standards pertaining to a number of areas such as workplace design and safety. There are also standards developed for specific domains such as the Association for the Advancement of Medical Instrumentation (www.aami.org/standards) or web design (https://18f.gsa.gov/tags/web-design-standards/).

Importantly, there are no short-cuts and no substitutions for user-testing, at the early formative stages as well as with prototypes and close-to-final versions at the summative stage. The reason government, consumer, military, and other organizations worldwide require HFE approaches is that they must be utilized in the representative contexts of use, for the potential range of users, for the target types of tasks, and with the type of products/devices/technologies available. We provide resources in the recommended readings sections for more details about these processes, as well as the methods briefly discussed next.

METHODS

Design is an iterative process, with HFE contributions in each stage of development and evaluation of a given product or system. In all cases, however, the earlier in the design process that HFE principles are incorporated, the easier and more cost-effective the recommendations will be. Field studies, incident reports, task analyses, surveys, literature reviews, simulations, and laboratory experiments are methods that can be used to answer questions regarding how the user interacts with the end product or system.

There are different ways to measure user-system interaction to assess the success of a particular design and/or provide guidance for design iterations and the development of training materials. Measurements can be physical, psychophysiological, cognitive, behavioral, team, environmental, and macroergonomic (Stanton, 2004) and can be informative to a diverse set of research questions:

- Physical: *What operating surface height is least likely to create surgeon fatigue?* (van Veelen, 2004).
- Psychophysiological: *How stressed/aroused is an air traffic controller when completing tasks necessary for adequate performance?* (Collet, Averty, & Dittmar, 2009).
- Cognitive: *What is a person's thought process when interacting with a novel user interface?* (Polson, Lewis, Rieman, & Wharton, 1992).
- Behavioral: *To what extent does a user trust an automated system to be successful in a critical situation?* (Miller, Johns, Mok, Gowda, Sirkin, Lee, & Ju, 2016).
- Team: *What is the level of team coordination necessary for maintaining situation awareness during an unstaffed air vehicle reconnaissance mission?* (Gorman, Cooke, Pederson, Connor, & DeJoode, 2005).
- Environmental: *How does the illumination and arrangement of lighting impact labor productivity in the workplace?* (Turekova, Kozik, Bagalova, & Neovesky, 2014).
- Macroergonomic: *Does a company's adoption of a quality management standard have the intended impact on the work life of employees?* (Heras-Saizarbitoria, Cilleruelo, & Allur, 2012).

The appropriateness of the methods and measurements used will depend on the design stage, the cost (time and financial) of the methodology, and the practical value of the findings. We will illustrate with specific examples in the application section. See also Sanders and McCormick (1993), Salvendy (2012), Meister (1985), and Wickens, Hollands, Banbury, & Parasuraman (2015) for details on different methodologies and the contexts that determine their appropriateness.

APPLICATION

The remaining chapters in this book provide in-depth reviews of different domains. The goal in this chapter is to demonstrate the breadth of the HFE field. We also aim to show the range of considerations and provide insight into the relevance of HFE throughout any given system. As such, we exemplify the complexities within two systems: education and home health care.

We have adapted the basic framework of the IEA model (www.iea.cc/whats/index.html) to illustrate the fundamental components of HFE that must be considered in the human-system interaction context, replacing "jobs" with "people" to emphasize the centrality of the human (and individual characteristics) for the design process. The primary categories of consideration are:

- the people involved,
- the task being accomplished,
- the product being used,
- the environmental context, and
- the characteristics of the organization.

We use this framework to illustrate the applications of HFE in education and home health care. We provide graphical representations for each system containing a list of considerations within various system dimensions (see Figures 1.1 and 1.2) and also provide research examples in the text. As is evidenced by the other practice domains covered in this book, these two example systems are only illustrative and not comprehensive. For every domain, there will be HFE considerations at the level of the person, the task, the product, the environment, and the organization.

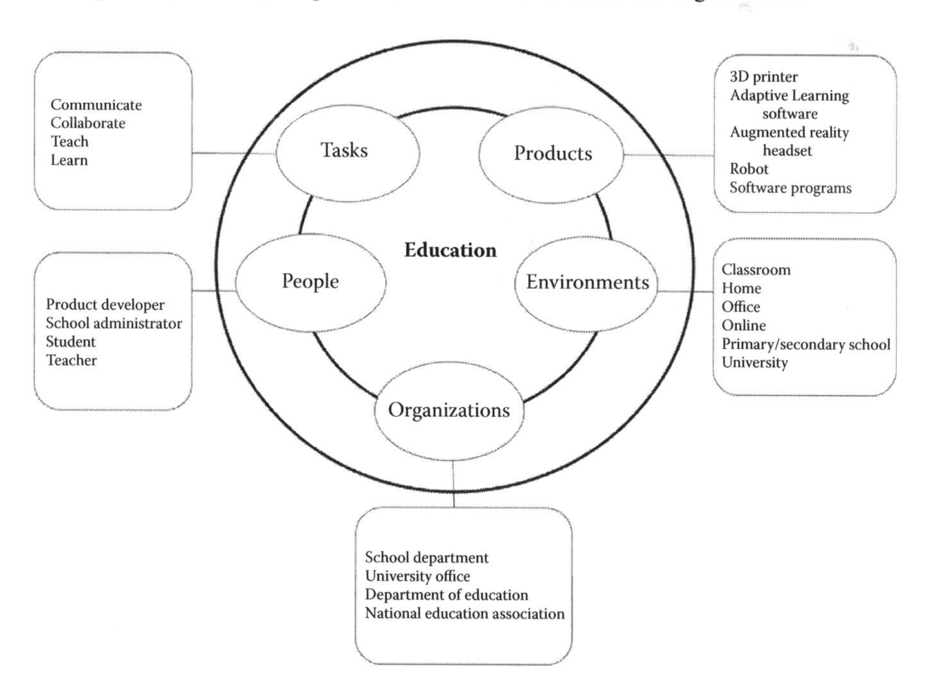

FIGURE 1.1 Illustrative examples of tasks, products, jobs, environments, and organizational considerations for the education system.

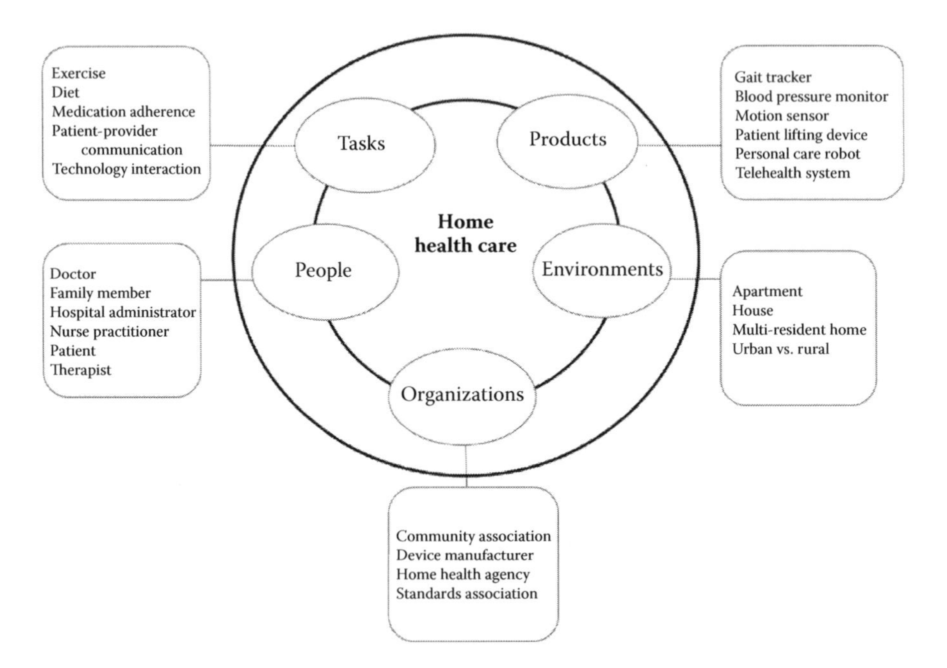

FIGURE 1.2 Illustrative examples of tasks, products, jobs, environments, and organizational considerations for the home health care system.

Education

A major part of HFE as a discipline involves understanding how to train people effectively. Historically, the focus of such training has been on specific jobs in the workplace; training pilots, soldiers, employees, and so on. One type of training relationship that is almost universally vital is the training that occurs between an instructor and a learner in an education system. An illustration of the range of people, tasks, products, environments, and organizations in this system is presented in Figure 1.1. Although depicted as separable, these components are often highly interactive such that changes in one could have a significant impact on other components or, in some cases, on the system as a whole. For example, moving the learning environment from the classroom to the home will change the instruction method, the teacher, and might introduce a new software program for the learner and teacher to interact with.

The transfer of information from one person to another is the core task of the education system. HFE research in this domain can be applied to help facilitate the transfer of information in such a way that it will be comprehended and remembered. On the side of the instructor or teacher, HFE research can inform education task-relevant questions regarding the types of instructions that best facilitate learning (Fonseca & Chi, 2011; Margulieux, 2012; Williams, 2013). From the student's perspective, when completing the task of studying, one might want to know which strategies lead to higher test scores (Elliot, McGregor, & Gable, 1999).

The different types of people involved in the education system create a plethora of user-centered design considerations. What is "best" will not be the same for every student. People vary in their goals, motivation, abilities, and rates of learning (Daneman & Carpenter, 1980), especially when bearing in mind that a person can choose to be a "student" well beyond the typical formal education ages of childhood and young adulthood (Cody, Dunn, Hoppin, & Wendt, 1999). Similarly, the methods used for assessing performance must be carefully selected, as they may not equally reflect knowledge of a given area across different students (Leonard & Jiang, 1999). There are individual differences in instructors as well. For example, pre- and post-tenure university faculty members may allocate their time very differently (Link, Swann, & Bozeman, 2008). It is also important to recognize that these roles of instructor and student exist within, and are influenced by, a larger network of people, which can include parents, peers, and administrators, among others.

Emerging technologies such as massive open online courses (MOOCs) and robots are good examples of what the future of education may look like (Liyanagunawardena, Adams, & Williams, 2013; Mubin, Stevens, Shahid, Al Mahmud, & Dong, 2013). These products have the potential to enhance learning or enable learning opportunities that may not otherwise have been possible (Cooper & Harwin, 1999). As interacting with technology continues to become a normal component of the education process, core HFE considerations such as interface design become increasingly pertinent (Yousef, Chatti, & Schroeder, 2014). New challenges emerge as well; for example, deciding if a robot designed to enhance childhood education should be assigned the role of the instructor or the student (Hood, Lemaignan, & Dillenbourg, 2015; Mubin, Stevens, Shahid, Al Mahmud, & Dong, 2013). Another challenge is ensuring that these technology-based education products are externally valid, such that the learning that occurs using these systems leads to performance gains outside of these specific situations (Chen, Davisy, Hauff, & Houben, 2016).

In addition to lesson formats, teaching styles, and performance metrics, a fundamental component of learning is the educational environment. Although a MOOC may be completed remotely and a robot may reduce the burden on the teacher, being present in a classroom with a teacher and peers may enable participation and provide a level of engagement and feedback that is not possible using these technologies (Koutropoulos & Zaharias, 2015). So even if technology usage in these domains may have benefits in some regards, it is critical to consider the negative potential outcomes as well.

Zooming out even further, all of these micro-level interactions are influenced by macro-level forces on the organization, such as school departments and university administration offices. In the USA, the Department of Education establishes policies, enforces educational laws, and makes decisions that affect elementary and secondary education. There are also local-level school boards and parent-teacher associations that influence education at specific schools. Likewise, for post-secondary education, the university system, governing boards, chancellors, provosts, deans, and department chairs are all part of the educational system that influences college education. Viewing the education system under an HFE lens enables understanding of how these different factors act individually and interactively influence the domain as a whole.

Home Health Care

The broad domain of health care is perhaps the fastest growing area wherein HFE is having an impact. This topic has been the focus of national academy reviews including *To Err is Human* (2000), *Patient Safety* (2004), *Preventing Medication Errors* (2007), and *Consumer Health Information Technology* (2011). All are available free from the National Academies Press site (www.nap.edu). These topics illustrate both the importance of the issues as well as the recognition that attention to HFE considerations in health care is critical to the health and quality of life of individuals.

The health care system is changing, with more focus on prevention and wellness as well as greater demands on patients and family members (e.g., less time in the hospital, more home health care). In addition, people are living longer, with multiple chronic conditions. Technology is permeating the health care system on both the provider side and the patient side. The range of relevant issues is illustrated in Figure 1.2. The system itself can best be described as a "care network" that may include the patient, family members (e.g., spouse, children, parents), primary care physician, specialist physicians, as well as home health providers, therapists, and more. Each individual in the network has particular goals as well as personal capabilities, motivations, limitations, and experience, all of which will contribute to the success of the overall system of care, which in turn operates in the context of a hospital system and higher-level healthcare policies and regulations. This was the topic of *Health Care Comes Home: The Human Factors* (National Research Council, 2011).

To illustrate this context, consider an older adult who has congestive heart failure and uses a telehealth system in her home to monitor her weight, activity, blood pressure, sleep, and medication adherence. A nurse practitioner receives alerts if any of the data appear outside of the range of what is normal for her. He can then communicate with the patient via a televideo system to provide guidance on behavior change and minimize unnecessary visit to urgent care facilities. The human factors issues in this context are well-described in Charness, Demiris, & Krupiski (2011).

The tasks in this context for the patient are multiple—some data are captured automatically, such as movements throughout the home, but the older adult must remember to step on the scale every morning (sometimes she does not hear the alerts), insert her arm into the blood pressure cuff properly, and so on. If she has questions, she must be able to initiate a video call to the nurse practitioner, but she is not very computer-savvy. Even devices that are purported by the device manufacturer to be easy to use can be complex. For example, we found that a medical device advertised as being "as easy as 1, 2, 3" actually required 52 steps—this was revealed via a detailed task analysis of what was required by the user (Rogers, Mykityshyn, Campbell, & Fisk, 2001).

On the nurse practitioner side, tasks include interpreting the data, which involves understanding the visualizations that are provided, and perhaps drilling down for more information. He must be able to initiate calls to the patient as needed and provide information that is specific to the patient's capacity for understanding, including both perceptual and cognitive limitations.

The general product here is the telehealth system but the specifics of the interface vary for the patient and the healthcare provider (Charness et al., 2011). The patient's system also includes peripheral devices, such as the scale and blood pressure monitor, whereas the provider's system is primarily a video display that presents data as well as serves as a communication tool for interacting with the patient.

An emerging technology in this context would be a personal care robot. Such a robot might reside in the home of the older adult and collect a range of health-related data and provide support for activities such as medication adherence, nutrition, activity monitoring, and social communication. Although initial indicators are that older adults are receptive to the use of robots for such tasks (Smarr et al., 2014), there remain a host of human-factor design issues that will have to be addressed before we will see widespread deployment of such technology.

One key person in the loop is the patient, whose goal is to maintain her health through monitoring of everyday activities and vital signs. She should adjust her behavior accordingly to enhance quality of life. Perhaps a less obvious issue on the patient side is the individual's confidence that she can perform the necessary tasks to maintain her health. This issue of self-efficacy as related to human-factor issues was highlighted in Mitzner, McBride, Barg-Walkow, & Rogers (2013). Thus, HFE design considerations may include information displays to increase motivation and provide guidance (i.e., persuasive technologies), as well as interface design and training.

Other people in the network include healthcare providers—in this case the nurse practitioner whose job is to provide healthcare at a distance—through interpretation of data but also through interactions with the patient and perhaps with somewhat impoverished information and in a stressful situation when the patient needs to know whether to change medication or go to a hospital. Home health care providers vary in experience and training, which can influence the type of care being provided (Beer, McBride, Mitzner, & Rogers, 2014).

The physical environment also raises HFE issues (see Stronge et al., 2007). The layout of the home may affect communication between the components of the system, alerts may be missed due to distance between rooms, and the heating/lighting/cooling may affect the reliability of the data. The nurse practitioner may work in a noisy environment and find it difficult to communicate with the patient.

For both patient and provider, HFE issues of interface design, instruction, and understanding of the potential stress of the situation are critical considerations. Their goals may be influenced by their previous history, family considerations, and even financial concerns. The physical environments on both ends of the telehealth system can enhance but may also degrade communication between the individuals. Overarching these components is the organization in which the care is taking place. For example, home health care providers work for companies that must follow regulations and meet standards of care.

These examples represent only a very small fraction of the HFE issues in the context of health care. There has been tremendous growth in this area—HFES started hosting a mid-year meeting in 2012 that focuses solely on this topic. The program tracks illustrate the range of topics in this domain: Clinical and Consumer Health-Care Information Technology; Health-Care Environments; Medical and Drug-Delivery Devices; and Patient Safety Research and Initiatives (www.hfes.org/web/HFESMeetings/2016HealthCareSymposium.html). For excellent reviews of HFE issues in this domain see Chapter 7 in the present book, as well as the review chapters in the volume edited by Morrow (2013), *Reviews of Human Factors and Ergonomics, Volume 8: Health-Care Human Factors/Ergonomics.*

FUTURE TRENDS

The future is bright for the field of HFE because the need for the knowledge, skills, and abilities of people in this discipline is growing. As the complexity of systems continues to increase, the variety of technologies grows, and population demographics change—but human needs must be met. We must ensure that the goal of designing for human use is achieved.

Futurists try to envision the state of the world at some time point, with a focus on preferred scenarios and the potential ideal situations (Datar, 1995). The value of this exercise is to identify potential and to determine what research and development needs to occur for that future ideal scenario to be realized. However, this is a challenge in part because, as Sardar (2010) expressed in his first law of future studies: "Almost all the problems we face nowadays are complex, interconnected, contradictory, located in an uncertain environment and embedded in landscapes that are rapidly changing" (p. 183). Envisioning scenarios for the future of HFE is no different, and in fact quite epitomizes this law.

There is a strong base of knowledge in the field of HFE, as evidenced by the research literature as well as national and international standards for design practices. However, it is also the case that HFE influence is based on *processes.* A task analysis is needed to fully understand what a user is trying to accomplish. A person analysis is necessary to identify the capabilities, limitations, experiences, attitudes, and other individual differences of the target user groups. Usability testing will be required to identify design limitations as well as training and instructional needs. The demands of the job and the role of the organizational climate will need to be specified, and the characteristics of the environment must be determined in terms of variables that might influence task performance. All of these activities require HFE knowledge and training. Consequently, it will be critical to: (a) continue research in HFE to advance the fundamental knowledge base and provide guidance for design and instruction; and (b) ensure that future practitioners are receiving the necessary education to provide HFE input throughout the development process.

On both the engineering and the science side there is a future wherein HFE will have to be integral, for successful growth, development, and life quality for society. For example, the National Academy of Engineering poses grand challenges (www.engineeringchallenges.org) that include topics such as personalized learning, virtual reality, health informatics, and cybersecurity—all of which will have fundamental HFE challenges and will thus benefit from HFE involvement from the beginning. The Director of the National Science Foundation, France Cordova, proposed ten big ideas in science (www.nsf.gov/about/congress/reports/nsf_big_ideas.pdf) that include science and engineering through diversity; the human-technology frontier; and navigating the arctic. One can easily imagine the HFE issues relevant to these areas and the need for HFE scientists to be involved in these initiatives.

CONCLUSION

The goal of this book is to provide an introduction to HFE for academics, professionals, and practitioners in the Gulf Cooperation Council (GCC; Bahrain, Kuwait, Oman, Qatar, Saudi Arabia, and the United Arab Emirates). Concordantly, the goal for this chapter is to provide a broad overview of the discipline and a general introduction to the range of topics and issues that are represented in HFE. The following chapters will provide in-depth discussions of the theory and research in the field as well as specific applications. A particular strength of the collection is the breadth of topic areas that are covered as well as the diversity of the backgrounds of the authors.

Perhaps most important to recognize is that HFE permeates all aspects of society. Designing for human use should be a fundamental consideration for all design efforts—from a game designed for children to a medical device that can enhance the lives of older adults. Certainly, there are things in our everyday environments that are not well-designed—that lead to frustration, error, accidents, and inefficiencies. However, there are many things that are designed with careful consideration of HFE principles and considerations—their ease of use may go unnoticed because of the elegance and simplicity of the design that completely matches the goals, abilities, and limitations of the user. This is what the HFE field strives for.

AUTHOR NOTE

The authors were supported in part by the National Institutes of Health (National Institute on Aging) through: (1) Grant P01 AG17211, the Center for Research and Education on Aging and Technology Enhancement (CREATE; www.create-center.org), and (2) the Ruth L. Kirschstein National Research Service Award Institutional Research Training Grant (T32AG000175).

KEY TERMS

Environment. The aggregate of surrounding things, conditions, or influences; surroundings.

Human factors and ergonomics. The scientific discipline concerned with the understanding of interactions among humans and other elements of a system, and the profession that applies theory, principles, data, and methods to design to optimize human well-being and overall system performance.

Job. A piece of work, especially a specific task done as part of the routine of one's occupation or for an agreed price.

Organization. A social unit of people that is structured and managed to meet a need or to pursue collective goals.

Product. An article or substance that is manufactured or refined for sale.

System. An entity that exists to carry out some purpose—composed of humans, machines, and other things that work together (interact) to accomplish some goal that these same components could not produce independently.

Task. A definite piece of work assigned to, falling to, or expected of a person; duty.

REFERENCES

Beer, J. M., McBride, S. E., Mitzner, T. L., and Rogers, W. A. (2014). Understanding challenges in the front lines of home health care: A human-systems approach. *Applied Ergonomics*, 45, 1687–1699.

Chapanis, A. (1999). *The Chapanis chronicles: 50 years of human factors research, education, and design*. Santa Barbara, CA: Aegean Publishing.

Chapanis, A., Garner, W. R., and Morgan, C. T. (1949). *Applied experimental psychology: Human factors in engineering design*. New York: Wiley.

Charness, N., Demiris, G., and Krupinski, E. (2011). *Designing telehealth for an aging population: A human factors perspective*. Boca Raton, FL: CRC Press.

Chen, G., Davisy, D., Hauff, C., and Houben, G. J. (2016). Learning transfer: Does it take place in MOOCs? In *Proceedings of the Third (2016) ACM Conference on Learning@ Scale*. ACM.

Collet, C., Averty, P., and Dittmar, A. (2009). Autonomic nervous system and subjective ratings of strain in air-traffic control. *Applied ergonomics*, 40(1), 23–32.

Cooke, N. J. and Durso, F. (2007). *Stories of modern technology failures and cognitive engineering successes*. Boca Raton, FL: CRC Press.

Cooper, D. K. and Harwin, K. D. (1999). Robots in the classroom-tools for accessible education. *Assistive technology on the threshold of the new millennium*, 6, 448.

Dalipi, F., Imran, A. S., Idrizi, F., and Aliu, H. (2017). An Analysis of Learner Experience with MOOCs in Mobile and Desktop Learning Environment. In *Advances in Human Factors, Business Management, Training and Education* (pp. 393–402). Orlando, FL: Springer International Publishing.

Daneman, M. and Carpenter, P. A. (1980). Individual differences in working memory and reading. *Journal of Verbal Learning and Verbal Behavior*, 19(4), 450–466.

Datar, J. (1995). *What futures studies is, and is not*. Retrieved from www.futures.hawaii.edu /publications/futures-studies/WhatFSis1995.pdf.

Elliot, A. J., McGregor, H. A., and Gable, S. (1999). Achievement goals, study strategies, and exam performance: A mediational analysis. *Journal of Educational Psychology*, 91(3), 549.

"Environment" (n.d.). *Dictionary.com Unabridged*. Retrieved July 27, 2017 from Dictionary.com website www.dictionary.com/browse/environment.

Fitts, P. M. and Jones, R. E. (1947). Analysis of factors contributing to 460 pilot-error experiences in operating aircraft controls. *Aero Medical Laboratories, Wright-Patterson Air Force Base*.

Fonseca, B. and Chi, M. T. H. (2011). The self-explanation effect: A constructive learning activity. In: Mayer, R. and Alexander, P. (eds), *The Handbook of research on learning and instruction* (pp. 270–321). New York, USA: Routledge Press.

Gilbreth, F. B. and Gilbreth, L. M. (1919). *Applied motion study: A collection of papers on the efficient method to industrial preparedness*. New York: Sturgis & Walton Company.

Gilbreth, F. B. (1911). *Motion study: A method for increasing the efficiency of the workman*. New York: D. Van Nostrand Company.

Gorman, J. C., Cooke, N. J., Pederson, H. K., Connor, O. O., and DeJoode, J. A. (2005). Coordinated awareness of situation by teams (CAST): Measuring team situation awareness of a communication glitch. In *Proceedings of the Human Factors and Ergonomics Society* (pp. 274–277). Thousand Oaks, CA: SAGE Publications.

Heras-Saizarbitoria, I., Cilleruelo, E., and Allur, E. (2014). ISO 9001 and the Quality of Working Life: An Empirical Study in a Peripheral Service Industry to the Standard's Home Market. *Human Factors and Ergonomics in Manufacturing & Service Industries*, 24(4), 403–414.

Hood, D., Lemaignan, S., and Dillenbourg, P. (2015, March). When children teach a robot to write: An autonomous teachable humanoid which uses simulated handwriting. In *Proceedings of the Tenth Annual ACM/IEEE International Conference on Human-Robot Interaction* (pp. 83–90). ACM.

"Job" (n.d.). *Dictionary.com Unabridged*. Retrieved July 27, 2017 from Dictionary.com website www.dictionary.com/browse/job.

Khan, M. I. (2010). *Industrial Ergonomics, 1/e*. PHI Learning Pvt. Ltd.

Koutropoulos, A. and Zaharias, P. (2015). Down the Rabbit Hole: An initial typology of issues around the development of MOOCs. *Current Issues in Emerging eLearning*, 2(1), 4.

Langan-Fox, J., Canty, J., and Sankey, M. (2009). Human Factors Issues in Air Traffic Control under Free Flight. In *Proceedings of the 45th annual human factors and ergonomics society of Australia conference 2009* (pp. 37–44).

Liyanagunawardena, T. R., Adams, A. A., and Williams, S. A. (2013). MOOCs: A systematic study of the published literature 2008-2012. *The International Review of Research in Open and Distributed Learning*, 14(3), 202–227.

Margulieux, L. E., Guzdial, M., and Catrambone, R. (2012, September). Subgoal-labeled instructional material improves performance and transfer in learning to develop mobile applications. In *Proceedings of the ninth annual international conference on International computing education research* (pp. 71–78). ACM.

Marmaras, N., Poulakakis, G., and Papakostopoulos, V. (1999). Ergonomic design in ancient Greece. *Applied Ergonomics*, 30(4), 361–368.

Miller, D., Johns, M., Mok, B., Gowda, N., Sirkin, D., Lee, K., and Ju, W. (2016, September). Behavioral Measurement of Trust in Automation: The Trust Fall. In *Proceedings of the Human Factors and Ergonomics Society Annual Meeting* (Vol. 60, No. 1, pp. 1849–1853). Thousand Oaks, CA: SAGE Publications.

Mitzner, T. L., McBride, S. E., Barg-Walkow, L. H., and Rogers, W. A. (2013). Self-management of wellness and illness in an aging population. In D. G. Morrow (ed.), *Reviews of Human Factors and Ergonomics* (Vol. 8, pp. 278–333). Santa Monica, CA: HFES.

Mubin, O., Stevens, C. J., Shahid, S., Al Mahmud, A., and Dong, J. J. (2013). A review of the applicability of robots in education. *Journal of Technology in Education and Learning*, 1, 209–0015.

National Research Council (US). Committee on the Role of Human Factors in Home Health Care. (2011). *Health Care Comes Home: The Human Factors*. National Academies Press.

Newstead, S. E. (1992). A study of two "quick-and-easy" methods of assessing individual differences in student learning. *British Journal of Educational Psychology*, 62(3), 299–312.

Norman, D. A. (1988). *The Psychology of Everyday Things*. New York: Basic Books.

Norman, D. A. (1990). *The Design of Everyday Things*. New York: Doubleday.

"Organization." *BusinessDictionary.com*. Retrieved July 24, 2017, from BusinessDictionary.com website: www.businessdictionary.com/definition/organization.html.

Polson, P. G., Lewis, C., Rieman, J., and Wharton, C. (1992). Cognitive walkthroughs: a method for theory-based evaluation of user interfaces. *International Journal of Man-Machine Studies*, 36(5), 741–773.

"Product." *OxfordDictionaries.com*. Retrieved July 27, 2017, from OxfordDictionaries.com website: https://en.oxforddictionaries.com/definition/product.

Rogers, W. A., Mykityshyn, A. L., Campbell, R. H., and Fisk, A. D. (2001). Analysis of a "simple" medical device. *Ergonomics in Design*, 9, 6–14.

Sanders, M. S. and McCormick, E. J. (1993). *Human Factors in engineering and design*. McGraw-Hill, New York.

Salvendy, G. (2012). *Handbook of human factors and ergonomics*. Hoboken, NJ: John Wiley & Sons.

Sardar, Z. (2010). The namesake: Futures; future studies; futurology; futuristic; foresight – what's in a name? *Futures*, 42, 177–184.

Smarr, C.-A., Mitzner, T. L., Beer, J. M., Prakash, A. Chen, T. L., Kemp, C. C., and Rogers, W. A. (2014). Domestic robots for older adults: Attitudes, preferences, and potential. *International Journal of Social Robotics*, 6(2), 229–247.

Stanton, N. A., Hedge, A., Brookhuis, K., Salas, E., and Hendrick, H. W. (eds) (2004). *Handbook of Human Factors and Ergonomics Methods*. Boca Raton, FL: CRC Press.

Stronge, A. J., Rogers, W. A., and Fisk, A. D. (2007). Human factors considerations in implementing telemedicine systems to accommodate older adults. *Journal of Telemedicine and Telecare*, 13, 1–3.

"Task" (n.d.). *Dictionary.com Unabridged*. Retrieved July 27, 2017 from Dictionary.com website: www.dictionary.com/browse/task.

Taylor, F. W. (1914). *The Principles of Scientific Management*. New York: Harper.

van Veelen, M. A. (2004). Ergonomics in minimally invasive surgery. *Minimally Invasive Therapy & Allied Technology*, 13(3), 131–132.

Turekova, I., Kozik, T., Bagalova, T., and Neovesky, J. (2014). Workplace lighting as an element influencing the working process. *Advances in Physical Ergonomics and Human Factors: art I*, 14, 34.

Williams, J. J. (2013, June). Improving learning in MOOCs with cognitive science. In *AIED 2013 Workshops Proceedings Volume* (p. 49).

ADDITIONAL RECOMMENDED READINGS

Bailey, R. W. (1982). *Human performance engineering: A guide for system designers*. Upper Saddle River, NJ: Prentice Hall Professional Technical Reference.

Fisk, A. D., Rogers, W. A., Charness, N., Czaja, S. J., and Sharit, J. (2009). *Designing for Older Adults: Principles and Creative Human Factors Approaches* (2nd ed.). Boca Raton, FL: CRC Press.

Norman, D. A. (2013). *The Design of Everyday Things: Revised and Expanded Edition*. New York: Basic books.

Wickens, C. D., Hollands, J. G., Banbury, S., and Parasuraman, R. (2015). *Engineering Psychology & Human Performance*. Upper Saddle River, NJ: Prentice Hall.

Case Study 1
The ErgoWELL Program

Samantha J. Horseman and Steven Seay

INTRODUCTION

The ErgoWELL program is aligned with Saudi Aramco's corporate objectives of intensifying their focus on safety and preparing their workforce for the future. Systematic ergonomics is an injury-prevention process that is regarded as best practice within the industry, analyzing the human factor* interface systems between the end-user, with task management, equipment, workspace design, the environment, and the organization at large (see Figure CS1.1). It is well documented that one size does not fit all. Therefore, to meet the requirements of 90–95% of end-users that fall within the outliers of the normative curve, this initiative provides solutions for injury prevention and injury management. ErgoWELL is founded on a population health management and health and productivity management (workforce and workplace), and systematic ergonomic framework (environment), model. Placing human factors and ergonomics at its core, ErgoWELL is able to address the underlying root causes and lifestyle factors that contribute to injuries on the job with successful outcomes.

Similar to in the West, the GCC is also concerned about the growing cost of injury management, absenteeism, and loss of productivity due to the increasing rate of preventable musculoskeletal (MSK) injuries in the workplace. The bottom line is that prevention is better than a cure. Globally, the trend in ergonomics now is to also invest in preventive strategies that maintain the health and well-being of the workforce (Horseman et al., 2016). Research indicates that this approach is more efficient and cost-effective (Horseman et al., 2013). Aligning with the company's safety culture provides opportunities to enhance awareness across multiple stakeholders. Furthermore, such a program enhances self-accountability and ensure that safety is a number-one priority.

* Human factor is defined as within the context of human factors ergonomics: "the scientific discipline concerned with the understanding of interactions among humans and other elements of a system, and the profession that applies theory, principles, data and methods to design in order to optimize human well-being and overall system performance" (International Ergonomics Association).

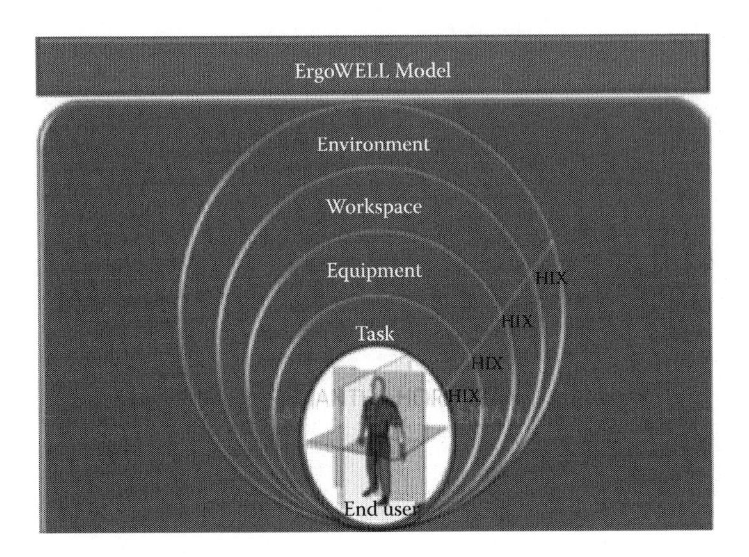

FIGURE CS1.1 ErgoWELL model is designing from the end user as the inner (core) system. HfX = human factor across each system.

HUMAN FACTORS AND ERGONOMICS IMPLICATIONS

Human factors and ergonomics (HFE) is a science. It seeks to fit workplace conditions and job demands to the capabilities and limitations of the workforce. The goal is to achieve the optimal fit between the work and worker so that the safety and health of the workforce is maintained while productivity is optimized. Three core principles are essential for an effective ErgoWELL program. These include:

1. **The design of work and equipment to suit the 90–95 percentile range of workers** through understanding the science of demographics (population health) and anthropometrics.
2. **Allowing "people" to do the work (i.e., cognitive processing, analytical tasks) that they do well**, while "machines" do what they do well (i.e., heavy, repetitive, monotonous tasks).
3. **Using a participatory team approach** including engineer, designer, project manager, safety officer, end-user, etc.

In applying these principles, the ErgoWELL methodology consists of four key elements (Figure CS1.2).

- Phase 1: a survey and awareness workshop is conducted to evaluate and measure current ergonomic issues and collate significant data points.
- Phase 2: an observational walkthrough of the workspace and risk analysis is completed.

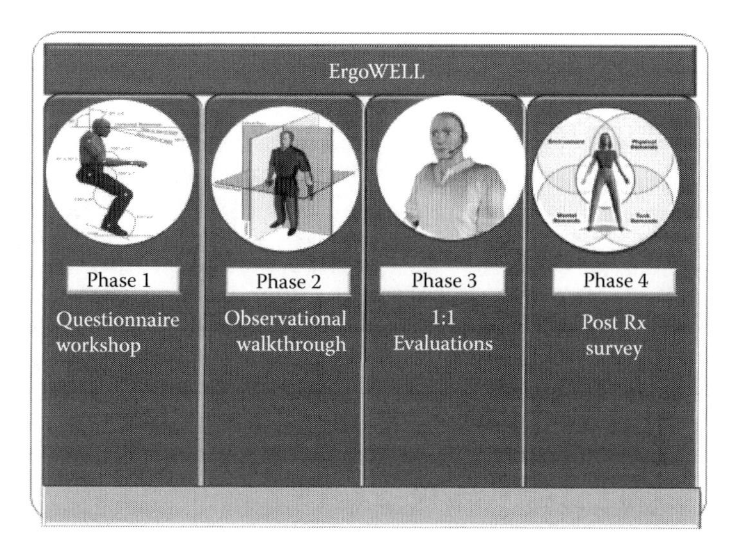

FIGURE CS1.2 The ErgoWELL Program Deliverables.

- Phase 3: after identifying low-, medium-, and high-risk cases, specific individual evaluations are completed with a four-points-of-contact (POC)* risk analysis evaluation. In specialized cases, the rapid upper-limb assessment (RULA) and rapid entire-body assessment (REBA) are deployed to assist in quantifying high risk. Ergonomic modifications and preventive practices are reinforced. High risk employees are also fitted with a wearable technology – known as 'Lumo'†.
- *Phase 4*: After 1 year the survey is conducted again to measure the effectiveness of the program. Also another reevaluation of baseline measures and outcomes of the program are conducted.

GOAL & APPLICATION

The objectives of ErgoWELL is to deploy a systematic participatory injury prevention program that has the 'human factor' interfaced with work task factors, workspace/design factors, environment factors, and organizational dimensions. In addition to integrating health and well-being that addresses the underlying factors of Musculoskeletal (MSK) injuries. Factors, such as lifestyle behaviors: smoking, obesity, poor nutrition, physical inactivity, and stress are major contributors. Thus, addressing the root cause of human factors that contribute towards injuries in the workplace.

The challenges that emerged were making sure that the Departments deploying the ErgoWELL program committed to aligning their safety officers to the program

* POC: A point of contact is a term referenced where the human interfaces with another system. E.g. where the eyes to connect to display screen equipment (DSE), and the hands connect with an input device (i.e., lever, mouse, keyboard, etc).

† Lumo – biofeedback wearable feedback device https://www.lumobodytech.com/

for the entire year. This was a critical success factor in both the initial and post survey. Departments that had committed safety professionals that worked with the ErgoWELL team also had successful outcomes. The outcomes of the ErgoWELL are compelling and demonstrates the success of adopting a systematic participatory methodology. In a one year follow up (n = 1145) there was a significant reduction in work related MSK injuries. In addition there was a reduction in medical leave taken due to MSK, initially there was a total of 22% absenteeism rate that was reduced to zero. These results are highly significant ($p < 0.001$). An impressive outcome was the culture shift towards self-accountability behaviors and investing in preventive practices throughout the day. Pre-program investing in preventive practices and healthy lifestyle factors was 40%, following the ErgoWELL intervention this improved to 96% adoption of daily preventive practices. The lessons learned, for a successful launch and deployment include—multiple stakeholders need to commit both time and resources with the ErgoWELL team. High quality programs were successful when there was a dynamic and active leader championing the program. Additionally, developing and sustaining a culture of self-accountability, safety, and well- being takes time. Therefore, applying a robust continuous quality improvement process is a critical success factor.

REFERENCES

Horseman, S.J, Dhubaib, K. T., Sall, M., Antony, A., Burgess, P., Hayman, S., and Birrer, R. (2013). Health & Productivity Management Strategy Concept Paper; A Proposed Equation for Presenteeism. *Journal of Health & Productivity*. 7, 1,

Horseman, S.J., Sullivan, S., Mattson, B.W., and Seay, S.A. (2016). The Value Proposition for Healthy Human Capital. *International Journal of Health & Productivity*. 1, April Issue.

2 Macroergonomics
An Overview

Jussi Kantola and Ari Sivula

CONTENTS

INTRODUCTION

We live in a society in which employees' well-being has a direct impact on a company's results (de Weerd et al., 2014). According to the International Labor Organization (ILO), 2.34 million people die each year from work-related accidents or diseases, and 317 million suffer from work-related injuries (ILO, 2012). The ILO also estimates that four percent of the world's gross domestic product is lost due to accidents and work-related diseases (ILO, 2012). The main cost components are medical costs, productivity losses, and human costs (Lebeau and Duguay, 2014). In short, deaths are counted in millions of cases, accidents and illnesses in hundreds of millions of cases, and financial losses in billions of USD/EUR every year. It is clear that such vast problems cannot be solved easily. Holistically, it is important to view well-being issues in the workplace from both detailed and systemic perspectives. Together, these views can provide more sustainable solutions than local safety fixes.

"Lebeau and Duguay, 2014" be changed to "Lebeau and Duguay, 2013". Please check if correct.

Ergonomics (or the study of human factors) is the scientific discipline concerned with the understanding of interactions between humans and the other elements of a system, and the profession that applies theory, principles, data, and methods to design, in order to optimize human well-being and overall system performance (International Ergonomics Association, 2016). Good ergonomics is also good economics (Hendricks, 1996; de Weerd et al., 2014), since employee productivity increases with well-designed methods, tools, and processes. Results are profitable when working conditions are appropriately designed, analyzed, and managed.

Macroergonomics, in turn, is the field of science that concentrates on designing overall work systems (WS) by providing the knowledge and methods necessary for the improvement of work systems and, thus, developing the effectiveness and performance of companies (Hendricks, 1996). The process typically involves designing, analyzing, developing, and improving work systems (Hendrick and Kleiner, 2009). Macroergonomics is highly relevant in terms of solving the serious occupational problems reported by the ILO. Macro-level problems cannot be solved by micro-level solutions, but micro-level problems can be solved with macro-level solutions. We can also consider the link to double-loop learning, proposed by Argyris and Schön (1978). According to them, double-loop learning means resolving incompatible organizational norms by setting new priorities and the weighing of norms, or by restricting norms altogether. In other words, accessing the system level facilitates the finding of more permanent solutions to local and repetitive occupational problems. That is why macroergonomics is so important when it comes to attempting to improve employee well-being and the profitability of businesses. In the next section, the fundamentals of macroergonomics are introduced.

FUNDAMENTALS

In the 1980s, it was noticed and reported that it is important for organizational design to begin with an understanding of the user's role in the overall system performance, and for systems to exist to serve their users (Kleiner, 2008). A sub-discipline of human factors focusing on organizational design and management factors associated with systems (Hendrick, 1991) was born. Soon, the new discipline was named macroergonomics by Hendrick (1991). In contrast with (micro)ergonomics, which deals with human-machine, human-task, human-interface, usability, and safety-type issues in the workplace (Putkonen, 2010), macroergonomics is a system-level approach to designing the interaction between humans and technology and to understanding how activities and processes are organized to produce products and services for customers. This is called work system design, and with the help of this new system-level approach clear performance gains were reported. Macroergonomics aims at a holistic understanding of how work is designed. The concept of socio-technical systems (STS) (Trist and Bamworth, 1951) in organizational development provides a theoretical foundation for macroergonomics. STS explores how people and technology interact in the workplace (Walker et al., 2008; Majchrzak and Borys, 2001). The more complex work systems become, the more important it is to understand the interaction between humans and technology. The main concepts of STS are humans, tasks, structure, technology (hardware and software), and processes, as well as regulations, policies, and culture (c.f. Mohr and Amelsworth, 2016). According to macroergonomics, socio-technical work systems consist of five elements, namely technological, personnel, organization, external environment, and internal environment:

- **Technology** is needed to carry out the goals set for the work system. With the help of technology, tasks and problems can be solved. Some examples of technology are AIT (advanced information technology), ICT (information and communication technology), and advanced manufacturing technologies.
- **People** (personnel) occupy a core role in work systems, since only people can achieve goals with the help of technology. Important aspects regarding people include competence, demographics, psychosocial factors, and teams.
- **Organizational** design specifies how technology and people are organized and managed together in work systems. Contemporary areas include lean processes and digitalization.
- **The external environment** is something that cannot be influenced from inside work systems. However, the external world can be analyzed, for example, with PEST (political, economic, social, and technological) or PESTEL (PEST plus environmental and legal) frameworks.
- **The internal environment** includes considerations such as the psychosocial and physical environment and cultures that are not part of the organizational system design.

Different human factors and ergonomic theories and methods can be applied to various levels of the macroergonomic design process. A macroergonomic system may include several interconnected work systems. As Figure 2.1 demonstrates, the right-hand model illustrates a work system that typically consists of people (who are doing the work), organization (how work is organized), and technology (how work is done). The left-hand model illustrates macroergonomics as a holistic approach with several

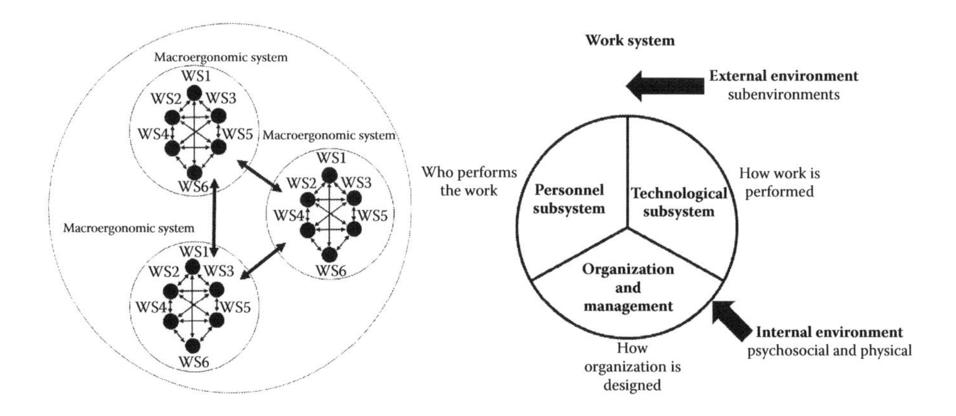

FIGURE 2.1 A macroergonomic system and a work system (adapted from Kleiner, 2008).

work systems. Thus, the design process occurs from the bottom up; microergonomic design → work system design → macroergonomics.

Macroergonomics is a top-down socio-technical approach to work system design. The design of work is done from the top down, in such a way that macroergonomic design is considered first, work system design second, and microergonomic design last. Therefore, the design is carried through from macroergonomic system-level design to the microergonomic design of human-technology interaction and interfaces (Hendrick, 2007). Interfaces are areas in which interactions between humans and machines and software occur. Interfaces are typically designed to support goal-oriented human being and activity. The next section describes known macroergonomic methods and models.

METHODS

We are currently living at a time where work systems are becoming increasingly complex due to the development of technology, globalization, digitalization, and urbanization. The development of socio-technical systems started prior to World War II, and the term macroergonomics was coined in the late 1970s, as Figure 2.2 illustrates.

Waterson et al. (2015) describe the current decade as the age of complex socio-technical systems, where complexity is part of macroergonomic work systems. According to several macroergonomic field and laboratory research cases, significant performance improvements have been achieved by analyzing and adjusting a variety of factors or work system designs (Kleiner, 2008). Typically, work systems are not

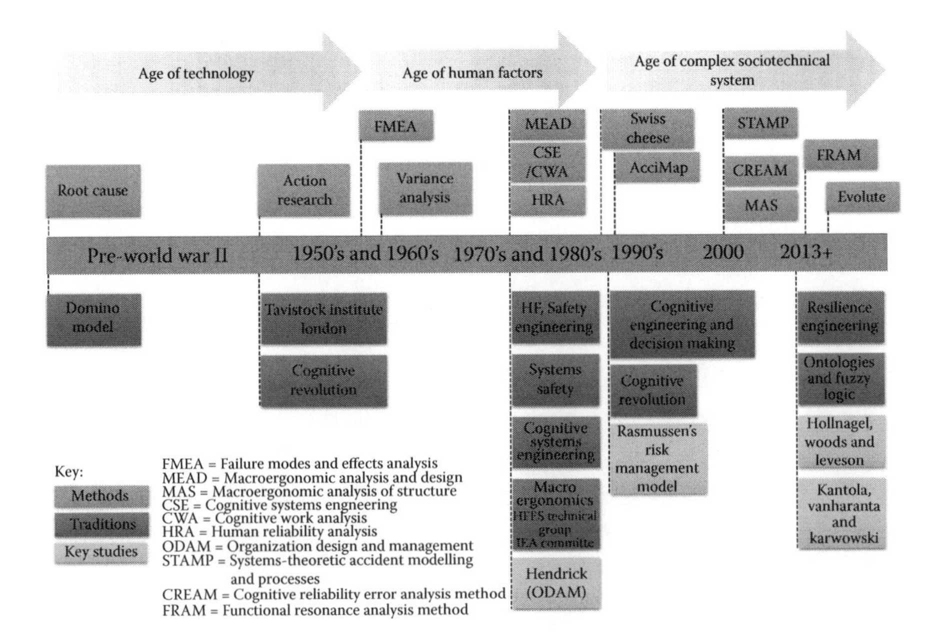

FIGURE 2.2 A timeline of the development of methods for socio-technical systems and safety (adapted from Waterson et al., 2015).

"automatically" designed in an optimal or compatible manner, and therefore TOP (technology-organization-people) adjustments within work systems are required. Additionally, external environment-TOP-internal environmental adjustments, as well as inter-work system adjustments, are needed in order to optimize the whole system. The goal is to improve the performance of work systems by analyzing, adjusting, and aligning the design. There is no one way of doing this, but instead many different methods that can be used, depending on the need. In this section, some known methods, as well as one newer ontology-based method, are described, namely MEAD, MAS, FRAM, and Evolute. The methods were selected based on their different approaches. Macroergonomic methods need to be holistically understood in the context of the management of organizations and their resources. Table 2.1 summarizes a generic description of each method and their potential strengths and weaknesses, as well as potential application areas. These methods are described in the next sections.

Macroergonomic Analysis and Design

Macroergonomic analysis and design (MEAD) begins with an analysis of environmental and organizational subsystems, which can be categorized into production, processes, functions, and tasks (Kleiner, 2002). It is an iterative ten-step process to improve work system design (Kleiner, 2006):

1. scanning the environmental and organizational design subsystem
2. defining the type of production system and setting performance expectations
3. defining unit operations and work process
4. identifying variances
5. creating the variance matrix
6. creating the key variance control table and role network
7. performing function allocation and joint design
8. understanding perceptions of roles and responsibilities
9. designing or redesigning support subsystems and interfaces
10. implementing, iterating, and improving

The MEAD methodology offers several points of view of an organization. The first step in the MEAD method is to scan organizational, internal, and external subsystems. It is useful for determining the nature and extent of any variance between the organization's public and private identities during the scanning; therefore, it allows for an assessment of formal statements regarding the company's mission, vision, and principles (Kleiner, 2006). Defining the type of production system helps to determine optimal levels of complexity, centralization, and formalization (Kleiner, 2006). In the third step, unit operations are defined as groupings of conversion steps that, together, form a complete piece of work and that are bounded from other steps by territorial, technological, or temporal boundaries (Kleiner, 2006). In steps four and five, key variances are variances that significantly impact performance criteria, or that may interact with other variances to exert a compound effect (Kleiner, 2006). The sixth step is performed to learn how existing variances are controlled, and whether the personnel responsible for variance control require additional intervention.

TABLE 2.1
Macroergonomic Methods Use Different Approaches to Analyze Human-System Integration (c.f. Waterson et al., 2015)

Name	Approach	Strengths	Challenges	Application Areas
MEAD - Macroergonomic Analysis and Design (Kleiner 2008)	Ten-step iterative process, integrates STS theory and ergonomics into the analysis and design of the micro-ergonomic and work system's environmental and organizational characteristics and subsystem interfaces (Kleiner, 2003)	Traditional and well-known method and easy to adapt because of its clear ten-step approach	Application of the ten steps may be a challenge in some cases	Complex systems, healthcare, construction (Kleiner, 2006, 2008)
MAS: macroergonomic analysis of structure (Hendrick, 2005)	Seven-step process studying the structure of the compatibility of work systems, according to their socio-technical characteristics	Empirically developed analytical model	Managers need to have a holistic view of the socio-technical characteristics of a company to provide a comprehensive analysis	Generic model, which can be adapted widely to any industry to provide a holistic view of its socio-technical characteristics
FRAM: functional resonance analysis method (Hollnagel, 2012)	Four-step process based on analyzing central functions according to six dimensions (Hollnagel, 2012)	Provides an overall understanding of how STS works (Hollnagel, 2016)	The analysis can be difficult and time-consuming (Hollnagel, 2016)	Generic to any field, including aviation, the energy sector, transportation (Hollnagel, 2016)
Evolute approach (Kantola, 2015)	Four-step process to develop organizational resources in a co-evolutionary manner (Kantola, 2015)	Research-based tools are comprehensive and fast to use	Development of ontology is slow	Generic to any field or object

The provisional allocations made to humans, machines, both, or neither in steps seven and eight provide important information as to how workers perceive their roles (Kleiner, 2006). The goal of step nine is to determine the extent to which a given support system impacts the socio-technical production system, the nature of the variance, the extent to which the variance is controlled, and the extent to which tasks should be taken into account when redesigning operating roles in the supporting subsystem units (Kleiner, 2006). In the final step, changes are implemented by the management of an organization because, commonly, the macroergonomic team does not have enough authority to implement the required changes in the organization (Kleiner, 2006).

Macroergonomic Analysis of Structure

Macroergonomic analysis of structure (MAS) uses predefined analytical models for technological, personnel, and environmental subsystems and attempts to structure the work system accordingly (Haro and Kleiner, 2008). The MAS analysis can be used to correct inconsistencies in an organization by comparing the results of the analysis with the existing organizational structure (Hendrick, 2005). The external environment factors of MAS models are divided into five categories that have an effect on an organization. Hendrick (2005) highlights that these factors belong to socioeconomic, educational, political, cultural, and legal categories. The socioeconomic category refers to stability, the nature of the competition, and the availability of materials. Education refers to educating human resources. The political category refers, for instance, to government attitudes towards companies and controlling market prices. The cultural category refers, for example, to social status, values, and attitudes towards work. The legal category refers, for instance, to legal controls and restrictions. These subsystems are assigned weights on a scale of 1 (low) to 5 (high) (Haro and Kleiner, 2008). After assigning the weights, the results are combined to get an overview of the current work system situation and the results of the MAS analysis.

Functional Resonance Analysis Method

The functional resonance analysis method (FRAM) provides an analysis of the variability of individual, technical, and organizational performance (Hollnagel et al., 2008). Moreover, FRAM provides a way of describing outcomes using the idea of resonance arising from the variability of everyday performance (Hollnagel, 2012). FRAM is used for the investigation of fault finding and the improvement of work systems (Hollnagel, 2013). According to Hollnagel (2012), functions and activities of FRAM are analyzed from six viewpoints, which are illustrated in Figure 2.3.

FRAM points to a single function/activity of an organization and can be combined to create a map of several activities or functions. Time describes the point at which the activity takes place and control highlights the regulations of the function or activity. Control is the supervision or adjustment of a function to get a desirable result or output. Input and output provide information on what comes in and out of the activity or function. Precondition sheds light on the state of the function or

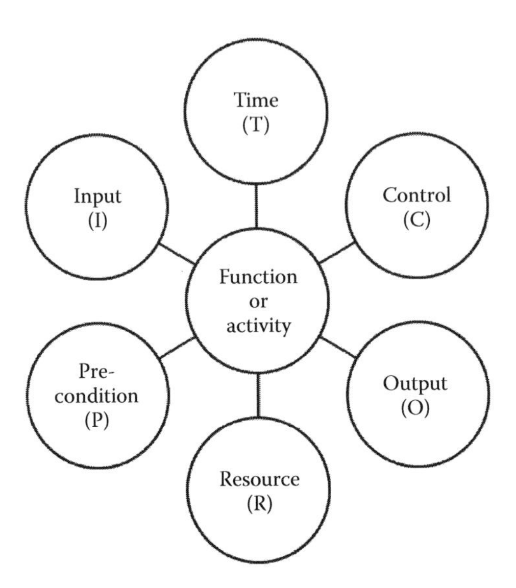

FIGURE 2.3 The FRAM approach (adapted from Hollnagel, 2012).

activity, and the tasks which are required to be fulfilled before the function or activity can initiate. Resources are the tools required for the function or activity to work properly. Thus, FRAM provides a detailed view of a specific function or activity of an organization. FRAM can be used via FRAM Model Visualizer (FMV) software. For more information on FRAM, see Hollnagel (2016).

The Evolute Approach

The evolute approach has partly evolved from the computer integrated manufacturing, organizations and people (CIMOP) method (Karwowski and Kantola, 2005). The CIMOP method was developed for the evaluation of system design quality from the points of view of technology integration, organization, information systems, and people subsystems. The evolute approach is based on four propositions: (1) model, (2) involve, (3) compute and (4) manage (Kantola, 2015).

1. Ontology engineering is applied to **model** the content and structure of organizational resources.
2. Knowledge input is **involved** and collected online from individuals and stakeholders.
3. The fuzzy logic-based evolute system **computes** the meaning of organizational resources, both today and in the future.
4. The computed meaning is examined and **managed** in current and future contexts for decision-making.

Ontology refers to the specification of the conceptualization of a domain (Gruber, 1993). Fuzzy logic is the precise logic of imprecise things (Zadeh, 1965, 1973). Fuzzy logic allows reasoning using natural language and fuzzy rules. The evolute system

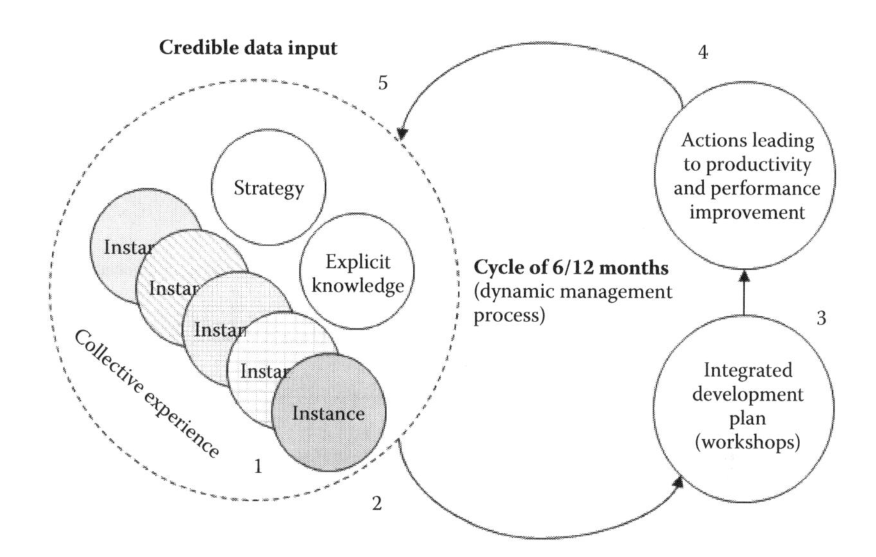

FIGURE 2.4 The evolute approach as a management process.

supports online domain ontology-based applications, which target groups can use with natural language (Kantola, 2015). The evolute system computes and visualizes the perceived current reality and the future vision of objects—for example, organizational subsystems. The evolute system can therefore be used to design and develop work system components (i.e., people, organizations, technology, cultures, and processes). Figure 2.4 illustrates the process of the evolute approach.

The process of human-technology-organization co-evolution is a continuous cyclic process, as Figure 2.4 represents. The steps are: (1) collecting instances within a set time window; (2) validating the perceived impact of previous development plans and fitting together instances, focus areas coming from the an organization's strategy and local conditions, and explicit knowledge about the management objects being developed; (3) making targeted development plans for the objects in wor shops; (4) taking action according to the plan; and (5) returning to step 1 and starting a new development cycle in six or 12 months. According to the process, all objects in an organization can be managed in a similar way and, therefore, the process is generic (Kantola, 2015). For more information on the evolute approach, see Evolute LLC (2016) and Kantola (2015).

APPLICATIONS

Construction Industries

Haro and Kleiner (2008) studied the adaption of MEAD and MAS methodologies in the construction industry. Together, MAS and MEAD represent the formalization of staple methods in macroergonomics and can be used to organize existing tools and methods, such as those that exist in system safety, and help to differentiate macroergonomics from other approaches (Haro and Kleiner, 2008). In the construction industry, a lot of deadly injuries happen due to falls, electrocutions, being struck by objects, or being caught in or between objects (OSHA, 2016). Such accidents are frequently repeated, making them systematic accidents (i.e., something that results from the design of the work system). By definition, macroergonomics is a system-level approach to designing how humans and technology work together and how activities and processes are organized. The design is realized on human-technology-task interfaces. Since the construction industry is a high-energy industry in terms of kinetic, potential, thermal, chemical, and electrical energy, the interfaces should be very carefully designed. Thus, the construction industry requires systematic change to decrease injuries and, as a result, expenses due to injuries. Haro and Klainer (2008) provide a MEAD input table, which can be utilized by the construction industry to adapt the MEAD method. This procedure is illustrated in Table 2.2.

The MEAD analysis can be utilized on multiple levels; these levels could be, for example, organizational, regulatory, national, and international. The MAS analysis extends the value of the MEAD analysis. Thus, the analysis is deeper when both methods are combined. Kleiner and Smith-Jackson (2005) have implemented an evaluation of the construction industry (socio-technical system) and recognized these components:

- technical subsystem (manner in which work is performed)
- heavy machinery (equipment, power tools, hand tools, methods, and procedures)
- personnel subsystem (sociocultural and socioeconomic characteristics of the construction workers, including selection and training)
- external environment (political, economic, technological, educational, and cultural forces)
- internal environment (physical and cultural job site)
- organizational and management structure (formal or informal)

These components form part of the factors presented by Hendrick (2005) in his research and were presented in the fundamental section.

Air Traffic Management System

The FRAM methodology can be used to analyze incidents and/or raise the everyday performance of the organization. Carvalho (2011) utilized FRAM to analyze the mid-air collision that happened on September 29, 2006 at 16:56 Brazilian time on a clear afternoon between flight GLO1907 (a commercial aircraft Boeing 737-800) and flight N600XL (an executive jet EMBRAERE-145), in order to investigate the key resilience characteristics of the air traffic management (ATM) system. The functions of the human-centered ATM system are illustrated in Figure 2.5.

The human-centered ATM system presented is based on the FRAM framework in Figure 2.5. The ATM system consists of two independent FRAM modules, which include the pilot flying and the air traffic control (ATC) monitoring. The FRAM methodology describes these as activities or functions. The pilots should monitor the aircraft's situation to fly according to clearances, plans, and airspace rules, and

TABLE 2.2
The MEAD Methodology in the Construction Industry (Haro and Klainer, 2008)

MEAD Step	Input Needed in the Construction Industry
Scanning analysis	• company documented safety information and employee perceptions (climate, culture) • suppliers, contractors, sub-contractors, inputs, processes, outputs, customers' users, feedback, internal controls, and outcomes (specific to the construction project) • environmental expectation of the system (regulations) and the system's expectation of the environment (regulatory support). Accounts for differences in metrics based on geography • levels of organizational complexity, centralization, and formalization (specific to construction project) with respect to the company, contractors, and sub-contractors
System type and performance analysis	• task descriptions and types • checkpoints • levels of organizational complexity, centralization, and formalization • human–machine information
Technical work process and unit operations	• trade schedule, shift work, unions • breakdown by functions and tasks
Variance data	• identify safety variances at the process and task level • differentiate between input and throughput variances
Construct variance matrix	• establish relationship between variances • identify key variances
Variance control table and role network	• construct key variance control table to include hazards, risks • construct role network to include owner, contractors, sub-contractors, etc. • evaluate effectiveness • specify organizational design dimensions
Function allocation and joint design	• perform functions with results from a system safety analysis tool • design technological, personnel, and organizational change to be managed with system safety tool recommendation/gap checklist
Roles and responsibilities	• evaluate role and responsibility perception of safety from identified stakeholders • provide safety training support
Design/redesign	• based on joint analysis
Implement, iterate, and improve	• based on follow-up analysis; may generate a reevaluation of the system

inform the controller of the flight's situation to close the control loop (Carvalho, 2011). The pilot, therefore, gives flight information to ATC monitoring and gets feedback on clearances and instructions; this is output in one function and input in another (O and I). The pilot and ATC monitoring take decisions based on the flight situation. Control is implemented through co-operation between the pilot and ATC monitoring, which includes cognitive functions, active monitoring, and risk

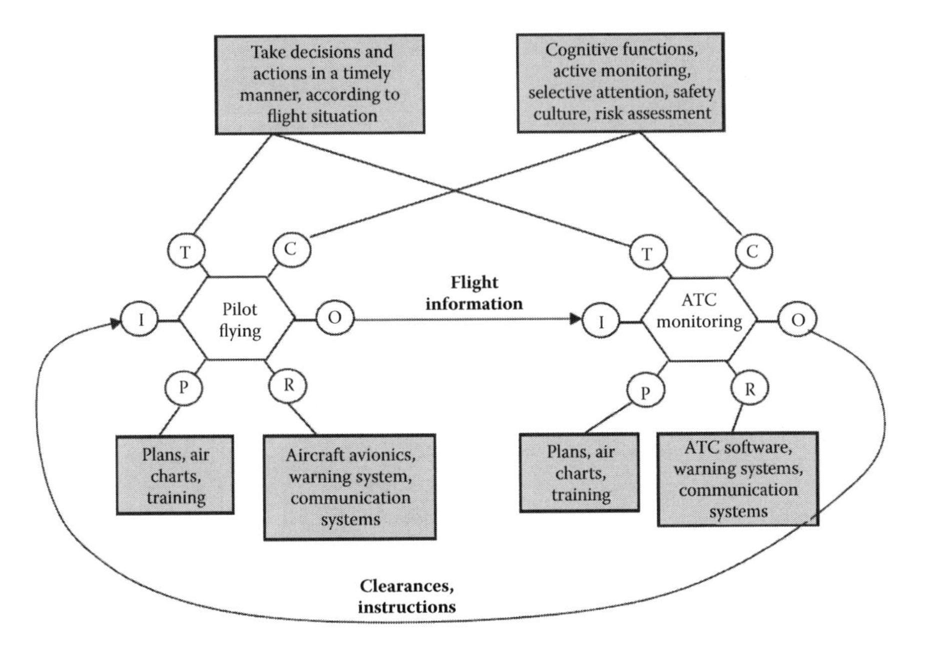

FIGURE 2.5 An example of the FRAM diagram: human-centered ATM system (Carvalho, 2011).

assessment. The FRAM methodology identifies these as time and control (T and C) aspects of the function. The pilot preconditions (P) include plans, air charts, and training. The pilot's resources (R) are aircraft avionics and the warning and communication systems. On the other hand, ATC monitoring also has preconditions (P), which include plans, air charts, and training, but from a different perspective. ATC monitoring resources (R) are ATC software, warning systems, and communication systems.

Organizational Resources

The evolute system provides a way of computing current and anticipated future states and an organization's current and future resources using ontologies and fuzzy logic (Kantola, 2015). The ontology repository provides conceptual frameworks for the resources that are under development. Organizational resource ontologies can be created through research (doctoral research, for example) or by domain experts. Pursoid 2.0 is a tool on the evolute system developed by Vanharanta (2015), which provides a systematic approach to analyzing and developing an organization's innovation resources. The ontology of Pursoid 2.0 is presented in Figure 2.6.

Pursoid 2.0 (Vanharanta, 2015) explicitly specifies concepts of innovation competence. Using the Pursoid 2.0 ontology on the evolute system provides managers with a way of visualizing and analyzing the state of an abstract innovation competence object within an organization. When all stakeholders provide their view of the object at hand, a comprehensive picture of an abstract object becomes available. Figure 2.7

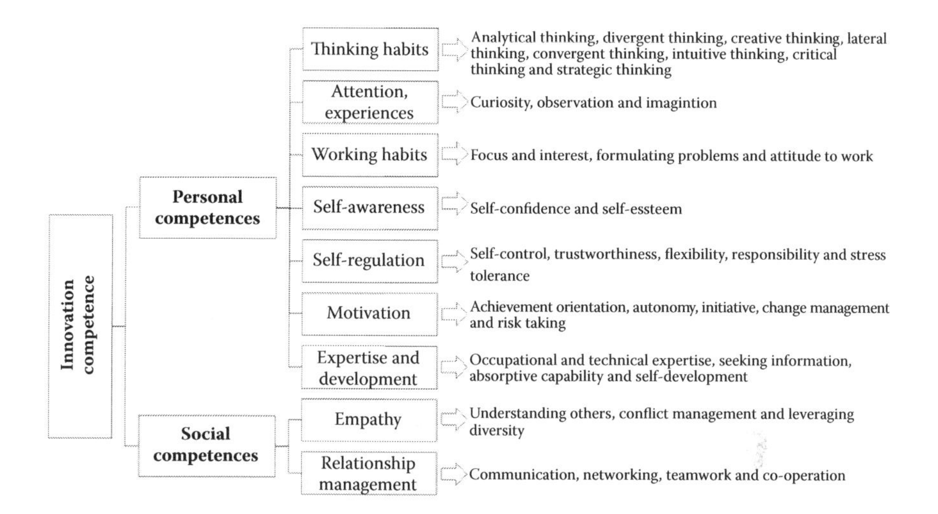

FIGURE 2.6 Innovation competence of human resources (Pursoid 2.0 ontology) (Vanharanta, 2015).

illustrates what this visualization looks like. On the left-hand side, the concepts of the ontology are shown, and the right-hand side shows the computed meaning of the concept. The darker upper bars represent the current state, while the lighter lower bars represent the anticipated future. The gap between the future and the present is called creative tension (Senge, 2006). Creative tension shows the need to develop a certain concept from its current state.

In the context of macroergonomics, objects that can be explicitly specified as ontologies and used for work system analyses according to the evolute approach, like in Figure 2.7, can be technology, organization, people, or the whole work system. The evolute approach to macroergonomics reveals those areas of the object in need of further attention and development; for example, better integration and interaction. The development action to be taken is decided within an organization according to the organization's strategy. The cycle is repeated every six or 12 months.

FUTURE TRENDS

Digitization

The digitization of work is providing a larger and shared information space, allowing leaps forward in the productive forces available to the organization (Boes, 2015). Cloud working and crowd sourcing, as well as digital platforms for work and services (Baums, 2015), will disruptively change what is happening inside and outside the organization. Therefore, digitization will have an effect on work system design and will provide new ways of applying existing methods, as well as developing existing methods further and developing completely new methods in macroergonomics. As these changes have taken place, much research has been conducted in the field of macroergonomics, and further continuous research is required due to the broad

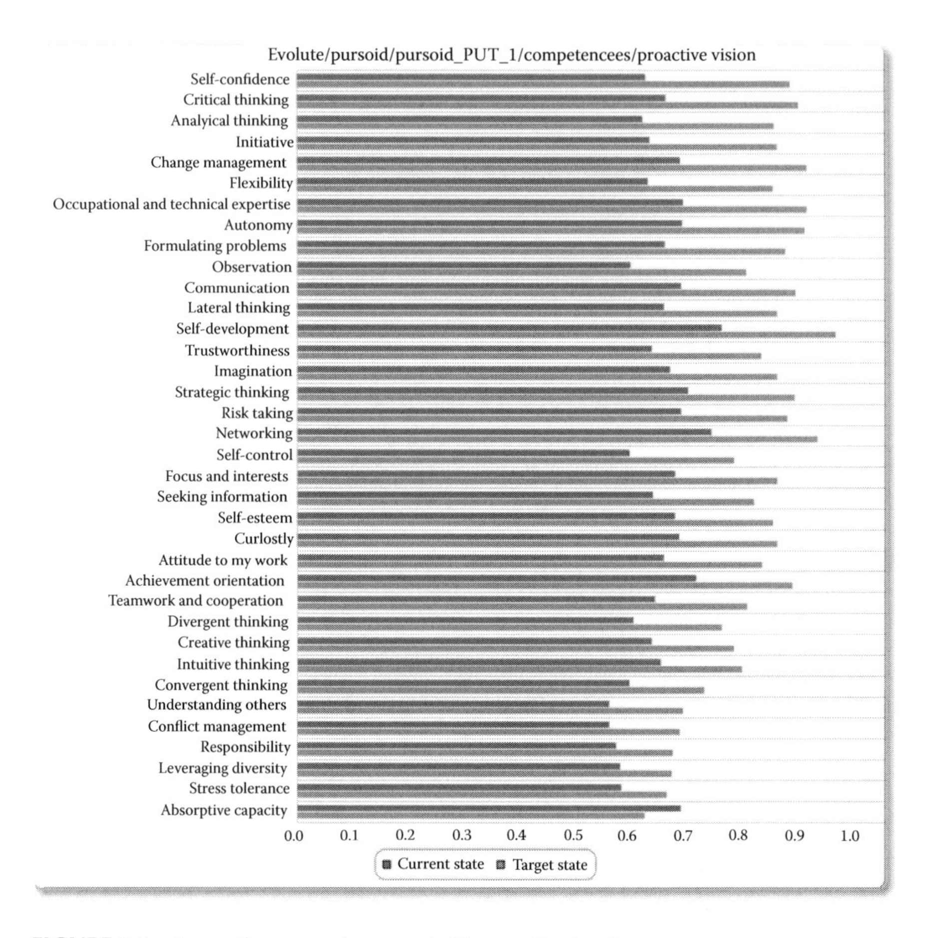

FIGURE 2.7 Innovation competence sorted by creative tension.

scope of the industries and trends that are changing how work will be designed and performed in the future. Existing methods will require further application and validation in these new digitized industries.

Evolving Work Roles

Due to digitization, platformization, automation, robotization, and AI, work roles are evolving (i.e., content of work, ways of working, and how work is organized between humans and technology). With these evolving shifts, major trends will impact work roles in terms of the physical space in which work will be carried out, as well as cognitive and physical ways of performing work. New work roles will provide opportunities for new innovative ways of managing and leading organizational resources to create value. Increased servitization, service-based logic (Vargo and Lusch, 2008), and value-creating networks are providing frameworks for how work systems and service systems will expand over the traditional border of organizations and how they will be designed in the future.

System of Systems Perspective

The study of systems of systems will be important as people learn to better understand causality and the behavior of complex systems (Marek et al., 2014), such as value-creating networks. In complex work and service systems that interact with their environment, many system elements are conceptual in nature (Kantola, 2015). Such conceptual elements reside in peoples' minds and cannot be directly observed or measured with variables, unlike technical system elements. Therefore, understanding contingency and the situational aspects of work system design and behavior will become increasingly important.

These future-oriented viewpoints indicate that the importance of considering factors beyond the human, machine, or human-machine interface will be growing (Dul et al., 2012; Wilson, 2014). Thus, methods that are able to consider conceptual elements in work system design beyond traditional borders and measurable work system elements will have increasing importance in the future.

CONCLUSIONS

This chapter described macroergonomics and macroergonomic methods. Also, cases in which each described method was applied were explained. This chapter described four methods, namely the MEAD, MAS, FRAM, and evolute approaches. These methods are commonly used and were most recently founded in the field of macroergonomics. There are, of course, many other methods, and new ones are being proposed by researchers and practitioners. Each method provides its own point of view for managing work system design and there is no single right way to proceed.

When a new work system is designed or an old one is improved, it is very important to look at the design from macroergonomic perspectives and to properly apply suitable methods. With the help of macroergonomic methods, work systems and their performance can be better understood, developed, and improved upon in an organization. Depending on the case, it is worth considering which tools are suitable for the job.

An expected outcome of a successful macroergonomic project will be an organization that is safer and performs well. Of course, the project goals have to be set clearly and selected methods must be used properly. A macroergonomic project is an investment that must be feasible. Therefore, anybody who proposes a macroergonomics project should also be able to show the feasibility of the project. Typically, the anticipated outcomes are greater than the investment.

KEY TERMS

Work system. A system that has personnel, technological, and organizational subsystems interacting with each other in the context of external and internal environments.

Work system design. A top-down approach to plan the roles and interactions of work system elements.

Macroergonomics. A field of science that concentrates on designing overall work systems by providing the knowledge and methods to improve work systems and, thus, develops the effectiveness and performance of companies (Hendricks, 1996).

Macroergonomics methods. Different kinds of approaches and practices to better understand, develop, and improve the performance of work systems in organizations

Microergonomics. Sides and views of ergonomics that deal with the design of human-technology interaction and interfaces (Hendrick, 2007).

Socio-technical system (STS) theory. A concept in organizational development that provides the theoretical foundation for macroergonomics (Trist and Bamworth, 1951). STS explores how people and technology interact in the workplace (Walker et al., 2008; Majchrzak and Borys, 2001).

REFERENCES

Argyris, C. and Schön, D. A. (1978). *Organizational Learning: A Theory of Action Perspective*. Reading, Addison-Wesley.

Baums, A. (2015). Industry 4.0: We Don't Need A New Industrial Policy But A Better Regulatory Framework, www.socialeurope.eu/2015/10/industry-4-0-we-dont-need-a-new-industrial-policy-but-a-better-regulatory-framework/, accessed March 21, 2017.

Boes, A. (2015). Digitization: New Work Concepts Are Revolutionizing The World Of Work, www.socialeurope.eu/2015/11/digitization-new-work-concepts-revolutionizing-world-work/, accessed March 21, 2017.

Carvalho, P. V. R. (2011). The use of Functional Resonance Analysis Method (FRAM) in a mid-air collision to understand some characteristics of the air traffic management system resilience. *Reliability Engineering & System Safety* 96(11), pp. 1482–1498.

de Weerd, M., Tierney, R., van Duuren-Stuurman, B., and Bertranou, E. (2014). Estimating the cost of accidents and ill-health at work: A review of methodologies. European Risk Observatory, European Agency for Safety and Health at Work, p. 60.

Dul, J., Bruder, R., Buckle, P., Carayon, P., Falzon, P., Marras, W. S., Wilson, J. R., and van der Doelen, B. (2012). A strategy for human factors/ergonomics: developing the discipline and profession. *Ergonomics* 55(4), pp. 377–395.

Evolute LLC. (2017). Evolute LLC home, www.evolutellc.com/, accessed February 23, 2017.

Gruber, T. R. (1993). A Translation Approach to Portable Ontologies. *Knowledge Acquisition* 5(2) pp. 199–220.

Haro, E. and Kleiner, B. M. (2008). Macroergonomics as an Organizing Process for Systems Safety. *Applied Ergonomics* 39, pp. 450–458.

Hendrick, H. W. (1996). The ergonomics of economics is the economics of ergonomics. *Proceedings of the Human Factors and Ergonomics Society Annual Meeting*. Vol. 40. No. 1. SAGE Publications.

Hendrick, H. W. (1996). *Good Ergonomics is Good Economics*. Santa Monica, CA: Human Factors and Ergonomics Society.

Hendrick, H. W. (2005). Macroergonomic Analysis of Structure (MAS). In Stanton, N., Hedge, A., Brookhuis, K., Salas, E., and Hendrick, H. (eds), *Handbook of Human Factors and Ergonomics Methods*, pp. 89.1–89.9.

Hendrick, H. W. (2007). Macroergonomics: The Analysis and Design of Work Systems. *Reviews of Human Factors and Ergonomics* 3(1), pp. 44–78.

Hendrick, H. W. and Kleiner, B. (2009). *Macroergonomics: Theory, Methods, and Applications*. Boca Raton, FL: CRC Press.

Hollnagel, E. (2012). *FRAM: The Functional Resonance Analysis Method*. Farnham: Ashgate Publishing Limited.

Hollnagel, E. (2012). A Bird's Eye View of Resilience Engineering. Presentation at Loughborough University.

Hollnagel, E., Pruchnicki, S., Woltjer, R., and Etcher, S. (2008). Analysis of Comair flight 5191 with the functional resonance accident model. *Proceedings of the 8th International Symposium of the Australian Aviation Psychology Association*, p. 8.

Hollnagel, E. (2013). An Application of the Functional Resonance Analysis Method (FRAM) to Risk Assessment of Organizational Change. Swedish Radiation Safety Authority.

Hollnagel, E. (2016). The Functional Resonance Analysis Method: FRAM Model Visualizer (FMV), http://functionalresonance.com/FMV/index.html, accessed December 8, 2016.

International Ergonomics Association. (2016). Definition and Domains of Ergonomics, www.iea.cc/whats/, accessed October 22, 2016.

International Labor Organization (2012). Estimating the Economic Costs of Occupational Injuries and Illnesses in Developing Countries: Essential Information for Decision-Makers Facts on Safe Work, p. 66.

Kantola, J., Vanharanta, H., and Karwowski, W. (2006). The Evolute System: A Co-Evolutionary Human Resource Development Methodology. In Karwowski, W. (ed.), *International Encyclopedia of Ergonomics and Human Factors* (2nd Edition). Boca Raton, FL: CRC Press.

Kantola, J. (2015). *Organizational Resource Management: Theories, Methodologies, and Applications*. Boca Raton, FL: CRC Press, p. 136.

Karwowski, W. and Kantola, J. (2005). The CIMOP System. In Stanton, N., Hedge, A., Brookhuis, K., Salas, E. Hendrick, H. (eds), *Handbook of Human Factors and Ergonomics Methods*. CRC Press.

Kleiner, B. M. (2002). Macroergonomic Analysis and Design (MEAD) of Work System Processes. *Proceedings of the Human Factors and Ergonomics Society Annual Meeting* 46(15), pp. 1365–1369.

Kleiner, B. M. (2006). Macroergonomics: Analysis and Design of Worksystems. *Applied Ergonomics* 37, pp. 81–89.

Kleiner, B.M. (2008). Macroergonomics: Work System Analysis and Design. *Human Factors* 50(3), pp. 461–467.

Kleiner, B. M. and Smith-Jackson, T. (2005). A Socio-Technical Approach to Construction Safety and Health in the United States. Paper presented at the Factors in Organizational Design and Management VIII, Madison, WI.

Lebeau, M. and Duguay, P. (2013). The Costs of Occupational Injuries: A Review of the Literature, REPORT R-787, The Institut de recherche Robert-Sauvé en santé et en sécurité du travail (IRSST), p. 75.

Majchrzak, A. and Borys, B. (2001). Generating Testable Socio-Technical Systems Theory. *Journal of Engineering and Technology Management* 18(3–4), pp. 219–240.

Marek, T., Karwowski, W., Frankowicz, M., Kantola, J., and Zgaga, P. (eds) (2014). *Human Factors of a Global Society: A System of Systems Perspective*. Boca Raton, FL: CRC Press, p. 1177.

Mohr, B. J. and van Amelsvoort, P. (2016). Co-Creating Humane and Innovative Organizations: Evolutions in the Practice of Socio-technical System Design, Global STS-D Network, p. 370.

OSHA (2016). OSHA statistics, www.osha.gov/oshstats/commonstats.html, accessed December 19, 2016.

Perrow, C. (1967). A Framework for the Comparative Analysis of Organizations. *American Sociological Review* 32(2), pp. 194–208.

Putkonen, A. (2010). Macroergonomic approach applied to work system modelling in product development contexts. Academic dissertation. *Acta Universitatis Ouluensis C360.* University of Oulu.

Senge, P.M. (2006). The Fifth Discipline: The Art and Practice of the Learning Organization (2nd edition). New York: Doubleday.

Trist, E. and Bamforth, K. (1951). Some social and psychological consequences of the long-wall method of coal getting. *Human Relations* 4, pp. 3–38, 14.

Vanharanta, H. (2015). Pursoid: Innovation Competence of Human Resources. In Kantola, J., *Organizational Resource Management: Theories, Methodologies, and Applications.* Boca Raton, FL: CRC Press.

Vargo, S. and Lusch, R. F. (2008). Service-dominant logic: continuing the evolution. *Journal of the Academy of Marketing Science* 36(1), pp. 1–10.

Walker, G. H., Stanton, N. A., Salmon, P. M., and Jenkins, P. D. (2008). A Review of Sociotechnical Systems Theory: A Classic Concept for New Command and Control Paradigms. *Theoretical Issues in Ergonomics Science* 9(6), pp. 479–499.

Waterson, P., Robertson, M. M., Cooke, N. J., Militello, L., Roth, E., and Stanton, N. A. (2015). Defining the methodological challenges and opportunities for an effective science of sociotechnical systems and safety. *Ergonomics* 58(4), pp. 565–599.

Wilson, J. R. (2014). Fundamentals of systems ergonomics/human factors. *Applied Ergonomics* 45(1), pp. 5–13.

Zadeh, L. (1965). Fuzzy Sets. *Information and Control* 8(3), pp. 338–353.

Zadeh, L. (1973). Outline of a New Approach to the Analysis of Complex Systems and Decision Processes. *IEEE Transactions on Systems, Man, and Cybernetics* 1(1), pp. 28–44.

Case Study 2
Macroergonomics: A Prospective for Action in Arid Environments

Ali M. Al-Hemoud

INTRODUCTION

Macroergonomics is a holistic systematic approach to optimize the best fit between workers and their workplace. It explores cognitive and physical effort performed in the workplace, that is, it integrates all workplace demands and energizers into one macro, top-down or bottom-up approach for the effective implementation of ergonomic interventions designed to improve "workplace health" in order to achieve harmonization of work productivity, work quality, social responsibility, and worker satisfaction in any organization. As such, one can optimize the task and the hierarchical surroundings with the workforce capabilities at the cognitive, emotional, and biomechanical levels. This approach was utilized to assess macroergonomics compatibility and worker performance in Kuwait oil companies.

MACROERGONOMICS IN KUWAIT

Nine macroergonomics workplace factors were studied to quantify their impact upon worker performance in two oil companies in Kuwait—a government-owned and a private company. It was hypothesized that the more harmonized the macroergonomics factors within an enterprise system, the better the work performance outcome measures; that is, achieving a balance between work output by simultaneous optimization of macroergonomics factors. The following nine macroergonomics factors were analyzed: (1) organizational (i.e., time management, work responsibilities, task meaningfulness, autonomy); (2) individual growth (i.e., skill development, skill utilization); (3) technological (i.e., hardware and software, work flow, information technology procedures, expertise); (4) economic (i.e., work pay and benefits, job security, bonuses, incentives, promotion); (5) social and communication (i.e., conflict, support, praise, feedback, knowledge of results); (6) cognitive (i.e., information processing, memory-related, sensory-related, cognitive processing); (7) physical conditions (i.e., physical hazards, immediate danger, architectural design, chemical environment, biological environment); (8) muscular activity (i.e., strength, endurance, sudden handling, fixed body positions, effort); and (9) experienced demands (i.e., achievement, satisfaction, perceived risk, and perceived benefit).

A macroergonomics factor instrument was designed and tested for validity and reliability. The psychometric properties—i.e., questionnaire validity and reliability of the instrument—were examined on a random sample of seven employees from the two selected companies. Analysis through Cronbach's alpha test showed a range of "good" to "excellent" (0.761 to 0.950), and the reliability coefficients were considered excellent (0.855) for the nine macroergonomics factors (Al-Hemoud and Behbehani, 2017). For the private company, four senior supervisors were randomly selected from four lines of service operations: maintenance, crude/chemical handling, wellhead maintenance, and slickline. From the government-owned company, samples of three senior safety supervisors of production operations were selected from the department of health, safety, and environment (HSE).

For the private company, results revealed that the nine macroergonomics factors were perceived as average, specifically for maintenance and crude/chemical handling operations, while they were considered weak for the wellhead maintenance and slickline operations on three factors (i.e., individual growth, economics, and physical environment). Surprisingly, for the government-owned company seven of the nine macroergonomics factors (except for economics and social/communication) were considered weak. Overall workload, job safety risk factors, and work stress were perceived as acting demands on the workers for both government and private companies.

MACROERGONOMICS IN THE GCC VS. INTERNATIONALLY

The application of macroergonomics in the GCC and industrialized enterprises in other parts of the world is shown schematically in the two figures below. The first class of relationship is characterized by a slow rise or fall around the extremes and a steep rise or fall in the middle (Figure CS2.1). This case typically occurs in organizations in the GCC countries that have performance quality and productivity problems and a low quality of work life. For instance, as physical work and environmental

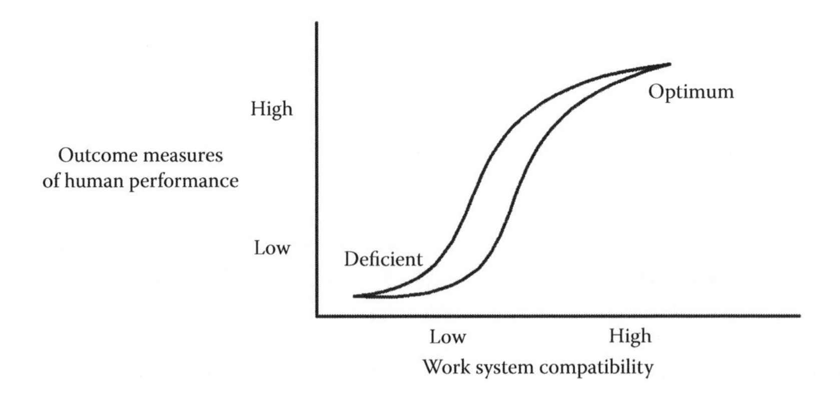

FIGURE CS2.1 Macroergonomics application in the GCC.

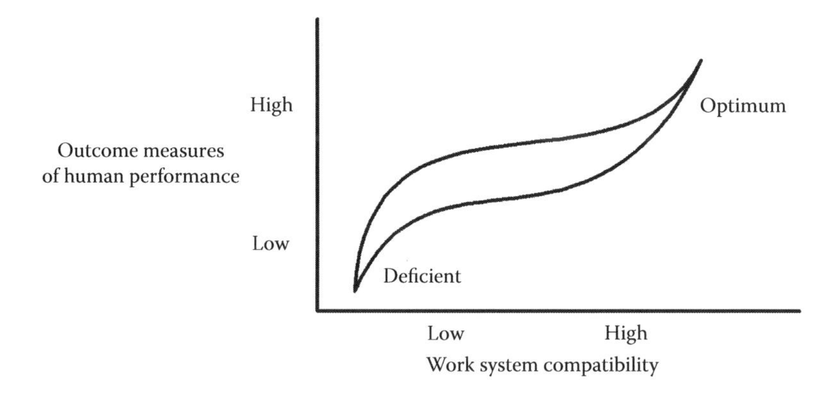

FIGURE CS2.2 Macroergonomics application in industrialized nations.

conditions are redesigned, work system compatibility increases; however, performance measures improve only slightly as the level of organization health is poor; in other words, problems are not fixed holistically but in pieces. In this case, to improve organizational health, more dramatic changes are required to realize substantial results. As dramatic changes occur, compatibility increases and all outcome measures are greatly enhanced due to the improvement in macroergonomics factors, thus accounting for the steep slope in the middle of the curve.

In Figure CS2.2, the relationship can be described by a steep rise or fall around the extremes and a slow rise or fall around the middle. For example, this case may occur in an organization that is already fundamentally healthy in terms of quality of work life with nearly optimum macroergonomics factor compatibility. In this case, the employees have high morale and job satisfaction; yet, the productivity and performance quality levels are not being met efficiently. As system compatibility is increased as a result of some obvious changes at the process level (e.g., work team operational changes), dramatic results are seen in performance productivity and quality. Thus, increasing system compatibility through enhancement of macroergonomics factors in the moderate region of the S-curve in Figure CS2.2 will yield moderate results in terms of productivity and performance quality levels as the process is now functioning successfully.

IMPLICATIONS OF MACROERGONOMICS IN KUWAIT'S OIL SECTOR AND BEYOND

The major conclusion extracted from this study is that the oil sector organizations in Kuwait are not managed as macro-ergo enterprises where there is disharmony in the ergonomics system, work-compatibility system, and the work output interface interaction. It was concluded that organizations in Kuwait and other GCC countries are performed in a traditional mode, where organizations focus on the improvement of a few selected measures of employee performance at a time (i.e., work productivity, work output, or work quality). This is quite evident in the independent work improvement philosophies implemented in industry in the form of isolated programs, that is, with no apparent link between the various programs with respect to their contribution to total employee performance.

These gaps in the enterprise-wide system have a significant impact on employee performance, work productivity, and work quality. Strategic improvement actions include reengineering ergonomic work conditions, human capital training investment, and job enrichment rather than job enlargement and development of proactive key performance indicators to synchronize and integrate all macroergonomics factors in the organization.

REFERENCES

Al-Hemoud, A. and Behbehani, W. (2017). Workplace environmental demands and energizers at two Kuwait oil companies. *International Journal of Environmental Sciences and Technology* 14(5), 983–992.

3 Human Factors in Safety Management

Safety Culture, Safety Leadership, and Non-Technical Skills

Rhona Flin and Cakil Agnew

CONTENTS

INTRODUCTION

Human factors/ergonomics (HFE) aspects of safety management are wide-ranging and it is now recognized that these are essential for effective risk control, as well as performance efficiencies and worker well-being. Traditionally, safety management

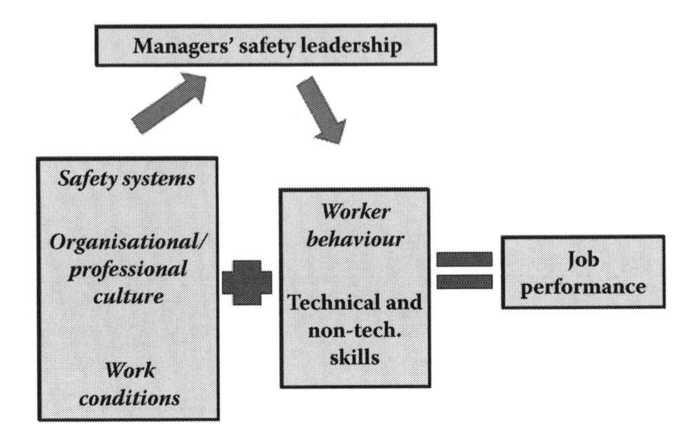

FIGURE 3.1 Three components of safe job performance discussed in this chapter.

was mainly concerned with engineering barriers (e.g., blast walls, guard rails), personal protection (e.g., hard hats), and devising rules and regulations to govern workers' and managers' actions. As accident investigations became more sophisticated, it was clear that these techniques alone did not provide sufficient protection and that human and organizational factors had to be considered and managed (Reason, 1997; CSB, 2016). The scope of human factors science applied to safety encompasses ergonomic issues of equipment design, usability and the layout of working environments, processes related to safety management systems (SMS), organizational issues (such as cultural aspects), and psychological and physiological effects on human performance for individuals and work teams. As the subject is extensive, in this chapter we have focused on just three of these safety topic areas that would be of particular interest to the Gulf region: (i) safety culture, (ii) managers' safety leadership, and (iii) worker behaviors relating to non-technical skills (see Figure 3.1, which indicates how these are related). We discuss these in the context of our research in two sectors: health care and the oil and gas industry.

FUNDAMENTALS

Safety Culture

Analysis of major industrial accidents in the 1980s began to shift regulatory and research focus from failures of equipment and of individual workers to an examination of the underlying culture of the organization. This began with the Chernobyl nuclear power plant disaster in 1986, where investigators concluded that aspects of the organizational culture had contributed to the accident (IAEA, 1986). Nowadays, culture frequently features as a causal factor. For example, the review of the crash of an RAF Nimrod aircraft in 2006 with 14 deaths is subtitled: "A failure in leadership, culture and priorities" (Haddon-Cave, 2009). Similarly, a recent report from the American National Academy of Sciences (NAS, 2016) emphasized the importance of managing the organizational culture on offshore oil and gas installations in order to enhance safety.

The term "safety culture" is often used interchangeably with "safety climate," but in the academic literature these are typically regarded as two distinct concepts (Cox & Flin, 1998). Safety culture is defined not only as encompassing safety-related attitudes, behaviors, and perceptions, but also covering deeply rooted values and assumptions that individuals hold about the organization (Pettita, Probst, Barbaranelli, & Ghezzi, 2017). On the other hand, safety climate is proposed to be a surface manifestation (Schein, 1990), a "snapshot" of the existing culture (Mearns, Flin, Gordon, & Fleming 1998). The distinction between the concepts can be crucial: recent research suggests that certain dimensions of safety culture (autocratic and bureaucratic) can undermine the effects of safety climate on safety outcomes (Petitta et al., 2017). Given the limited space in this chapter, we use the term safety culture to cover both concepts.

There are different approaches to studying organizational culture and safety outcomes. The most common is to talk of an over-arching safety culture that essentially reflects managerial and worker attitudes related to the control of risk and the prioritization of safety.

> The safety culture of an organization is the product of individual and group values, attitudes, perceptions, competencies and patterns of behaviour that determine the commitment to, and the style and proficiency of, an organisation's health and safety management. (ACSNI, 1993, p. 23.)

The main dimensions of organizational safety culture typically include management commitment to safety, work practices, relative prioritization of safety, adherence to safety rules, risk management, and reporting of errors and incidents.

A second approach emphasizes subcomponents that are types of culture. For instance, Reason (1997, p. 195) suggested that a safety culture had elements which were: an "informed" culture—knowing about all the factors that influence the safety of the system; a "reporting" culture—that encourages telling about incidents; a "just" culture—where employees believe they are treated fairly and will not be inappropriately blamed for errors; a "flexible" culture that favors a flatter structure; and a "learning" culture that is willing to draw appropriate conclusions and act on them. Dekker (2016) provides a detailed account of why a just culture, which he describes as a culture of trust, learning, and accountability, is particularly important for safety management. Thirdly, safety culture maturity models (based on Westrum, 1995) have been devised, which characterize a staged evolution from a pathological culture that does not pay attention to safety, through to a very safety-conscious "generative" culture (Fleming, 2000; Goncalves et al., 2010; Parker et al., 2006).

Whichever framework is adopted, the safety culture is of interest because it essentially influences what becomes the normal workplace behaviors in relation to safety, such as taking risks, following rules, speaking up about safety concerns, and reporting accidents and errors. Essentially, organizations need to maintain a culture that makes it "easy to do the right thing, and hard to do the wrong thing" for safety. The safety culture is normally measured by questionnaires that assesses workforce perceptions of the dimensions (e.g., supervisor support for safety) and their associated behaviours (e.g., willingness to report incidents).

Across industries, safety culture has been shown to be a robust predictor of both workers' safety behaviours and objective safety outcomes, such as injury and accident rates. Worksites with more positive cultures show lower accident rates, workers who perceive their supervisor/manager to be more committed to safety engage in more safety-related behaviour and fewer risk-taking behaviours (Clarke, 2010; Probst & Estrada, 2010; Zohar, 2014). In a meta-analysis based on different industrial settings, safety climate was shown to influence employees' safety behaviors through its effects on safety knowledge and motivation (Christian et al., 2009). The motivational mechanism linking culture to behavior is likely to be a function of expectations, that is, whether workers expect to be rewarded or reprimanded for particular actions related to safety and production (Zohar, 2014). Again, this indicated the important role of managers and supervisors in creating expectations that will affect workers' behavior choices. Later, the roles of safety climate acting as a source for employees' safety motivation and knowledge, as well as a predictor of employees' safety behaviors, were validated across cultures in both English-speaking and non-English-speaking countries (Barbarabelli, Pettita, & Probst, 2015).

Following the Piper Alpha accident in 1988, studies of safety culture were conducted on offshore installations in the North Sea (Mearns et al., 1998, 2001; Rundmo, 1992). These portrayed the key cultural features showing that management and supervisor commitment to safety were particularly important for building safer norms of behavior. A study comparing offshore installations across the Norwegian and UK sectors showed that there were more differences among companies than between the two nationalities (Mearns et al., 2004).

In healthcare, interest in measuring safety culture emerged after high rates of adverse events suffered by patients were revealed (Vincent, 2010); therefore, the focus has usually been on features of the safety culture in hospitals that protects patients from errors and harm. Fan and colleagues (2016) found an association between safety culture scores and surgical site infections. A study of intensive care units revealed an association between poorer safety culture scores and increased length of stay for patients, as well as a link between less-favorable perceptions of management and higher mortality rates (Huang et al., 2010).

The relationships between safety culture and healthcare workers' well-being have also been assessed. In a study with nurses, negative associations were found between unit-level safety culture scores and workers' back injuries, as well as with patient urinary tract infections and medication errors (Hofmann & Mark, 2006). Similarly, Gimeno and colleagues (2005) found that safety culture was related to self-reported work-related injuries. Blood and body fluid exposure incidents for workers were lower when senior management support, safety feedback, and training were perceived favorably (Gershon et al., 2000). Zohar and colleagues (2007) showed both group- and hospital-level culture as predictors of future safety behaviors.

As mentioned above, one of the most influential safety culture dimensions is leadership quality (Nahrgang, Morgeson, & Hoffman, 2011), with managerial commitment to safety emerging as the most robust predictor of future incidents (Beus et al., 2010). We consider managers' and supervisors' safety leadership in the next section.

SAFETY LEADERSHIP

For effective safety management, leadership is important at every level of management, from team leaders, to site managers, to top-level managers. Most research on managerial leadership concerns productivity (Yukl, 2013), but there is now an increasing interest in the relationship of leadership styles to safety outcomes (e.g., Agnew et al., 2014a; Hofmann & Morgeson, 2004). Particular styles of leadership are associated with better safety behaviors by workers (e.g., compliance to rules) and more favorable organizational safety performance, such as decreased accident rates.

The model most often applied to the study of managers' safety leadership is the transactional/transformational model (Bass, 1998). The transactional component involves the leader offering incentives and/or punishments that are contingent on the subordinate's performance meeting agreed standards. Bass argued that this transactional relationship, at best, produces expected performance levels, because it only appeals to individual goals and aspirations. While all leaders use the transactional component, he showed that leaders of the highest-performing teams also display transformational behaviors. Transformational leaders are charismatic, inspiring, stimulating, and considerate. They provide followers with a sense of purpose; portray an image of success, self-confidence, and self-belief; and articulate shared goals and question traditional assumptions, while taking into account the needs of subordinates. Clarke (2013) conducted a meta-analysis on transactional/transformational leadership and safety outcomes. The findings demonstrated the crucial role of both leadership styles to predict the safety behaviors of workers. For example, active transactional leadership such as anticipating problems and taking proactive actions were strongly associated with workers' compliance with the organization's safety rules and regulations. On the other hand, transformational style was a better predictor of safety participation behaviors of the workers. The related theory of authentic leadership (Avolio & Gardner, 2005) has also been applied in safety research (e.g., Nielsen et al., 2013) indicating that this style can also be effective.

Although leadership practices as a predictor of safety outcomes are well documented, Yukl (2013) pointed out that the level of management has to be taken into account. Top managers may have more of an influence on rule-related behaviors whereas supervisors may be better at encouraging voluntary activities that are related to safety (Clarke & Ward, 2006). Zohar and Luria (2005) identified supervisors as better than senior managers at influencing workers' safety behaviors.

Supervisory safety practices have been found to decrease the number of minor accidents and positively influence workers' safety climate perceptions. Transformational leadership behaviors of supervisors were related to fewer occupational injuries (Zohar, 2002). The literature on supervisors and safety emphasizes the importance of good communication, the need to build trust and to care about team members, and setting and reinforcing safety standards, especially when there are strong production or cost-reduction goals (Hofmann & Morgenson, 2004). Therefore, relying solely on written safety rules might not ensure an increase in voluntary safety activities of workers. Rather, supporting and providing training for supervisors on specific leader behaviors to improve their leadership styles might yield more desirable safety-related outcomes.

Safety leadership behaviors of supervisors (Fleming, 1996) and managers (O'Dea & Flin, 1991) on offshore installations have been investigated showing the importance of both transactional and transformational behaviors. More recently, Nielsen and colleagues (2016) in a time-lagged study of Norwegian offshore workers found that constructive leadership emerged as the only significant predictor of subsequent psychological safety climate.

An investigation into patient safety in a healthcare organization in England (Healthcare Commission, 2009) revealed that failure of the senior management's leadership was one of the factors contributing to high mortality rates. In a study of surgery in the USA, the team leader's behaviours were shown to influence team members' willingness to speak up (Edmondson, 2003).

While both safety culture and safety leadership can create supportive conditions for safe working practices, it is also necessary to consider the skills of the workforce and how these can relate to job performance, errors, and accidents (as shown in Figure 3.1).

NON-TECHNICAL SKILLS

The term "non-technical skills" (NTS) was first used by the Joint Aviation Authorities (now called the European Aviation Safety Agency) in relation to airline pilots' behavior on the flight deck. NTS can be defined as "the cognitive, social and personal resource skills that complement technical skills, and contribute to safe and efficient task performance" (Flin et al., 2008, p. 1). It is not only in aviation where these skills contribute to workplace safety; studies of accidents in other industries reveal similar patterns. Today, Crew Resource Management (CRM, i.e., non-technical) skills training is used as part of safety management and skills development in the maritime industry, rail, nuclear power production, mining, and the emergency services.

In essence, non-technical skills enhance workers' technical skills. As Figure 3.2 shows, poor NTS can increase the chance of error, which in turn can

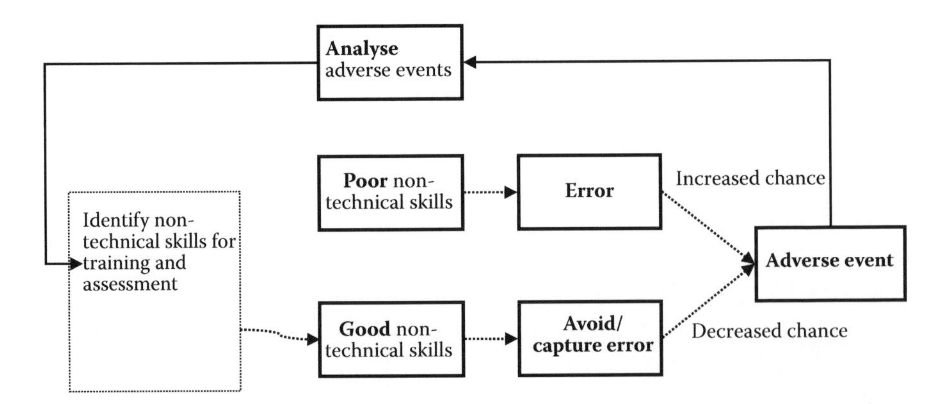

FIGURE 3.2 Relationship between non-technical skills and adverse events (Reprinted from Flin, R. et al., *Safety at the Sharp End: A Guide to Non-Technical Skills*, Farnham, Ashgate, 2008. With permission of Taylor & Francis.)

increase the chance of an adverse event. Good NTS (e.g., high vigilance, clear communication, and team coordination) can reduce the likelihood of error and consequently of accidents.

The aviation industry had realized by 1980, from a series of accidents with no primary technical failure, that maintaining high standards of safety was going to require attention to pilots' behaviors that could diminish or enhance flight safety (Kanki et al., 2010). From interviews, experiments, and accident analysis the behaviors that contributed to accidents or were effective in preventing them were extracted. A key source of information was the cockpit voice recorder, which enabled analysis of pilots' conversations prior to an accident. The identified behaviors were classified into categories of non-technical skills and a special training course for pilots was devised called Crew Resource Management. This was designed to increase pilots' understanding of the importance of particular behaviors for safety and to provide opportunities to practise non-technical skills in exercises and simulated flights (CAA, 2016). Non-technical skills are assessed alongside technical skills as part of licensing requirements for pilots (Flin, 2018).

The main categories of non-technical skills are similar, although not identical, for operational jobs in higher-risk work settings. Each category can be subdivided into constituent elements and, for each element, examples of good and poor behaviours (behavioural markers) can be specified. A typical set of non-technical skills (described in Flin et al., 2008) is shown below in Table 3.1.

Many higher-risk work domains (e.g., ships, mines, hospitals, railways) have adopted a non-technical skills approach and introduced CRM training. In healthcare, studies are beginning to show that the non-technical skills of clinical staff are related to patient outcomes (Hull et al., 2012). Analyses of the Deepwater Horizon drilling rig accident have indicated specific failures in non-technical skills (Reader & O'Connor, 2014; Roberts et al., 2015). A new report on the accident from the Chemical Safety Board in the USA (CSB, 2016) recognizes that there is a "need for development and use of non-technical skills, including communication, teamwork, and decision making by the operator, drilling contractor and other well services providers" (p.24). It should be noted that safety culture and safety leadership are important for the maintenance of non-technical skills at the worksite (see McCulloch et al., 2009).

METHODS

Many sources of information are available on human factors methods for safety management, for example, books on the design of safe work environments (McLeod, 2015); accident analysis methods (Weigmann & Shappell, 2003; Gordon et al., 2003); human error (Reason, 1997); data-gathering techniques (Crandall et al., 2006; Stanton et al., 2013); risk management (Glendon & Clarke, 2015). Specialist journals publish reports of studies using human factors techniques to study safety issues (e.g., *Human Factors; Ergonomics; Safety Science; Journal of Safety Research; Accident Analysis and Prevention, Journal of Loss Prevention in the Process Industries; BMJ Quality and Safety).* Many organizations concerned with safety management have websites with human factors advice, for example:

TABLE 3.1
Examples of Categories and Elements in a Generic Non-Technical Skills Framework

Categories	Definitions	Typical Elements
Situation awareness	Developing a dynamic awareness of the situation during a task, based on assembling data from the environment, understanding what it means and anticipating future developments	• Gathering information • Comprehending (forming a mental picture) • Anticipating (thinking ahead)
Decision-making	Determining possible courses of action (options) to deal with the assessed situation; reaching a judgement in order to choose an appropriate course of action; implementing the chosen option and reviewing its effect	• Generating one or more options • Evaluating options • Selecting and implementing options • Reviewing
Teamwork	Skills for working in a team context to ensure that the team has an acceptable shared picture of the situation and can complete tasks effectively	• Co-ordinating actions • Resolving conflicts • Sharing information • Helping others
Leadership[a]	Leading the team and providing direction, demonstrating high standards of practice and care, and being considerate about the needs of individual team members	• Setting and maintaining standards • Monitoring progress • Supporting others • Allocating tasks
Managing personal resources (e.g., stress and fatigue)	Skills for diagnosing one's state of mental and physical fitness for the task; taking action to maintain the necessary level of fitness or to find an alternative solution	• Identifying causes of stress and fatigue • Recognizing effects • Implementing coping strategies

[a] Leadership within the discussion of non-technical skills refers to the behaviors of the leader who is co-located with his or her team during task execution. Safety leadership in the previous section refers to the leadership style of managers and supervisors who are not directly engaged in the workers' task execution.

Civil Aviation Authority (CAA, UK), www.caa.co.uk/Safety-initiatives-and-resources/Working-with-industry/Human-factors/Human-factors/.

Clinical Human Factors Group, www.chfg.org.

Energy Institute (UK), www.energyinst.org/technical/human-and-organisational-factors.

Eurocontrol (Air Traffic Management), www.eurocontrol.int/tags/human-factors.

Health and Safety Executive (HSE, UK), www.hse.gov.uk/humanfactors/index.htm.

Methods of Measuring Safety Culture

Assessing the state of the safety culture requires a baseline assessment of the current level of relevant cultural factors in the workplace, so that interventions can be targeted and any subsequent improvements can be assessed (Antonsen, 2009). The measurement is normally achieved with a questionnaire survey asking workers and managers about their attitudes to safety and perceptions of how safety is prioritized and managed in their work unit or across the organization. It may also ask respondents to report on their behaviors (e.g., reporting incidents) and to say how many injuries or accidents they have suffered or witnessed. There are many safety culture questionnaires available, generic instruments such as the HSE safety climate tool (website given above) or bespoke questionnaires designed for a specific sector, e.g., healthcare (Jackson, Sarac [Agnew], & Flin, 2010; Waterson, 2015). To determine if safety culture has an effect on safety behaviors and accidents, different types of outcome data can be collected, e.g., (i) near-miss and accident incident records; (ii) self-reports of incidents and injuries; (iii) workers' safety behaviors (self-reported or observed).

The nuclear power industry has advocated the measurement and management of safety culture for 30 years following the Chernobyl accident, and it provides guidance on performing safety culture assessments (IAEA, 2016). The focus of this report is on using such assessments as a learning opportunity for organizational growth and development rather than as a fault-finding or "find and fix" exercise. The guidance advises having engagement with all levels of the organization and using techniques such as document reviews, questionnaires, interviews, observations, and focus groups. It emphasizes the need to use multiple measurements and qualitative, as well as quantitative, methods of gathering data. Similarly, the level of safety culture maturity can be assessed by questionnaires or card sorting and discussion tasks (see the Energy Institute, Hearts and Minds toolkit). Nowadays, many oil and gas companies conduct regular safety culture surveys as part of their safety management system (e.g., Tharaldsen et al., 2008).

Methods of Measuring Safety Leadership

Safety leadership is also usually measured by questionnaires, either completed by the leader and/or by those directly reporting to that leader. For example, *Perceptions of supervisory behaviours for safety* by Zohar and Luria (2005) was designed to identify how supervisors prioritize safety over productivity using self-report items in a questionnaire. There are standard leadership questionnaires that can be purchased from psychometric test suppliers, such as the Multifactor Leadership Questionnaire (MLQ) designed to assess transactional and transformational leadership (available from www.mindgarden.com). One method of assessing managers' commitment to safety and safety behaviors, originally developed for the multinational company Shell, involves the leader completing a self-rating questionnaire and asking several of his or her team to complete an "upward" rating (Bryden et al., 2006), which means that the staff rate their direct boss. These scores are fed back to each manager as a personal report and aggregated scores are presented for group discussion. (This is now part of the Energy Institute Hearts and Minds Toolkit.)

Methods of Identifying and Measuring Non-Technical Skills

While the main skill categories are similar across professions, the component elements and examples of good and poor behaviors need to be specified for a given profession and task set. Analysis of incidents and interviews with experienced workers, as well as observation of behavior during routine and non-routine work, can reveal which workplace behaviors positively or negatively influence job performance and adverse events. These are all forms of task analysis (see Stanton et al., 2013) and where cognitive skills play a major part, e.g., control room operators, then cognitive task analysis can be used (Crandall et al., 2006). Having identified the skills and related behaviors, these need to be refined and organized into a concise, hierarchical structure or taxonomy. This is usually achieved using panels of subject matter experts. This skill set then forms the basis of NTS (CRM) training and related assessment methods (e.g., behavior rating systems).

Pilots are regularly assessed on their non-technical skills, using behavior rating systems such as NOTECHS (CAA, 2016; Flin, 2018; Flin et al., 2003). In healthcare, the Anaesthetists' Non-Technical Skills (ANTS) system was developed from data on anaesthetists' behaviour gathered from a literature review, observations, interviews, surveys, and incident analysis (Flin et al., 2010). There are also similar rating tools for surgeons (NOTSS) and for scrub nurses (SPLINTS) and anaesthetic assistants (ANTS-AP); see Flin et al. (2015). For papers and copies of these rating tools, see www.abdn.ac.uk/iprcs.

APPLICATIONS

Safety Culture

The main application of the safety culture concept has been in the use of diagnostic tools (described above) to measure the level of culture and identify strengths and weaknesses. For example, Agnew et al. (2013) conducted a study of 1,866 healthcare staff to provide a baseline assessment of safety culture in Scottish hospitals. Their findings illustrated the links between safety perceptions and safety outcomes both for worker and patient injuries and focus groups with frontline staff were conducted to get a deeper understanding of the significant factors. More importantly, both the qualitative and the quantitative data were used as a tool to generate discussions with the management through an interactive workshop designed as a feedback mechanism (Agnew & Flin, 2014b). The main aim of the workshop was to discuss the project findings, present the available tools to assess safety culture, and formulate recommendations for improving safety culture. This was based on an approach devised in air traffic management (Kirwan, 2008; Mearns et al., 2013).

Safety culture measurement can also be carried out across companies using a benchmarking approach (Mearns et al., 2001) so that organizations can learn from each other. Hudson (2007) discusses techniques that were used in a major operating company to enhance safety culture. The nuclear power industry (IAEA, 2016) recommends safety culture assessments as a learning opportunity for organizational development rather than as a fault-finding or "find and fix" exercise.

Safety leadership training and non-technical skills training are both designed to shift the norms of behavior, and thus drive the workplace culture in a safer direction.

Safety Leadership

The main applications of safety leadership research are diagnostic tools to determine leadership style and training programs for leaders. Part of the Hearts and Minds behavioral safety toolkit (mentioned above) is a training guide titled "Improving Supervision," designed to provide a step-by-step guide to identify the areas of concerns with the supervisors' leadership styles and the ways to improve people's safety behaviors and performance (https://heartsandminds.energyinst.org/). Similarly, IOGP (2013) produced guidance for shaping the safety culture through effective safety leadership. Recently, there have been improved efforts to train leaders to manage safety in health care organizations. The World Health Organization developed the *Leadership Competencies Framework on Patient Safety and Quality of Care* identifying key leader competencies to ensure safe and quality patient care (www.who.int). In the USA, the Institute for Healthcare Improvement offers a program called *High Impact Leadership* in order to help leaders to build safer health care organizations (Swensen, McMullen, & Kabcenell, 2013).

Non-Technical Skills

CRM training programs to enhance non-technical skills are very well established in the aviation industry (CAA, 2016). Other sectors are beginning to adopt this method. It was recommended for offshore oil and gas production operations following the Piper Alpha accident (Flin, 1995) but was not endorsed until the Deepwater Horizon rig blowout 22 years later. Now, courses on well operations CRM (WOCRM) have been developed for rig crews (IOGP, 2014). Other sectors of the energy industry, such as refineries and pipelines, have been advised to adopt a CRM approach (Energy Institute, 2014).

In healthcare, CRM is a recent innovation. Ab-initio courses in the medical and nursing schools are teaching students the importance of non-technical skills for patient safety. In some universities there are now psychologists employed to lecture on human factors and patient safety to health care students and these topics are being embedded throughout the curriculum. There are also courses on non-technical skills for qualified medical staff, such as surgeons (see Flin et al., 2015).

For recent guidance on the training and assessment of non-technical skills, see Thomas (2017).

FUTURE TRENDS

Product Safety Culture

New applications of safety culture have been emerging, relating to the safety of consumers rather than of workers. The concept of product safety culture has appeared in the manufacturing sector following a series of product failures that injured consumers, e.g., faults in cars, children's toys, and medical implants. Companies are now recognizing that organizational culture may influence their design, manufacturing, or service practices in ways that can ultimately affect the well-being of the product users. The empirical literature on this topic is, as yet limited, although there are studies of product safety culture in manufacturing (e.g., Zhu et al., 2016) and in the food industries, where the concept of food safety culture is now being discussed (e.g., Jesperson et al., 2016). At this stage it is unclear whether the components of the organizational safety culture that are protective for workers are the same as those that ensure the safety of consumers and product users (Suhanyiova et al., 2017).

Managers' Safety Leadership

Expressing some concerns about the limitations of a safety culture approach, Kirwan (2008), a senior psychologist at Eurocontrol, proposed that more attention needed to be paid to the knowledge and skills of senior managers in relation to organizational safety. He coined the term "safety intelligence" and subsequently sponsored research to explore the concept (Fruhen et al., 2014a). A resulting White Paper (Eurocontrol, 2013) explains how senior managers can become more intelligent safety leaders.

A related approach has been to examine "chronic unease" in senior managers. This concept first appeared in the high-reliability literature and refers to managers retaining a sufficient level of concern about the safety of their work sites and not being complacent about ever-present risks. Fruhen and colleagues (2014b, 2016) identified five key attributes of chronic unease: pessimism, propensity to worry, vigilance, requisite imagination, and flexible thinking. Interviews with senior managers from oil and gas companies found that chronic unease was described as having positive effects on safety.

Non-Technical Skills and the Safety Management System

The aviation industry continues to review its CRM (NTS) training and evaluation and recent guidance from the European Aviation Safety Agency (EASA, 2015) emphasizes that CRM methods should be evidence-based and embedded within the organization's safety management system, for example, by using safety data to inform CRM training requirements (see Flin, 2018). A new component that should be included in pilots' CRM training is on how to cope with the effects of a startle response that can occur after a sudden "threat" in the environment and cause a loss of concentration. This was recommended following the findings of the investigation report into the Air France (AF447) fatal accident in 2009 when the pilots lost control of a large passenger aircraft flying between Rio and Paris (BEA, 2012).

CONCLUSION

Organizations striving to improve their safety performance need to adopt a wide-ranging human factors/ergonomics approach. This chapter focuses on psychological research on workplace safety culture and associated safety behaviors, describing human-factor measurement tools and training techniques, applied in healthcare and the oil and gas industry. These are being used to measure safety culture and to address two key aspects of behavior that influence the culture, namely, managers' safety leadership and workers' non-technical skills.

KEY TERMS

Crew Resource Management (CRM). Training which was developed in aviation, designed to encourage aircrews to use all available resources—equipment, people, and information—in order to enhance flight safety.

Non-technical skills (NTS). The cognitive, social, and personal resource skills that complement technical skills, and contribute to safe and efficient task performance.

Safety culture. The product of individual and group values, attitudes, perceptions, competencies, and patterns of behavior that determine the commitment to, and the style and proficiency of, an organization's health and safety management.

Safety leadership. The behaviors of managers and supervisors that maintain, improve, and promote the state of workplace safety.

REFERENCES

ACSNI (1993). *Study Group on Human Factors. Third Report: Organising for Safety.* Sheffield: HSE Books.

Agnew, C., Flin, R., and Mearns, K. (2013). Patient safety climate and worker safety behaviours in acute hospitals in Scotland. *Journal of Safety Research* 45, 95–101.

Agnew, C. and Flin, R. (2014a). Senior charge nurses' leadership behaviours in relation to hospital ward safety: A mixed method study. *International Journal of Nursing Studies* 51, 768–780.

Agnew, C. and Flin, R. (2014b). Safety culture in Practice: Assessment, Evaluation, and Feedback. In P. Waterson (ed.), *Patient Safety Culture: Theory, Methods and Application,* (pp. 207–228). Ashgate, Aldershot.

Antonsen, S. (2009). *Safety Culture: Theory, Method and Improvement.* London: CRC Press.

Avolio, B. and Gardner, W. (2005). Authentic leadership development. Getting to the root of positive forms of leadership. *Leadership Quarterly* 16, 315–338.

Barbaranelli, C., Petitta, L., and Probst, T. (2015). Does safety climate predict safety performance in Italy and the USA? Cross-cultural validation of a theoretical model of safety climate. *Accident Analysis & Prevention* 77, 35–44.

Bass, B. (1998). *Transformational Leadership.* Mahwah, NJ: Lawrence Erlbaum.

BEA (2012). *Final Report. Accident on 1st June 2009 to the Airbus A330-203 operated by Air France flight AF 447 - Rio de Janeiro - Paris.* Paris: Bureau D'Enquetes et Analyses.

Beus, J. M., Payne, S. C., Bergman, M. E., and Arthur, W. (2010). Safety climate and injuries: An examination of theoretical and empirical relationships. *Journal of Applied Psychology* 95, 773–727.

Bryden, R., Flin, R., Hudson, P. et al. (2006). Holding up the leadership mirror then changing the reflection. *Proceedings of Society of Petroleum Engineers Health, Safety, Environment conference, Calgary, March. (SPE 98700).* Texas: SPE.

CAA (2016). *Flight-crew Human Factors Handbook.* CAP 737. Gatwick: Civil Aviation Authority, www.caa.co.uk.

Christian, M., Bradley, J., Wallace, J., and Burke, M. (2009). Workplace safety: A meta-analysis of the roles of person and situation factors. *Journal of Applied Psychology* 94, 1103–1127.

Clarke, S. (2010). An integrative model of safety climate: Linking psychological climate and work attitudes to individual safety outcomes using meta-analysis. *Journal of Occupational and Organizational Psychology* 83, 553–578.

Clarke, S. (2013). Safety leadership: A meta-analytic review of transformational and transactional leadership styles as antecedents of safety behaviors. *Journal of Occupational and Organizational Psychology* 86 (1), 22–49.

Clarke, S. and Ward, K. (2006). The role of leader influence tactics and safety climate in engaging employees' safety participation. *Risk Analysis* 26, 1175–1185.

Cox, S. and Flin, R. (1998). Safety culture: Philosopher's stone or man of straw? *Work & Stress* 12, 189–201.

Crandall, B., Klein, G., and Hoffman, R. (2006). *Working Minds: A Practitioner's Guide to Cognitive Task Analysis*. Cambridge, MA: Bradford.

CSB (2016). *Report into Macondo Blowout. Vol. 4 Human and Organisational Factors.* Washington, DC: Chemical Safety Board.

Cullen, D. (1990). *Report of the Public Inquiry into the Piper Alpha Disaster.* London: HSE Books.

Dekker, S. (2016). *Just Culture. Restoring Trust and Accountability in your Organisation.* (3rd ed.). Oxford: CRC Press.

EASA (2015). *CRM Requirements.* Part-FCL and EU-OPS. Cologne: European Aviation Safety Agency.

Edmondson, A. (2003). Speaking up in the operating room: How team leaders promote learning in interdisciplinary action teams. *Journal of Management Studies* 40, 1419–1452.

Energy Institute (2014). *Guidance on crew resource management (CRM) and non-technical skills training programmes* (authored by Flin & Wilkinson). London: Energy Institute.

Eurocontrol (2013). *White Paper: Senior Managers Safety Leadership*, www.eurocontrol.int, accessed December 20, 2016.

Fan, C., Pawlik, T., Daniels, T. et al. (2016). Association of safety culture with surgical site infection outcomes. *Journal of the American College of Surgeons* 222, 2, 122–128.

Fleming, M., Flin, R., Mearns, K., and Gordon, R. (1996, January). The offshore supervisor's role in safety management: Law enforcer or risk manager. In *SPE Health, Safety and Environment in Oil and Gas Exploration and Production Conference.* Richardson, TX: Society of Petroleum Engineers.

Fleming, M. (2000). *Safety culture maturity model. HSE Offshore Technology Report, 49.* London: HSE.

Flin, R. (1995). Crew Resource Management for training teams in the offshore oil industry. *European Journal of Industrial Training* 9, 19, 23–27.

Please provide publisher location.

Flin, R. (2018). European pilots' non-technical skills. In B. Kanki, R. Chidester, and J. Anca (eds.), *Crew Resource Management* (3rd ed.). Elsevier.

Flin, R., Patey, R., Glavin, R., and Maran, N. (2010). Anaesthetists' non-technical skills. *British Journal of Anaesthesia* 105, 38–44.

Flin, R., Martin, L., Goeters, K., Hoermann, J., Amalberti, R., Valot, C., and Nijhuis, H. (2003). Development of the NOTECHS (Non-Technical Skills) system for assessing pilots' CRM skills. *Human Factors and Aerospace Safety* 3, 95–117.

Flin, R., O'Connor, P., and Crichton, M. (2008). *Safety at the Sharp End: A Guide to Non-Technical Skills.* Farnham: Ashgate.

Flin, R., Youngson, G., and Yule, S. (eds) (2015). *Enhancing Surgical Performance. A Primer on Non-Technical Skills.* London: CRC Press.

Fruhen, L. and Flin, R. (2016). "Chronic Unease" for safety in senior managers: An interview study of its components, behaviours and consequences. *Journal of Risk Research* 19, 5.

Fruhen, L., Flin, R., and McLeod, R. (2014b). Chronic unease for safety in managers: A conceptualisation. *Journal of Risk Research* 17, 969–979.
Fruhen, L., Mearns, K., Flin, R., and Kirwan, B. (2014a). Safety intelligence: An exploration of senior managers' characteristics. *Applied Ergonomics* 45, 967–975.
Gershon, R., Karkashian, C., Grosch, J. et al. (2000). Hospital safety climate and its relationship with safe work practices and workplace exposure incidents. *American Journal of Infection Control* 28, 211–221.
Gimeno, D., Felknor, S., Burau, K., and Delclos, G. (2005). Organisational and occupational risk factors associated with work related injuries among public hospital employees in Costa Rica. *Occupational and Environmental Medicine* 62, 337–343.
Glendon, A. and Clarke, S. (2015). *Human Safety and Risk Management* (3rd ed.). London: CRC Press.
Goncalves F. A., Silveira A. J., and Marinho, M. (2010). A safety culture maturity model for petrochemical companies in Brazil. *Safety Science* 48, 615–624.
Gordon, R., Flin, R., and Mearns, K. (2005). Designing and evaluating a Human Factors Investigation Tool (HFIT) for accident analysis. *Safety Science* 43, 147–171.
Haddon-Cave, C. (2009). *The Nimrod Review: An independent review into the broader issues surrounding the loss of the RAF Nimrod MR2 Aircraft XV230 in Afghanistan in 2006.* London: Stationery Office.
Healthcare Commission (2009). *Investigation into Mid-Staffordshire NHS foundation trust.* London: Healthcare Commission.
Hofmann, D. and Mark, B. (2006). An investigation of the relationship between safety climate and medication errors as well as other nurse and patient outcomes. *Personnel Psychology* 59, 847–869.
Hofmann, D. and Morgenson, F. (2004). The role of leadership in safety. In J. Barling and M. Frone (eds), *The Psychology of Workplace Safety.* Washington, DC: APA Books.
Huang, D., Clermont, G., Kong, L. et al. (2010). Intensive care unit safety culture and outcomes: A US multicenter study. *International Journal of Quality in Health Care* 22, 151–161.
Hudson, P. (2007). Implementing a safety culture in a multi-national. *Safety Science* 45, 697–722.
Hull, L., Arora, S., Aggarwal, R., Darzi, A., Vincent, C., and Sevdalis, N. (2012). The impact of nontechnical skills on technical performance in surgery: A systematic review. *Journal of the American College of Surgeons* 214, 2, 214–230.
IAEA (1986). Summary Report on the Post-Accident Review Meeting on the Chernobyl Accident. International Safety Advisory Group (Safety Series 75-INSAG-1). Vienna: International Atomic Energy Agency.
IAEA (2016). *Performing Safety Culture Self-Assessments:* Safety Reports Series No. 83 Paperback – 31 Oct 2016. Vienna: IAEA.
IOGP (2013). *Shaping Safety Culture through Safety Leadership. Report 452.* London: International Oil and Gas Producers, http://www.iogp.org/bookstore/product/crew-resource-management-for-well-operations-teams/, accessed March 14, 2018.
IOGP (2014). *Crew Resource Management for Well Operations Teams* (Report 501; Flin, R., Wilkinson, G., and Agnew, C.). London: International Oil and Gas Producers, www.ogp.org.uk/publications/wells-committee/wocrm-report/.
Jackson, J., Sarac [Agnew], C., and Flin, R. (2010). Hospital safety climate surveys: measurement issues. *Current Opinion in Critical Care* 16, 632–638.
Jespersen, L., Griffiths, M., Mclaurin, T., Chapman, B., and Wallace, C. (2016). Measurement of food safety culture using survey and maturity profiling tools. *Food Control* 66, 174–182.
Kanki, B., Helmreich, R., and Anca, J. (eds) (2010). *Crew Resource Management* (2nd ed.). San Diego: Academic Press.

Kirwan, B. (2008, May). From safety culture to safety intelligence. In T. Kao (ed.), *Probability Safety Assessment and Management Conference, PSAM 9* (pp. 18–23). Hong Kong: International Association for Probabilistic Safety Assessment and Management.

McCulloch, P., Mishra, A., Handa, A., Dale, T., Hirst, G., and Catchpole, K. (2009). The effects of aviation-style non-technical skills training on technical performance and outcome in the operating theatre. *Quality and Safety in Health Care* 18, 109–115.

McLeod, R. (2015). *Designing for Human Reliability. Human Factors Engineering in the Oil, Gas, and Process Industries.* Oxford: Gulf Publishing.

Mearns, K., Flin, R., Gordon, R., and Fleming, M. (1998). Measuring safety climate in the offshore oil industry. *Work & Stress* 12, 238–254.

Mearns, K., Kirwan, B., Reader, T. et al. (2013). Development of a methodology for understanding and enhancing safety culture in Air Traffic Management. *Safety Science* 53, 123–133.

Mearns, K., Rundmo, T., Flin, R., Gordon, R., and Fleming, M. (2004). Evaluation of psychosocial and organizational factors in offshore safety: A comparative study. *Journal of Risk Research* 7, 545–561.

Mearns, K., Whitaker, S., and Flin, R. (2001). Benchmarking safety climate in hazardous environments: A longitudinal, inter-organisational approach. *Risk Analysis* 21, 771–786.

NAS (2016). *Strengthening the Safety Culture of the Offshore Oil & Gas Industry.* Washington, DC: National Academy of Sciences.

Nahrgang, J., Morgeson, F., and Hofmann, D. (2011). Safety at work: A meta-analytic investigation of the link between job demands, job resources, burnout, engagement, and safety outcomes. *Journal of Applied Psychology* 96, 71–94.

Nielsen, M., Eid, J., Mearns, K., and Larsson, G. (2013). Authentic leadership and its relationship with risk perception and safety climate. *Leadership & Organization Development Journal* 34, 308–325.

Nielsen, M., Skogstad, A., Matthiesen, S., and Einarsen, S. (2016). The importance of a multidimensional and temporal design in research on leadership and workplace safety. *The Leadership Quarterly* 27, 142–155.

O'Dea, A. and Flin, R. (2001). Site managers and safety leadership in the offshore oil and gas industry. *Safety Science* 37, 39–57.

Parker, D., Lawrie, M., and Hudson, P. (2006). A framework for understanding the development of organisational safety culture. *Safety Science* 44, 551–562.

Petitta, L., Probst, T., Barbaranelli, C., and Ghezzi, V. (2017). Disentangling the roles of safety climate and safety culture: Multi-level effects on the relationship between supervisor enforcement and safety compliance. *Accident Analysis & Prevention* 99, 77–89.

Probst, T. and Estrada, R. (2010). Accident under-reporting among employees: Testing the moderating influence of psychological safety climate and supervisor enforcement of safety practices. *Accident Analysis & Prevention* 42, 1438–1444.

Reader, T. and O'Connor P. (2014). The Deepwater Horizon explosion: Non-technical skills, safety culture and system complexity. *Journal of Risk Research* 17, 405–424.

Reason, J. (1997). *Managing the Risks of Organizational Accidents.* Aldershot: Ashgate.

Roberts, R., Flin, R., and Cleland, J. (2015). "Everything was fine": An analysis of the drill crew's situation awareness on Deepwater Horizon. *Journal of Loss Prevention in the Process Industries* 38, 87–100.

Rundmo, T. (1992). Risk perception and safety on offshore petroleum platforms—Part II: Perceived risk, job stress and accidents. *Safety Science* 15, 53–68.

Schein, E. (1990). Organizational culture. *American Psychologist* 45, 109–119.

Stanton, N., Salmon, P., Rafferty, L. et al. (2013). *Human Factors Methods* (2nd ed.). Aldershot: Ashgate.

Suhanyiova, L., Flin, R., and Irwin, A. (2017). Safety systems in product safety culture. In Walls, L., Revie, M., and Bedford, T. (eds), *Risk, Reliability and Safety: Innovating Theory and Practice* (pp. 1803–1808). Proceedings of the European Safety and Reliability conference, Glasgow, 2016. London: CRC Press.

Swensen, S., Pugh, M., McMullan, C., and Kabcenell, A. (2013). *High-Impact Leadership: Improve Care, Improve the Health of Populations, and Reduce Costs.* IHI White Paper. Cambridge, MA: Institute for Healthcare Improvement (available at ihi.org).

Tharaldsen, J., Olsen, E., and Rundmo, T. (2008). A longitudinal study of safety climate on the Norwegian continental shelf. *Safety Science* 46, 427–439.

Thomas, M. (2017). *Training and Assessing Non-Technical Skills: A Practical Guide.* London: CRC Press.

Vincent, C. (2010). *Patient Safety* (2nd ed.). London: Wiley Blackwell.

Waterson, P. (ed.) (2014). *Patient Safety Culture. Theory, Methods, Application.* Aldershot: Ashgate.

Weigmann, D. and Shapell, S. (2003). *Human Error Approach to Aviation Accident Analysis: The Human Factors Analysis and Classification System.* Aldershot: Ashgate.

Westrum, R. (1995). Organisational dynamics and safety. In N. McDonald, N. Johnston, and R. Fuller (eds), *Applications of Psychology to the Aviation System.* Aldershot: Ashgate.

Yukl, G. (2013). *Leadership in Organizations* (8th ed.). Harlow: Pearson.

Zhu, A., von Zedtwitz, M., Assimakopoulos, D., and Fernandes, K. (2016). The impact of organizational culture on Concurrent Engineering, Design-for-Safety, and product safety performance. *International Journal of Production Economics* 176, 69–81.

Zohar, D. (2002). The effects of leadership dimensions, safety climate, and assigned priorities on minor injuries in work groups. *Journal of Organizational Behavior* 23, 75–92.

Zohar, D. (2014). Safety climate: Conceptualization, measurement, and improvement. In B. Schneider and K. Barbare (eds). *The Oxford Handbook of Organizational Climate and Culture* (pp. 317–334). Oxford: Oxford University Press.

Zohar, D., Livne, Y., Tenne-Gazit, O., Admi, H., and Donchin, Y. (2007). Healthcare climate: A framework for measuring and improving patient safety. *Critical Care Medicine* 35 (5), 1312–1317.

Zohar, D. and Luria, G. (2005). A multilevel model of safety climate: Cross-level relationships between organization and group-level climates. *Journal of Applied Psychology* 90, 616–628.

Case Study 3

Human Factors in Safety Management: A Cross-Cultural Comparison

Yousuf Mohamed Al Wardi

INTRODUCTION

It is generally accepted that human errors are the underlying cause of 70–80% of civil and military aviation accidents (Sarter and Alexander, 2000). Chappelow (2006) further emphasized that 40% of military accidents in the United Kingdom are attributed to aircrew human factors. Failure to consider the impact of human factors principles on human performance may lead to gross misjudgment of total system safety and therefore compromise operational performance. To mitigate accidents, the FAA acquisitions are a multi-disciplinary effort to generate and apply human factors considerations and human performance information to acquire safe, efficient, and effective operational systems.

NATIONAL CULTURE AND AIRCRAFT ACCIDENTS

It has been suggested that national cultures have different accident rates as a result of different underlying human factors causes attributable to culture (Soeters and Boer, 2000). National culture has been proposed to be classified based on four different dimensions: power distance, individualism, uncertainty avoidance, and masculinity (Hofstede, 2001). Each society exhibits different patterns for these four dimensions. For example, the USA is low in power distance (authority is questioned and power is distributed); individualistic (emphasize "I" vs. "we"); low in uncertainty avoidance (tolerates ambiguity); and masculine (assertive, achievement-oriented). The Republic of China (ROC) is high in power distance (it is accepted and expected that power is distributed unequally); collectivistic (groups are highly integrated with strong loyalty), high in uncertainty avoidance (rules are strict and there is less acceptance of ambiguity and a fear of failure); and there is a feminine culture (with cooperation, care for the others, and modesty). Oman is categorized as high in power distance, collectivistic, high in uncertainty avoidance, and with a masculine culture. These dimensions can affect aircrew interaction, which may consequently have an impact on flight safety.

The Human Factors Analysis and Classification System (HFACS) framework was applied to existing aircraft accidents investigation reports for the analysis of human factors causes of aircraft accidents in Oman (Al-Wardi, 2017). The pattern and frequencies of 18 individual categories of the HFACS framework were examined and the results were compared with US data (Wiegmann and Shappell, 2001) and ROC data (Li and Harris, 2005) to assess national cultural differences in aircraft accidents.

CROSS-CULTURAL COMPARISON

The HFACS framework showed some significant differences in the relative rates of occurrence of human factors causes implicated in aircraft accidents between Oman, Taiwan, and the USA (Figure CS3.1).

The four cultural dimensions determine, to some extent, the manner in which people lead their lives and interpret events. The culture in Oman consists of tightly packed families where an individual is expected to respect and show obedience to elders. There is a low tolerance of uncertainty and ambiguity, and a fear of failure. This is also reflected in the organizational structures, which tend to be pyramidal with a great deal of bureaucracy. On examining the results of the applied HFACS framework, it was concluded that a high number of accidents occur due to unsafe supervision and organizational influences, which is highly attributable to the type of culture Oman has, where subordinates are dependent on their superiors and are unlikely to question those in authority. The rest of the GCC has an almost identical

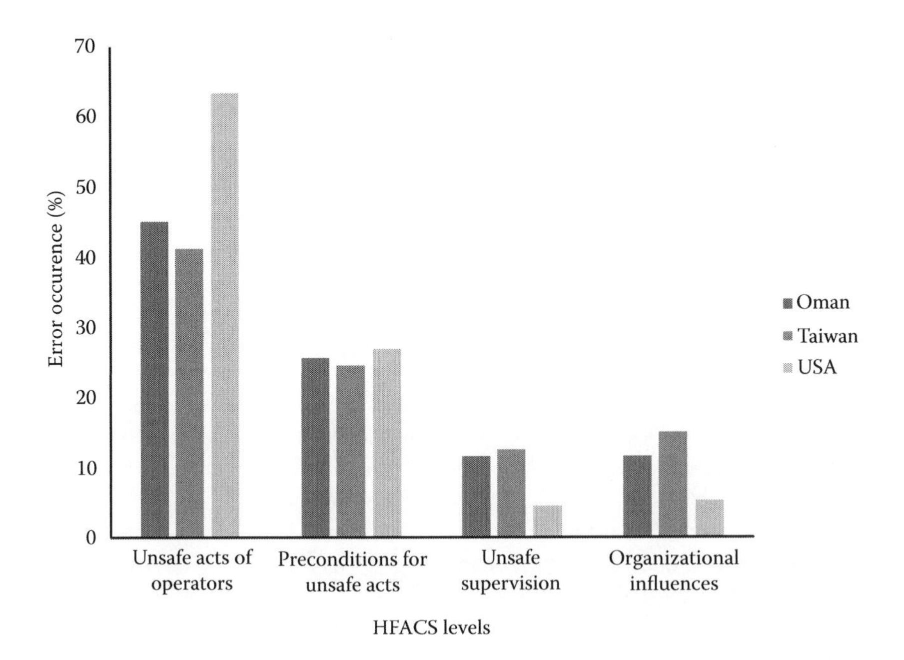

FIGURE CS3.1 The percentages of occurrence of HFACS levels between Oman, Taiwan, and the USA.

culture and it is therefore safe to say that the results can also be generalized to apply to the rest of the GCC.

The nature of the national cultures in both Oman and Taiwan were found to share similar dimensions, except in masculinity/femininity, and were quite the opposite to the USA's national cultural dimensions. Therefore, we found that the causal factors contributing to aircraft accidents in Oman and Taiwan were similar to a very large extent, whereas the causal factors in the USA differ completely. The USA has a higher percentage of accidents attributed to unsafe acts of operators, while, on the other hand, Oman and Taiwan exhibit a higher percentage of accidents as a result of unsafe supervision and organizational influences, which are directly attributable to the type of national culture these two countries have. In Oman and Taiwan, authority is focused on centralized decision-making. The same applies to the rest of the GCC, where the power of superiors is based on controlling uncertainties that are frequently managed by implementing new rules and regulations. On the contrary, in the USA, authority is less concentrated and decision-making is more distributed. Individuals are trained to work independently and, if necessary, act outside official procedures, which explains the rise in accidents as a result of unsafe acts of operators.

Oman has a lower number of occurrences of human errors at the "unsafe acts of operators" level in the HFACS system than the USA's sample of accidents. On the other hand, as mentioned previously, Oman exhibited a higher percentage of human errors at the management and supervisory levels. Organization and supervision are the most critical areas in order to avoid the formation of latent underlying error conditions. In HFACS, levels where co-ordination, consultation, and—if necessary—questioning superiors by subordinates are critical issues; Oman and Taiwan showed higher accidents than the USA.

SAFETY IMPLICATIONS

Flight safety is implemented by the utilization of proactive measures in the USA, which involves identifying threats and risks before they result in an accident. This is usually done by management. The HFACS USA data analysis illustrated the low frequency of human factors contributing to accidents at the managerial and supervisory levels. On the other hand, Oman, along with the rest of the GCC and Taiwan, heavily relies on reactive measures, which involves taking action after an accident has taken place.

This study revealed that national culture plays a significant role in aircraft accidents. Oman and other GCC states share similar if not identical national culture characteristics. These human factors methodologies may be used by all organizations, including surface transportation, health care, and energy industries to identify the human causes of accidents and provide tools to assist in the investigation process using target training and prevention efforts. Thus, it is suggested that these techniques are used by all safety departments to put into practice more guidance associated with human factors and ergonomics (HFE) in the accident/incident investigation process and develop a better accident database. Also, the importance and contribution of HFE in operational safety should be conveyed to all personnel in the organization. As was demonstrated, cultural differences in Eastern cultures (Oman, other GCC countries, Taiwan) and Western cultures (USA) have a direct impact on safety due to the relationships between the superiors and their subordinates. Therefore, developing a pro-active safety culture rather than a traditional reactive safety culture to facilitate reporting systems is critical in safety management.

REFERENCES

Al-Wardi, Y. (2017). Arabian, Asian, western: A cross-cultural comparison of aircraft accidents from human factor perspectives. *International Journal of Occupational Safety and Ergonomics* 23(3), 366–373.

Chappelow, J. W. (2006). Error and accidents. In Rainford, D. J. and Gradwell, D. P. (eds), *Ernsting's Aviation Medicine* (4th edition) (pp. 349–357). London: Hodder Arnold.

Hofstede, G. (2001). *Culture's Consequences: Comparing Values, Behaviours, Institutions, and Organizations Across Nations* (2nd edition). Thousand Oaks: Sage Publications.

Li, W. C. and Harris, D. (2005). HFACS analysis of ROC air force aviation accidents: Reliability analysis and cross-cultural comparison. *International Journal of Applied Aviation Studies* 5(1), 65–81.

Sarter, N. B. and Alexander, H. M. (2000). Error types and related error detection mechanisms in the aviation domain: An analysis of aviation safety reporting system incident reports. *The International Journal of Aviation Psychology* 10(2), 189–206.

Soeters, J. L. and Boer, P. C. (2000). Culture and flight safety in military aviation. *The International Journal of Aviation Psychology* 10(2), 111–133.

Wiegmann, D. A. and Shappell, S. A. (2001). Human error analysis of commercial aviation accidents: Application of the human factors analysis and classification system (HFACS). *Aviation, Space, and Environmental Medicine* 72(11), 1006–1016.

4 Information and Communication Technologies and Human Interaction

Catherine M. Burns and Carlos A. Lucena

CONTENTS

INTRODUCTION

Information and communication technologies (ICT) have evolved exponentially during the last couple of decades. In the 1980s and 1990s, most people experienced ICT through computers at the workplace. In the late 1990s and early 2000s, ICT changed in a dramatic way with the development of the smart phone. The smart phone made ICT a nearly-constant human companion. Networks and connectivity became widespread and this connectivity is enabling new technologies such as wearable devices, streaming media, and the "internet of things." ICT allows people great access to information and makes communication more affordable and accessible. ICT has enabled access to services, like banking, education, and tele-medicine. It has changed the work lives of many, reducing some kinds of work and increasing others. However, the impact of ICT is not always neutral or positive. Confidential personal information, such as our health and finances, are being exposed; at the same time there are large amounts of sophisticated technologies being developed to steal strategic private date. Furthermore, ICT has changed job roles, leading to an "always on" mentality. A significant and current topic of discussion is the prediction of automation and the anticipated widespread job losses.

The evolution of ICT now faces endless opportunities with the support of ubiquitous technologies. Although ICT are built in order to assist people in executing specific tasks, there are always human influences in the outcome of an applied technology. Therefore, the interaction of humans and ICT has to be taken into account if a task is to be done successfully. The most effective way to understand ICT, apply it effectively, and anticipate its effects is to always return to a discussion of how technology interacts with human capabilities and influences human work.

Historically, there are examples that show why ICT can go wrong if human factors/ergonomics (HFE) concepts are not followed properly. For instance, errors involving the handling of technology by humans can be found in the military, health, and finance fields. On September 26, 1983, the Russian lieutenant colonel Stanislav Petrov and his crew received information from their missile monitoring system that the USA had fired nuclear weapons. The radar system, in response to a false alarm, required Petrov to press the button and launch weapons back at the USA as a counterattack. Instead, Petrov waited to monitor the confusing signals showing three, four, and five strange images. Using his knowledge and skills, he recognized that an attack would fire hundreds of missiles, rather than just five. Petrov avoided a major international conflict by using his own judgment rather than using a faulty system (Cimbala, 2000). Petrov interpreted the information provided by the radar as a false alarm, and as a result did not pursue an offensive attack against US ships.

In the medical field, one of the most discussed cases involving software and health care ICT glitches is the Therac 25 radiation events (Leveson, 1995). The Therac 25 medical radiation device was involved in several cases related to massive overdoses of radiation applied to patients during cancer treatment, which killed at least three patients from 1985 to 1987. The controlling computer was placed in a room apart from the physical device where the patient's sessions took place and operated by trained professionals to release managed beams of radiation into specific areas on the skin. Bad design in software upgrades allowed operators to let laser beams act over patients without applying the proper filter. Operators were still correcting the filter setup to manage the beam while it was releasing at full capacity over the patient's skin. The resulting radiation was so strong that patients died a few days after the episode.

A bad ICT design incident that caused a massive economic catastrophe also occurred in 2012, involving trading technologies for the financial market. Software developed by the Knight Capital Group (Heusser, 2014) was released without the proper testing, resulting in financial losses of $440 million for clients. On August 1, 2012, during a period of only 30 minutes, Knight's trading software released false calculations for its traders, influencing enormous amount of bad deals and resulting in great losses. The programmer responsible for the problem had deployed a new version of the software to the production environment with an obsolete functionality that was not replaced by the new code.

The examples described above demonstrate how ICT products might fail if not designed with the user in mind. It is essential to understand people in order to design reliable products and experiences. A good example is illustrated by IDEO (a design company focusing on building meaningful user experiences), which partnered with the health industry company Medtronic to design products for patients and health care providers. IDEO designed and developed a device that nurses would use to enter data during specific procedures. Approaching this problem with the user mindset, IDEO delivered a device that enabled nurses to enter data using only one hand, leaving their second hand free to hold the patient's hand in order to provide comfort in difficult and stressful situations (Lanoue, 2016). Thus, using HFE principles, techniques and tools can be used to help design optimal user interactions.

FUNDAMENTALS

When people work with ICT, it is always in partnership. Well-designed ICT help to extend human capabilities by providing enhanced information processing, improved communication, and far greater access to information than would otherwise be possible. However, poorly designed ICT can interfere with successful work, interrupting workflow, becoming a barrier to communication, or making simple tasks more difficult. The first step in designing ICT successfully is to think carefully about how they can support humans in their work.

MATCHING HUMAN CAPABILITIES WITH ICT

There are certain tasks that humans are good at, while computers do a better job of other tasks. For instance, memory tasks are challenging for humans, while computing systems can easily store large amounts of data for future access. Humans can be very good at pattern recognition, particularly in the case of ambiguity or noise. Additionally, while humans may make errors computers will follow their programming exactly, only producing errors in the case of programming problems. For these reasons, computers are suited to calculating or processing large amounts of data, while humans may fatigue and suffer distractions without the proper support of ICT to accomplish tasks. Humans are currently better than ICT at complex tasks like driving, though this is changing quickly as advanced automation becomes more capable. Once automation can handle the ideal control of the vehicle and detection of various driving situations, it is very likely that automated driverless cars will be safer, more reliable, and more efficient than human drivers, who are prone to fatigue, distraction, and error. So while we want to be aware of human and computer capabilities, it is important to realize that with technological changes the fit of ICT to human capabilities is always evolving.

Human Information Processing

To better understand human capabilities, it is helpful to understand how humans process information. While there are many different models, most break down human information processing into four key blocks: sensing, perceiving, decision-making, and acting (Figure 4.1). Many models also include attention as a modifier, controlling and determining how the other processes work. Memory is also often distinctly identified, and sometimes broken apart into long-term and short-term memory. It is important to realize that information flows from sensing to acting, but that people are always observing their performance, and receiving feedback that allows them to adjust better to their environment. This feedback is shown in the arrows that move from right to left in Figure 4.1.

The human information processing model above lets us understand how people think and encourages us to ponder on the support needed for each process using ICT products. The model is somewhat simplified and may lead us to think that these processes occur linearly, when, in reality, all processes are occurring simultaneously. In ICT design, though, it is useful to think about whether a design supports perception/information acquisition, memory, decision-making, and action appropriately, and what problems might arise at each stage of processing. There are many good references on human information processing, such as the publication by Wickens and colleagues (2012).

Table 4.1 provides a list of common ICT technologies, in relation to how they support information processing. This table is a simplification, however, as most of these technologies may support more than one information processing step.

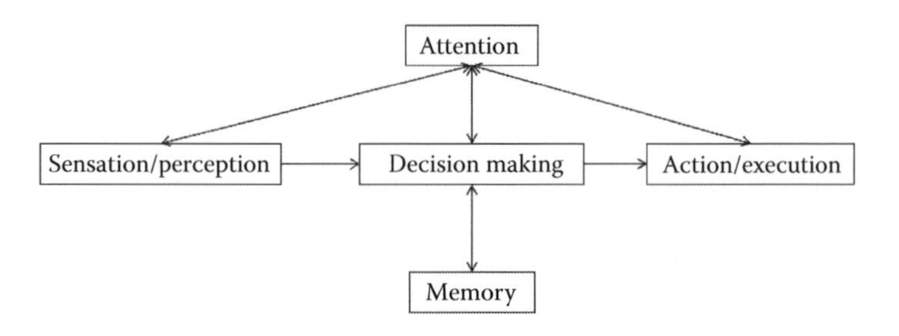

FIGURE 4.1 Simple human information processing model.

TABLE 4.1
Information Processing and ICT Examples

Information Processing Step	ICT Examples
Perception/information acquisition	News and travel websites, social media, data acquisition systems in industrial plants, sensor-driven data collection (e.g., satellite imagery or fitness trackers)
Decision-making	GPS systems for navigation, decision algorithms, automated financial trading, clinical decision support systems in health care, diagnosis systems in vehicles
Action/execution	Cruise control systems on cars, control systems for robots, autopilots on planes, stock purchase software, home automation
Attention	Notification systems, search engines
Memory	Calendar and workflow tools, education systems

When we examine ICT in this way, we can ask good questions about how it is helping people. For example:

- Is the human currently performing a task that is too complex, or presents risks that could be performed better by ICT?
- Is the ICT designed as well as possible to support human work? If the ICT is performing the task instead of the human, is the human appropriately aware of its actions? Can the human step in, if needed, and take over?

Types of Human Error

Understanding human interaction and error in particular is key to good design. While humans will inevitably make errors, errors also provide clues to how we can design better ICT. Human error can often be understood by considering whether it was automatic or intended, and whether the error involved taking an inappropriate action, or neglecting to take a needed action. This error categorization stems from the work of James Reason (1990). We can use these two categories to develop an error space that identifies four different kinds of human error, as in Table 4.2.

Slips and lapses are errors that are mainly unconscious. A slip is an error where the human unconsciously makes the wrong action, and a lapse is an error where a human neglects to take a needed action. These kinds of errors show up in highly practised behavior with highly skilled workers. Essentially, the user has progressed to the point where they can perform the task successfully with low levels of attention, and that is when these errors will tend to occur.

TABLE 4.2
Types of Human Error

	Not Intended, Unconscious	Intended
Action taken	Slip	Mistake
Action not taken	Lapse	Error of omission

Mistakes and errors of omission are arguably more serious. In both of these situations, the human has thought carefully about the situation, and for some reason has misunderstood the situation, and proceeds to make the error. When users make these kinds of errors they are either perceiving the wrong information or interpreting the information incorrectly, and therefore taking the wrong action or not taking the needed action. These errors can result from the user misunderstanding the information, holding an incorrect mental model of the situation, or not obtaining or attending to the information correctly.

When we design ICT, we want to design with the premise that our users will make errors. We want to anticipate what errors they might make and provide error recovery processes that will catch and correct those errors. By using early testing in the design process to observe what kinds of errors our users make, we can improve our designs so that we can eliminate slips, prompt to correct lapses, and design technologies that promote correct mental models and entice a user's attention to information correctly. Therefore, watching and studying human error can be one of the most informative ways of improving a design.

ICT Design Process

User-centered design (UCD; Norman & Draper, 1986) is a term used to describe the process of designing everyday products (physical or digital), with the focus during design being on the end user. UCD takes into consideration the end user on several stages of the design process, from planning to evaluation. The requirements of the design process are gathered by observing the needs of targeted users to actually interact with the proposed ICT solution, which evolve as the experience of the user is captured. The designer acts as a facilitator for the user's task or product, making sure that they interact with the product in an effortless matter.

Contextual design is a UCD process incorporating contextual inquiry (Beyer and Holtzblatt, 1997). Contextual inquiry is a technique that focuses on understanding the realism of user work. The designer spends time with the user in a master-apprentice role, while the user teaches the designer the complexities of their work. The designer should watch carefully for the context of the work—is it noisy, quiet, high-pressured? Are there particular artefacts and ways of organizing work that are already prevalent in the workspace? Contextual inquiry and design develops five different models of the user work: (1) the flow model, which shows the flow and coordination of information; (2) the sequence model, which shows the steps of certain activities and potential breakdowns; (3) the cultural model, which shows the influence of culture and community on the users; (4) the artifact model, which shows existing objects that users use to organize their work; and (5) the physical model, which shows the layout and environment where the work occurs. From these models, design needs emerge and are solidified through affinity diagrams. From those needs, storyboards and prototypes can be developed and evolved iteratively with users.

Ecological design (Burns and Hajdukiewicz, 2004) complements UCD in that the focus is on the work domain or environment of the user, rather than the on the user or their specific task(s). This approach was introduced specifically for complex dynamic systems such as nuclear power plants, aviation, and medicine. Ecological design looks at complexities that were discovered during work domain analysis and seeks to provide strong visual support for users to understand those relationships better. A fundamental premise of ecological design is that complex and effortful "knowledge-based" behavior can be moved to easier visual or perceptual "rule-based" behavior. This approach reduces user workload but can also help novice users to understand and develop more accurate mental models of how their work system operates. Due to their emphasis on the visual display of complexity, ecological designs often incorporate object displays or data visualizations in the design.

Design thinking, which is also based on the UCD process, was first used by architects and urban planners (Rowe, 1987). Although the term has its origins in architecture, it was widely disseminated for business purposes by David Kelley, who founded the company IDEO in 1991. It focuses on creative ways to look at a problem and try to find possible solutions. Design thinking processes have seven stages: define, research, ideate, prototype, choose, implement, and learn. It emphasizes exploring multiple possibilities by fast prototyping and testing selected ideas. Design thinking is being adopted by several companies around the world today and is used with multidisciplinary teams looking at specific problems through the perspective of different industries, areas, or activities.

METHOD

A systematic approach to ICT design requires a user needs assessment and iterative evaluation. We will provide an introduction to some core methods in each of these steps, but we caution that there are many different methods that can be used to suit different situations. Some other sources to consider are included in the references of this chapter.

User Needs Assessment

All design must begin with understanding the needs of users as well as possible. Designers spend time talking to users, watching them work, and conducting interviews to prompt users to consider their needs more explicitly. There are many good questions that can be used to initiate and develop a discussion with users (Crandall and Klein, 2006). Once time has been spent understanding a user's needs, these needs must be transformed into design requirements. Two possible approaches that may be used to help with this process are cognitive task analysis and cognitive work analysis.

Cognitive Task Analysis

Cognitive task analysis (CTA) seeks to understand how people work to develop good recommendations on how to support that work. While there are many variations and similarities on how to perform this analysis, we are presenting a general approach to breaking down tasks using hierarchical task analysis (Kirwan and Ainsworth, 1992). Some other approaches that can be useful include concept mapping (Novak and Canas, 2008) and the critical incident method (Klein, Calderwood, and Macgregor, 1989). Concept mapping can help you understand user mental models of complex decision-making. The critical incident method is a useful approach when trying to understand how experts solve complex time-pressured problems by conducting an interview after an incident occurs.

A CTA involves interviews and observations of how people work. The goal of CTA is to understand current working conditions, while having a systematic manner to discover improvements and design new work processes and to evaluate the success of your new design. The first step is to understand what tasks the user must perform. Users will perform routine tasks and tasks that occur under unexpected conditions like emergencies or new problems. The best CTA will look at both kinds of tasks. When we improve the workflow on routine tasks, we make the user's everyday experience better. However, we must also look at how our users work in less routine situations, to ensure that our product will support those unexpected situations effectively.

A CTA breaks down user tasks into smaller components and then uses those components to analyze the needs at various stages. If we consider a very simple user task of withdrawing funds from a bank, we could have a CTA as follows (Figure 4.2).

Table 4.3 demonstrates how a task analysis can be used. At each task step (column 1), the needs of the user can be identified (column 2). By identifying these needs, you are able to identify design ideas to assist with fulfilling user requirements (column 3). For more information on CTA, we suggest the following sources: Annett and Stanton (1998), Kirwan and Ainsworth (1992), Klein (2000), Lane, Stanton, and Harrison (2006), and Militello and Hutton (1998).

Cognitive Work Analysis

Cognitive work analysis (CWA) can be useful for complex tasks or situations where unexpected events can be particularly dangerous (Vicente, 1999). CWA has been useful for power plant displays, health care (Burns, Enomoto, and Momtahan, 2008), and financial decision-making support. It supplements interviews and observations by looking at the problem through several analyses that explore different perspectives. Table 4.4 explains the five analyses that CWA explores. For more information on CWA, we suggest Bisantz and Burns (2008).

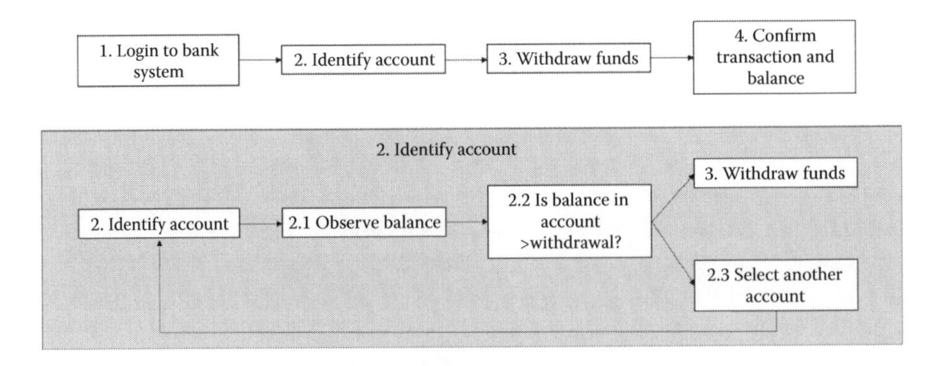

FIGURE 4.2 A simple hierarchical task analysis.

TABLE 4.3
Simple Connection between User Tasks, Needs, and Design Ideas

Task Step	User Needs	Design Idea
Login	Remember password	Incorporate password prompt or recovery
Identify account	Remember accounts. Associate accounts with purpose (e.g., saving, credit, foreign transactions)	Allow user to write personal description of accounts
Withdraw funds	Understand amount in account, execute withdrawal, receive funds in appropriate form (cash or transfer)	Clearly show balance and balance remaining after withdrawal
Confirm	Understand new balance in account	Show the transaction has been completed and provide guidance for further action

TABLE 4.4
The Five Analyses of Cognitive Work Analysis

View	Analysis	Elements
What are the fundamental complexities of the problem?	Work domain analysis	How does plant equipment work and what is the core process that is occurring? What is being managed and how? What are the financial laws governing these transactions?
How do people and the software process information in this work situation?	Control task analysis	What is the best human and computer allocation of information processing? How do experts and novices process information differently?
What are the different ways people approach their work?	Strategies analysis	How do people shift what they do under high workload? Missing information? Fatigue?
How do teams work together? How does the organization influence work?	Social and cooperation analysis	What are the rules, regulations, and culture? How do teams pass information? When are goals shared or disparate?
What skills should workers have?	Worker competency analysis	What physical and perceptual skills are required? What rules and heuristics are needed? What training is needed to make novices advance to being experts?

Iterative Evaluation

Following the concepts of UCD mentioned above, designers involved in developing products, digital interactions, or processes apply iterative cycles of evaluation to quickly deliver an interactive solution. Assessment of performance, error, and workload are key to evaluating solutions. The method of evaluation should be linked to the main goal of the product designed. As there are different characteristics of products, there should be adequate evaluation methods explored for each context. By using this approach, designers are able to collect feedback from user interactions and apply design improvements for the next iterative cycles. Table 4.5 demonstrates techniques to involve users in the development and evaluation of ICT products/ interactions (Preece et al., 2002).

When interviewing, observing, and/or collecting data from users, there are several key issues to consider while using the information processing model as your framework to guide in the design process.

- Perception and Information acquisition. How does the technology present information to the users? Is it well laid out and easily extracted? Are key relationships shown so that users can understand them? Is information hidden and possibly hard to find?

TABLE 4.5
Involving Users in the Design Process

Technique	Purpose	Stage of the Design Cycle
Background interviews and questionnaires	Collecting data related to the needs and expectations of users; evaluation of design alternatives, prototypes, and the final artefact	At the beginning of the design project
Sequence of work interviews and questionnaires	Collecting data related to the sequence of work to be performed with the artifact	Early in the design cycle
Focus groups	Include a wide range of stakeholders to discuss issues and requirements	Early in the design cycle
On-site observation	Collecting information concerning the environment in which the artefact will be used	Early in the design cycle
Cognitive walkthroughs, role-playing, and simulations	Evaluation of alternative designs and gaining additional information about user needs and expectations; prototype evaluation	Early and mid-point in the design cycle
Usability testing	Collecting quantitative data related to measurable usability criteria	Final stage of the design cycle
Interviews and questionnaires	Collecting qualitative data related to user satisfaction with the artefact	Final stage of the design cycle

Source: Adapted from Preece, J. et al., *Interaction Design: Beyond human computer interaction*, Chichester, West Sussex, John Wiley & Sons, 2002.

- Decision-making. Does the technology support decision-making? Does it show information in a context that makes decisions easier? Does the technology have automation that can help the user? Are the role and the actions of the automation clearly understandable?
- Action and execution. Is the task flow efficient? Can the user easily take action? How does the technology respond to errors? Are they easily corrected, or do they have serious consequences?
- Memory. Does the technology support and aid human memory? Or does it add more memory load on the user?

Designing to improve ICT incorporates these questions throughout the design process, as each hypothetical design is returned to users for feedback. There are many good sources to understand this iterative design process better (e.g., Lanoue, 2015; Preece et al., 2002).

APPLICATION

GPS Systems

Global positioning systems (GPS) are common these days and are present in cars, on smart phones, and in watches and other technologies. The overall goal of most consumer GPS systems is to track location and provide navigation. Using our information processing approach, how does the GPS system work to enable people? In Figure 4.3 below, we show the task of driving without a GPS system on the left, and with a GPS system on the right.

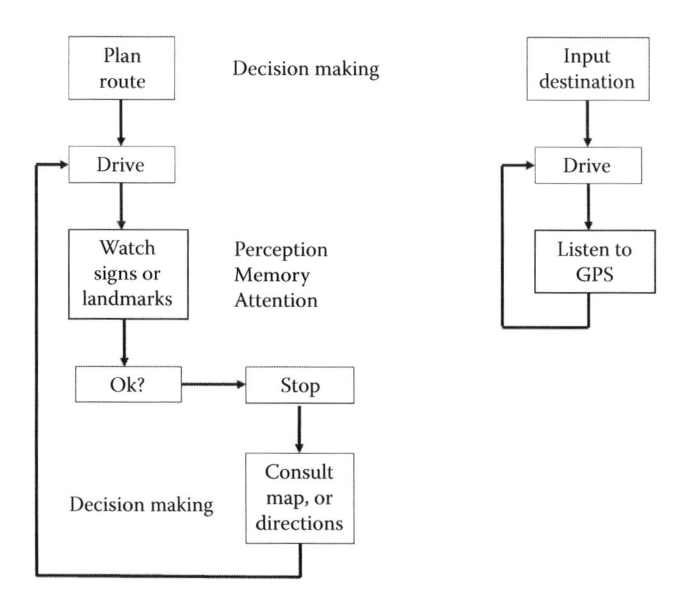

FIGURE 4.3 Driving with GPS (from BME 162 course material, copyright permission by Catherine Burns).

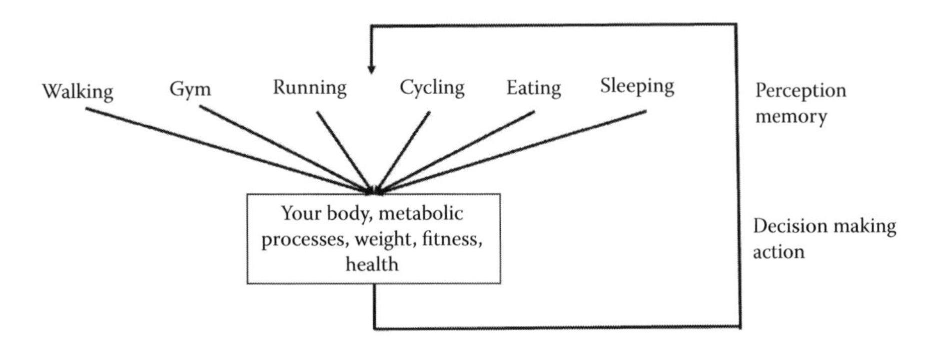

FIGURE 4.4 Information collection with fitness tracking.

It is immediately clear that the GPS technology replaces some difficult and sometimes dangerous human information processing tasks. First, the person no longer needs to plan their route and can instead input the destination and let the GPS system plan the route. The driver no longer needs to watch for road signs and landmarks, both demanding perception and memory tasks. This should free cognitive resources in order to focus on driving. If the driver goes off-route, the GPS system can guide the driver back, aiding the driver by choosing the route for the driver, providing decision-making support. Driving is a complex task for humans, and it is clear that the GPS system makes a strong contribution to aiding this task.

Are there any risks with GPS systems? One documented risk that has led to accidents is excessive trust in the system. As you can see in Figure 4.4, the GPS system has largely removed the human from the task of route planning and monitoring. This allows the driver to focus on driving but can also take their attention away from their surroundings. This had led to cases of drivers following poor GPS directions leading to inappropriate locations such as lakes or roads not appropriate for the vehicle.

Fitness Tracking

Fitness trackers are one of the early success stories of the "internet of things." These devices, usually worn on the wrist, contain a range of sensors to track activity and connect to software on a phone or computer to provide more detailed analytics, coaching, and health advice to consumers. These devices have replaced many perception tasks that are very difficult for people. You might be able to measure how long you have walked for, but counting your steps is a dull and error-prone task. It is very hard for people to monitor the calories they eat but these devices can integrate food databases that help people to look up this information. Further, tracking the quality of your sleep is nearly impossible. Clearly, these devices capture more information than humans could capture on their own.

Fitness tracking devices also aid decision-making. They can estimate the calories burned by exercise and compare that with calories consumed. They can recommend certain exercise levels that can help the consumer reach their goals. Combining sensors, timing, and contextual information they can provide analytics that help you to understand your patterns better, and how to develop better fitness practices.

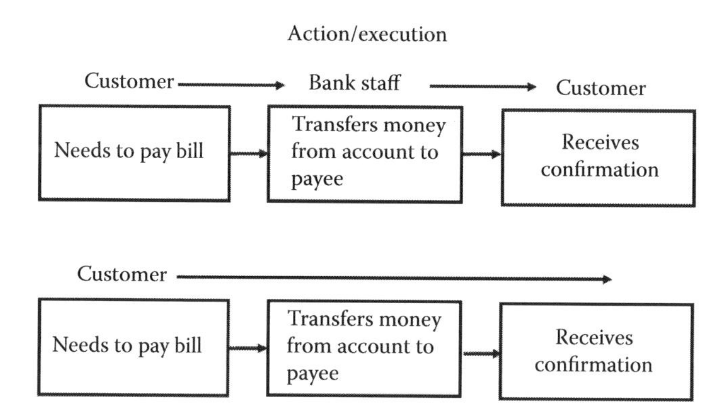

FIGURE 4.5 ICT enabling action in new ways.

Some of the devices can prompt for action by reminding you to move more. A weakness of these devices might be in execution; they cannot actually force the user to perform the activity, just persuade. (For more information on how these technologies can be designed to be more persuasive, consider Fogg, 2002; Mercer et al., 2015.)

Electronic Finance

In many cases, ICT has made executing actions much easier for people. Some of the most successful technologies are directly enabling this kind of support for people by helping them to communicate at a distance or carry out new kinds of interactions, or reducing the need for others to help them. Electronic finance is an example of this kind of interaction. Before ICT, people would seek an intermediary such as a bank teller to help them to conduct a transaction. Electronic banking systems have put this action into the capability of customers.

Action and execution ICT needs to be particularly well designed. People need to be aware of the actions that can be taken, and need good feedback after the action has occurred. These systems need error recovery mechanisms, and ways to show the user the effect of their actions. In the transaction shown in Figure 4.5 above, a good system will typically show the account balance before and after the transaction, and confirm the transaction before executing it.

Action and execution technologies are increasing in scope and ubiquity. Automated financial trading systems capitalize on the speed gains that are possible from fast connections to stock markets and employ complex algorithms to execute thousands of trades per day. The faster these systems execute, with less human feedback, the greater the risks for unexpected execution effects to occur (Minotra and Burns, online first).

FUTURE TRENDS

Emerging Technologies of ICT

According to Vicki Huff Eckert, PwC US, and Global New Business Leader (PwC, 2016), there is a shortlist of eight emerging technologies that will drive our future endeavors. All tools listed allow the exploration of eventual fits into ICT opportunities and challenges with human capabilities. As information and communication can be explored in all areas that human beings interact with, it is a matter of applying the UCD model in order to plan and deliver solutions that are useful to the specific user profile.

- "Internet of things." A wireless communication system that allows object to interact and exchange information.
- Augmented reality. Gathering and adding information to digital information
- Virtual reality. Simulated experiences that replicate real-world environments
- Blockchain. New technology using a continuously growing list of records (blocks), linked and secured using cryptography for secure communications.
- Artificial intelligence. Software algorithms designed to learn and apply human decision-making processes.
- 3D printing. Fast prototyping and modeling activity that allows physical representations of computer-designed elements.
- Drones. Devices capable of providing visual perspectives of physical spaces.
- Robots._Machines with embedded software that are able to aid human tasks.

Increasing Information

The amount of information that is available continues to increase with ICT. The world of data is continuing to grow and in many cases is exceeding the human capability to interpret that data. DNA can be sequenced far faster than humans can process the sequences. Human knowledge as represented by journal articles is expanding faster than students can keep track of. In many ways we are moving into the world of "big data." From a human factors perspective, this means increasing efforts to collect and interpret information for people. With large stores of data, data analytics becomes more important through its ability to summarize trends and extract meaning from data. Search engines begin to become less useful and people will turn to data collection programs or bots that can utilize machine learning to understand a person's search needs in more detail and provide a better filter on the world. Humans will need to learn how to integrate with machine learning systems more effectively. When each input begins to change the system, the user needs to become more careful about how they manage that interaction. Returning to the ideas of supporting human perception and filtering information appropriately, and supporting decision-making and putting that information in context, will continue to be the foundation of successful ICT in this space.

Increasing Connectivity

As more devices become connected, people will gain access to more information but may also need to consider new kinds of decision-making. Having your home devices controllable permits a wider range of options for the homeowner to consider, from ranges of accessibility to complex environmental control schemes. In each case, returning to an analysis of how the particular technology will help the human is crucial to designing for the success of the technology. Whenever the new technology is replacing a task that is challenging or effortful for humans, it is likely to be successful. The risk with some new technologies is that they may replace simple tasks with complex ones.

Increasing Automation

Automation is possibly the most influential new technology in the immediate future. It will become more pervasive and have a larger role in people's daily lives. From automated vehicles to automated industries, automation will greatly change the roles of humans. Already automation can be described using the information processing model in the same way we have used it here, looking at whether the automation replaces perception, decision, action, or all three of these functions. Depending on the type of the automation, the human factors question becomes how does automation interact with people? Can people understand its actions and does it communicate what it will do next?

CONCLUSION

The evolution of ICT seems endless. Although there are already trends that might guide the way ICT is used and developed, it will reshape itself as new technologies arise. The world population is connecting itself by the use of technology while communication is being delivered instantly and ubiquitously. Despite the automation of processes and solutions, humans are increasingly getting involved with technology even in traditional fields such as education and health. Therefore, it is increasingly relevant to value concepts and techniques that focus on that human interaction.

KEY TERMS

Human information processing. An approach that looks at how people receive information, make decisions, and take actions. This approach draws on concepts such as perception, attention, and memory to explain how information is used.

Human error. An unexpected consequence that can be tied to a human action or omission. From a human factors perspective, there are different types of human error, depending on whether the human's intention was right or wrong, or if the human took an action or neglected to take a required action.

User-centered design. A process that considers user needs and usability at the core of the design process. A user-centered design process involves users at several different stages of the process.

Mental model. A representation in someone's head of relationships in the world.

Storyboard. A presentation in the form of images in sequence. The sequence of images tell a storyline and the images help to provide a visualization of that story. Storyboards are used in motion picture planning and animation and can be used to show a user experience as part of a user-centered design process.

Affinity diagram. A tool used to organize large amounts of data or ideas. When making affinity diagrams, individual ideas or data are first listed in blocks or on paper, then organized to create groups of meaning.

Knowledge-based behavior. Analytical and deductive reasoning behavior exhibited by a user; used in cognitive work analysis.

Rule-based behavior. Behavior that is driven by heuristics and the recognition of similar situations from experience; used in cognitive work analysis.

Workload. The amount of work that has to be done by somebody. From a human factors perspective, workload is studied as physical or cognitive workload. Cognitive workload can be further differentiated as mental demand, temporal demand, frustration, and effort.

Cognitive walkthrough. A user-centered design method where users are presented with a prototype to evaluate. The evaluator guides the process by exposing the user to parts of the design as needed and asking the user to identify ways that the design can be improved.

ACKNOWLEDGMENTS

The writing of this chapter was supported by funding through NSERC Discovery Grant #132995 to Catherine M. Burns and a Science Without Borders Post-Doctoral Fellowship to Carlos Lucena.

REFERENCES

Annett, J. and Stanton, N. A. (1998). Research and developments in task analysis. *Ergonomics* 41(11), 1529–1536.

Beyer, H. and Holtzblatt, K. (1997). *Contextual design: Defining customer-centred systems*. Burlington, MA: Morgan Kaufman.

Bisantz, A. M. and Burns, C. M. (2008). *Applications of Cognitive Work Analysis*. Boca Raton, FL: CRC Press.

Burns, C. M., Enomoto, Y., and Momtahan, K. (2008). A cognitive work analysis of cardiac care nurses performing teletriage. In Bisantz, A. and Burns, C. M. (eds), *Applications of Cognitive Work Analysis*. Mahwah, NJ: Lawrence Erlbaum and Associates, pp. 149–174.

Burns, C. M. and Hajdukiewicz, J. R. (2004) *Ecological Interface Design*. Boca Raton, FL: CRC Press.

Cimbala, S. (2000). Year of Maximum Danger? The 1983 "War Scare" and US-Soviet Deterrence. *The Journal of Slavic Military Studies* 13, 1–24.

Crandall, B. and Klein, G. (2006). *Working minds: A practitioner's guide to cognitive task analysis*. Bradford, Bingley, UK.

Fogg, B. J. (2002). *Persuasive technology: Using computers to change what we think and do*. Burlington, MA: Morgan Kaufman.

Heusser, M. (2014). Software Testing Lesson Learned From Knight Capital Fiasco. Available at www.cio.com/article/2393212/agile-development/software-testing-lessons-learned-from-knight-capital-fiasco.html.

Kirwan, B. and Ainsworth, L. K. (1992). *A guide to task analysis: The task analysis working group*. Boca Raton, FL: CRC Press.

Klein, G. (2000). Cognitive task analysis of teams. In J. M. C. Schraagen, S. F. Chipman, and V. J. Shalin (eds), *Cognitive task analysis*. Mahwah, NJ: Lawrence Erlbaum Associates.

Klein, G. A., Calderwood, R., and D. Macgregor (1989). Critical decision method for eliciting knowledge. *IEEE Transactions on Systems, Man and Cybernetics* 19, 462–472.

Lane, R., Stanton, N., and Harrison, D. (2006) Hierarchical task analysis to medication administration errors. *Applied Ergonomics* 37 (5), 669–679.

Lanoue, S. (2015). IDEO's 6-step Human-Centered Design Process: How to make Things People Want. Available at www.usertesting.com/blog/2015/07/09/how-ideo-uses-customer-insights-to-design-innovative-products-users-love/.

Militello, L., and Hutton, R. (1998). Applied cognitive task analysis (ACTA): A practitioner's toolkit for understanding cognitive task demands. *Ergonomics* 41, 1618–1641.

Minotra, D. and Burns, C. M. (online first). Understanding safe performance in rapidly evolving systems: A risk management analysis of the 2010 US stock mark flash crash with Rasmussen's risk management framework. *Theoretical Issues in Ergonomics Science.*

Novak, J. D. and Canas, A. J. (2008). The theory underlying concept maps and how to construct and use them. IHMC Technical Report Cmap Tools 2006-01 Rev 01-2008. Florida Institute for Human and Machine Cognition, 2008. Available at http://cmap.ihmc.us/Publications/ResearchPapers/TheoryUnderlyingConceptMaps.pdf.

Preece, J., Rogers, Y., and Sharp, H. (2002). *Interaction Design: Beyond human-computer interaction*. Chichester, West Sussex: John Wiley & Sons.

Reason, J. (1990). *Human Error.* Cambridge: Cambridge University Press.

Wickens, C. D. (2002). *Elementary Signal Detection Theory.* New York: Oxford University Press.

Wickens, C. D., Hollands, J. G., Banbury, S., and Parasuraman, R. (2012). *Engineering Psychology and Human Performance*. Boca Raton, FL: CRC Press.

Vicente, K. J. (1999). *Cognitive Work Analysis: Toward safe, productive and healthy computer-based work*. Mahwah, NJ: Lawrence Erlbaum Associates.

Xing, L. and Amari, S. (2008). *Fault Tree Analysis*. Springer, pp. 595–620.

Case Study 4
The DubaiNow Story: An Omni-Channel, Coherent, and Delightful City Experience

Ali al-Azzawi

THE HISTORY OF DUBAINOW

DubaiNow aims to be the single point of access for online public and private sector services in the smart city of Dubai. Rather than a mobile app or a website, DubaiNow is an omni-channel platform that provides services through a coherent and connected experience for residents and visitors for a variety of segments, such as business, travel, and health. This case study briefly describes the customer-centric processes involved in platform design, starting from policy rationale, to data gathering and customer engagement techniques, design principles, prototyping, user testing, and eventually the early development stages.

Since the early 1990s, Dubai has been undergoing a digital transformation, initially through initiatives like e-Government (focusing on digitization), then m-Government (moving towards provision of services on mobile platforms), then Smart Dubai Government, and more recently in 2014, through the formal establishment of the Smart Dubai Office (SDO) by His Highness Sheikh Mohammed Al Maktoum, Vice President and Prime Minister of the UAE and Ruler of Dubai. With a clear mandate to orchestrate the transformation of Dubai into the smartest and happiest city on Earth, SDO has focused on establishing clear guidance and structure for both happiness and well-established EU smart city dimensions; economy, living, mobility, people and society, environment, and governance. These dimensions were positioned along four pillars, driving all projects and initiatives in SDO to provide *efficient*, *seamless*, *safe*, and *impactful* experiences within the city.

ICT EXPERIENCE DESIGN IN DUBAI

Although user experience (UX) and customer experience (CX) disciplines were not initially, formally, or widely established in the Dubai government, the DubaiNow project kick-started the successful establishment of the first government-based, in-house, purpose-built, formal CX lab in the region. Today this practice is more prevalent, with many staff at government agencies undertaking regular training, and practice in CX. However, this is still a growing trend, and there is much room for consolidating and enhancing the citywide CX sector in Dubai, and the UAE in general.

THE DUBAINOW PROJECT

As a precursor to the DubaiNow project, an assessment was made regarding the state of residents' customer experience in Dubai and their access to government services. The assessment found a myriad of websites (sometimes more than one per government entity), as well as at least one mobile app for each government entity, each with their own visual and interaction style, login credentials, and payment method where appropriate. Such diversity, incoherence, and fragmentation in the overall experience for residents certainly did not contribute to a happy city scenario. Further, such duplication of effort in design, development, and infrastructure is counter to the concept of a smart city, where resource efficiency is a key pillar. Therefore, in line with the overall goal of creating the "smartest and happiest city on Earth," a strategic goal for Dubai as a smart city was to provide a way to deliver government and other citywide services in the most seamless, convenient, and delightful way for Dubai's residents. The task was therefore outlined to create a platform much like Amazon.com, which provides a coherent omni-channel experience and, when done correctly, has the potential to provide a delightful holistic experience for residents interacting digitally with city services. This would be in positive contrast to the findings of the assessment of the pre-existing conditions outlined above.

Therefore, in line with modern design practice, the choice was made to start with mobile and then extend the project to other channels in later phases. The practical part of the project then started by examining a service catalog of over 1,600 government services, with a design brief of squeezing these services into a single mobile app. Such a mammoth task has many challenges, including those related to sociopolitical, design, logistical, and technical issues.

DESIGN IMPLICATIONS

The approach was to design a modular platform from the point of view of the user, designer, and developer. The rationale was that when users experienced repeated and coherent patterns of interaction that are based on user research, they would quickly "learn" how to use the user interface (UI), and this would lead to increasing the "intuitive" nature of the UI. Further, this would also reduce the designer's workload, due to the reduced number and types of UI "blocks" (eventually reduced to a total of 17 blocks for most government services), and the designers would then find it more efficient to focus on creating high-quality interactions for each block, without the excessive penalty on production-time. These advantages are also transferred to the developers, as they would focus on creating high-quality re-usable code, e.g., adherence to WCG 2.0 accessibility guidelines, and cross-platform/browser compatibility.

The developers also benefit from reduced lines of code, and therefore decrease the probability of bugs, minimizing consequent servicing requirements. Finally, such modularity provides marked and substantial opportunities for efficiency across the entities joining the platform, as they are now engaging at a functional block level rather than the code level, reducing their workload and allowing them to focus on their core-business (e.g., utility provision and innovating for new services).

Throughout the design process three primary design principles were used to guide the team. First, the tone should be welcoming, with "how can I help?" rather than "first tell me who you are, then I'll let you see what we have." The second principle was design for the common case and allow secondary space for special cases. The third principle was directly associated with the strategic goal of the project, which was to focus on providing a direct service, or specific information. Any UI or page suggestions that did not fit within these principles were either modified to fit or removed.

In total, the initial stages of the DubaiNow project conducted 180 hours of one-to-one interviews with 180 participants and used actual service interaction data and web analytics to drive the initial design efforts. The project also employed several standard user-centered design methods, such as card sorting, eye-tracking, one-to-one interviews, tree testing of information architecture, as well as standard usability testing with full prototypes. The project also delivered a complete design guidebook, showing design patterns and branding guidelines, along with standardized iconography, and a platform to host developer resources. Finally, the DubaiNow platform continues to host more than 57 services from 25 entities (to date), with a mixture of public and private sectors, e.g., Dubai Police; Dubai Culture, Roads, and Transport Authority; Dubai Electricity and Water Authority; telecom operators; Dubai Health Authority; petrol stations, as well as charity organizations. DubaiNow offers services varying in complexity and integration via "microapps" (an integrated UI within the platform UI structure). For example, users can simply look up school/university details, as well as hospital and doctors, or just call an emergency service, or pay a parking fee. The platform also offers business users an AI-based chat microapp to help entrepreneurs learn about the specifics of setting up a business. There are also more transactional microapps using built-in payment features to pay bills for utilities, such as electricity, water, and phone/internet connections, as well as toll fees, and to top up the citywide "Nol" smart card (for travel, museum access, restaurants, and micro purchases), or cashless refuelling for car owners. Importantly, having many services in a single platform also allows integration for context, such as the MyCar microapp, which pulls in data from the police and transport licensing authority to present services relevant to a car owner/operator, along with notifications and related actions. In this way, DubaiNow is well on its way towards providing Dubai residents with a truly omni-channel, coherent, and delightful city experience.

5 Human Factors in Cyber Security Defense

Prashanth Rajivan and Cleotilde Gonzalez

CONTENTS

INTRODUCTION

Most organizations, large or small, rely on some level of computing infrastructure to support their daily operations. Cyber attacks pose a major threat to the confidentiality, integrity, and availability of critical computing systems. Attacks often lead to extensive operational, financial, legal, or reputational loss for an organization (Freund & Jones, 2014).

In recent years, we have seen how cyber attacks have evolved from conventional scams (mass phishing emails) and isolated denial-of-service (DOS) and malware attacks (computer viruses) launched by independent groups of hackers, to sophisticated attacks that employ a combination of methods launched by nation-state operators and organized cybercrime groups. Cyber-criminal organizations operate through functional specializations and formal divisions of labor (Dishman, 2015). They have sufficient abilities and infrastructure to deploy targeted, strategic, stealthy, persistent, and multi-step attacks that have dire implications for national security (Liu, Chen, & Lin, 2013). Sophisticated multi-step attacks are likely to go unnoticed for several months, with popular examples including the Stuxnet attack on Iranian nuclear power plants, and the attacks on Sony and Target (Singer & Friedman, 2014).

To counter the growing number of sophisticated attacks, organizations need high-performing teams assisted by security tools that are well designed and integrated with their workflow. Operators conducting cyber defense tasks (i.e., "security analysts") are essential to the analysis of and response to emerging cyber threats. Although cyber analysts are armed with sophisticated and powerful tools, their inherent cognitive limitations, information processing boundaries, poor situational awareness (SA), and decision-making biases may decrease the efficacy of their defense efforts. Furthermore, defense processes are dynamic and involve an asymmetric interaction with adversaries. Attackers have an advantage over defenders. They exploit and use deception, can carefully study an attack target, and may acquire advanced knowledge about vulnerabilities to exploit. Security analysts are outnumbered by adversaries, and they are often overloaded with alerts from security sensors, including false alerts, which affect their situational awareness and defense performance. Human factors research can play a critical role in improving cyber defense strategies by addressing the behavioral factors that affect individual security analysts and by analyzing and characterizing attackers' behavior (Crossler et al., 2013).

Importantly, the increased complexity of cyber attacks requires not only the individual action of a security analyst but also the coordinated action of an organized team of analysts (i.e., a cyber defense team). Complex attacks such as multi-step attacks require the engagement of cyber analysts across multiple sub-networks, over different points in time and work shifts. Human factors research that addresses team processes and organizational factors, such as communication, organization, and leadership, are crucial to security defense performance. Recent advancements have started to address the human element of cybersecurity, but we are still in the early stages of understanding the organizational and team factors of cyber defense. In what follows, we present the fundamentals, methods, and applications of human factors research in cybersecurity.

FUNDAMENTALS

Cyber defense is primarily a cognitive task with limited physical demands. It is knowledge- and skill-intensive and requires extensive technical training in computer science and expertise in computer security. Globally, there is a crippling shortage of security defense professionals and the gap is expected to reach 1.8 million by 2022 (GISWS, 2017). Determining the characteristic factors of security expertise and its nuances in expert analysts is essential to developing effective training programs to meet the staggering demand for security professionals. In practice, security defense is performed in a dynamic and uncertain environment. Amidst uncertainty and complex workflows, security analysts are required to maintain superior levels of situational awareness to assess risk, detect emerging threats, and respond to attacks. Furthermore, security defense is a team task, with operators assuming different roles and working interdependently on different aspects of the task. Hence, teamwork and leadership are significant determinants of the success of security defense. Next, we present landmark findings in each of the three categories of human factors (expertise, SA, and teamwork) in the context of cyber security threat detection.

Security Analysts: Expertise

Expertise in cyber security is crucial for timely threat detection and response (Goodall, Lutters, & Komlodi, 2004). Expertise in a domain is developed through extensive training and explicit practice (Chi, Glaser, & Farr, 2014). Likewise, security analysts develop expertise through the pursuit of formal degrees, practical training, technical certifications, and significant work experience. Compared to novices, expert analysts are significantly better at using their specialized knowledge in interpreting security events, fusing different attributes of an event, and judging their priority by considering the operational requirements (Ben-Asher & Gonzalez, 2015).

Expertise in security defense is tied to specific roles and types of experience. For example, an expert analyst on threat detection in a particular operating system (e.g., Linux and Windows) may not be an expert in another system and would not be an expert in other critical functions such as risk management or forensic analyses. Similarly, expertise is also bound to a specific network environment because activities considered "normal" in one environment may be indicative of malicious activity in another. Situated expertise is acquired through repeated interactions within the specific environment. Therefore, experts in threat detection in one organization may appear as novices in another because they are unable to transfer their specific situated knowledge (Ben-Asher and Gonzalez, 2015; Goodall, Lutters, & Komlodi, 2004; Goodall, Lutters, & Komlodi, 2009). Thus, expertise in threat detection can be characterized as a combination of domain and situated knowledge.

Security defense involves making decisions based on partial information about plausible events. Expertise is necessary to build a mental representation of the situation, define the situation, and add constraints to make effective decisions (Chi, Glaser, & Farr, 2014). To help efficiently process large amounts of information and manage overload, expertise provides organized knowledge structures (Chase & Simon, 1973). Led by these structures, experts can perceive and process the connections between seemingly disconnected events and are more likely to demonstrate superior perceptual ability and discern meaningful patterns in security events than novices (Ben-Asher & Gonzalez, 2015). Experts are more sensitive to critical cues in the environment that could be a central characteristic of detection of cyber attacks; and, finally, experts are better than novices in knowing their limitations in a situation, which enables them to seek out help from others in the team (Chi, Glaser, & Farr, 2014). Thus, the expertise of security analysts is a fundamental human factors concept that deserves more attention in cyber security research.

Security Analysts: Situation Awareness

Situation awareness (SA) is defined as the three-step process of "the perception of the elements in the environment within a volume of time and space, the comprehension of their meaning, and the projection of their status in the near future" (Endsley, 1995, p. 97). In security analysis, perception involves paying attention to the status of a network and system nodes, being vigilant of critical events in the network, and perceiving patterns of deviant behaviors. Comprehension involves understanding how vulnerabilities in the network can be exploited and comprehending the full scope of attack vectors and the correlation of disparate events. Projection involves anticipating how current events in the network will unfold in the future. For example, after ascertaining the extent of a malware infection in a machine, an analyst would need to predict the likelihood of such an infection transferring to other connected nodes.

Good SA enables analysts to make effective defense decisions. However, achieving a high level of security SA is inherently challenging due to the complexity of computer networks, the non-physical environment, and information overload, all of which inhibit analysts from developing a unified picture of network events (Taylor, 1990). Monitoring the network traffic logs and user activity logs that track the emergence of threats is perceptually challenging, much like finding a needle in the proverbial haystack. Hence, better SA does not necessarily result from having more information (Marusich et al., 2017). Situation awareness in threat detection is susceptible to vigilance decrement because the cognitive resources necessary to perform signal/noise discrimination depletes with time (Sawyer et al., 2014).

lease check if
is should be
Marusich et al.,
016" since it
as in ref. list.

Security visualizations are designed to mitigate the limitations of human perception and may draw analysts' attention to salient cues indicative of malicious changes in the network (Goodall et al., 2005; D'Amico et al., 2016). Also, effective work scheduling and non-intrusive ways to monitor analyst performance have been proposed as potential solutions to minimize effects from vigilance decrement (McIntire et al., 2013). Workflow analysis of information processing in threat detection has found that security analysts iterate through different stages of analysis to maintain SA. For example, Paul and colleagues (2005, 2013) present a taxonomy of mental queries that analysts employ in different stages of workflow to comprehend a security event and improve their SA (D'Amico et al., 2005; Paul & Whitley, 2013). Another workflow analysis revealed that effective knowledge transfer during shift hand-off is essential to maintaining cyber SA (Gutzwiller, Hunt, & Lange, 2016). The main point is that the "awareness" resides neither with the analyst nor the technology alone but with the joint human-technology system (McNeese, Cooke, & Champion, 2011).

Please
"Paul a
leagues
2013)"
"D'Amic
2005" a
& Whitl
If not p
provide
publica
for this
citation

Teamwork among Security Analysts

Coordination, communication, shared situation awareness, and leadership are four elements of teamwork crucial to security analysts (Goodall, Lutters, & Komlodi, 2009; Jariwala et al., 2012). The Department of Homeland Security highlighted the lack of coordination among analysts as a major reason for ineffective and delayed responses to multiple cyber incidents (Moore, 2015). Security analysts are often organized loosely rather than arranged in a coherent, functioning team (Champion et al.,

2012). In general, any professional team should include people with diverse backgrounds who are identified by role and work together in an interdependent manner toward common objectives (Salas et al., 1992). Team effectiveness largely depends upon appropriate leadership, team structure, communication, collaboration, and the distribution of tasks (Hackman & Katz, 2010).

Teams of analysts need to manage uncertainty and information overload in an agile and adaptive way (Terreberry, 1968). Communication is key to forming relationships and sharing information, but the amount of communication necessary for effective team performance differs according to team composition, task type, and team maturity (Cooke et al., 2013). However, some amount of effective communication is critical. Communication is the conduit that transforms individual expertise and situational awareness into team-level knowledge and shared situation awareness (Cooke et al., 2013; Saner et al., 2009). Field studies with security analysts have found that communication and collaboration between security analysts is an integral aspect of security defense, particularly during a widespread security crisis (Goodall, Lutters, & Komlodi, 2009; Jariwala et al., 2012). Lab experiments on collaboration during threat detection have also found evidence that cooperation between security analysts during triage analysis augments signal detection performance, particularly in novel and complex security defense situations (Rajivan et al., 2013). Unfortunately, communication across the hierarchy of security analysts is often inefficient and largely one-directional (bottom-up; Staheli et al., 2016). Research suggests that tools for collaborative threat detection developed using human systems engineering principles would help in mitigating losses in communication between security analysts (Rajivan, 2014). What matters is for the team have good shared SA. In contrast to theories that view team cognition as the sum of the knowledge of individuals (Cannon-Bowers & Salas, 1993), shared SA has enhanced emergent properties and typically arises from interactions in teams where knowledge shared by team members is complementary (Cooke et al., 2013; Saner et al., 2009).

Leadership also aids security defense team development and performance (Buchler et al., in review). Typically, a leader is expected to develop team capabilities, facilitate problem solving, provide performance expectations, synchronize and integrate team member contributions, clarify team member roles, initiate meetings, and provide feedback (Salas et al., 1992; Webber, 2002). Field studies have shown that leadership is a significant predictor of defense performance. In one such study, two security teams—otherwise equivalent in skills, experience, and knowledge—performed extremely differently, primarily due to differences in leadership approach and amount of collaboration (Jariwala et al., 2012). In a subsequent study, it was found that functional specialization and adaptive leadership strategies were important predictors of security defense performance (Buchler et al., in review). However, the determinants of effective teamwork and leadership among security analysts are still an emerging area of research.

HUMAN FACTORS METHODS

To study human factors in cybersecurity, one should take a hybrid research approach that combines field studies, simulation, cognitive modeling, and laboratory

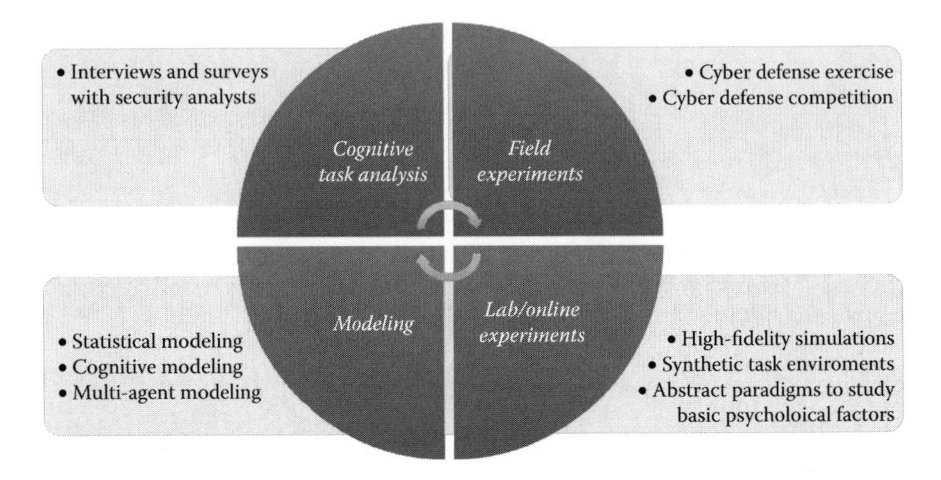

FIGURE 5.1 Hybrid methodology for studying human factors in cybersecurity.

experiments. Multiple methods should be used in an integrated mode such that results from one approach inform the other (Figure 5.1).

Cognitive Task Analysis

Cognitive task analysis (CTA) is a common method used in human factors research to represent the cognitive activities and demands of a task, critical decision points, operational policies, strategies, and tools used, both at the individual and team level (Crandall, Klein, & Hoffman, 2006). These cognitive activities of a task are often elicited via expert interviews. Findings from CTA typically inform all other research methods, including simulation development, measure development, experiment design, and cognitive modeling.

There is a significant body of work describing CTA in cyber security teams (D'Amico et al., 2005; Goodall, Lutters, & Komboldi, 2009; Thompson, Rantanen, & Yurcik, 2006). CTA of threat detection is primarily conducted using semi-structured interviews with a diverse number and type of security analysts (Goodall, Lutters, & Komboldi, 2004). Since access to security analysts and their experiences in naturalistic attack scenarios is often restricted, interviews are sometimes conducted using hypothetical threat scenarios (D'Amico et al., 2005). CTA also involves participatory observations of threat detection workflows. Observations and transcripts of interviews are examined to characterize the analytical processes of threat detection.

So far, CTA on security analysis has predominantly focused on cognitive processes involved in threat detection, but a CTA of analysts performing other critical functions such as risk management and forensics analyses has yet to be conducted.

Field Experiments

Field experiments study behavior in naturalistic environments, often while individuals do their work in the wild. Cyber security defense exercises and competitions provide ecologically valid environments for field experiments on security defense. Cyber defense *exercises* are training events used to *augment* personnel performance, and they often execute operation-centric missions. Examples include exercises conducted by The National Guard (CyberShield) and The National Security Agency. In contrast, cyber defense *competitions* are usually conducted for the purpose of *assessing* skills. Competitions often occur at the college level, where student teams trained in computer security participate. Examples include capture-the-flag style competitions (DEFCON, 2016; Childers et al., 2010), high-school-level security competitions (Chapman, Burket, & Brumley, 2014), and collegiate- (NCCDC, 2016) and professional-level competitions (SANS, 2016).

Cyber defense exercises and competitions employ simulated network environments and controlled attack scenarios and collect performance measures that objectively assess the efficacy of participating teams' defense and network management skills. Participants in these events assume roles that could be adversarial, defensive, or a combination of both. In all cases, there is a team leader assisted by team members who monitor networks for live intrusions and respond to already-discovered intrusions. These events emphasize a collaborative approach to security defense, which makes it relevant to the study of team processes. However, other defense activities such as forensics analyses and risk management are seldom conducted and are in need of study.

Large-scale human behavior data collection from defense exercises and competitions should be done with extreme care because of the noisiness of the data and other analytic challenges particular to big data (Granasen & Andersson, 2016; Henshel et al., 2016). We advocate for field studies driven by specific research questions and focused data collection plans.

Laboratory and Online Experiments

Behavioral experiments are conducted in a laboratory or with online communities using simulated environments to recreate tasks, workflows, and the cognitive processes that are involved during security analyses (Synthetic Task Environments; Cooke & Shope, 2004). Simulation approaches can be expensive and resource- and time-intensive, but they are a good way to study the basic human factors challenges that security analysts confront. Simulation approaches in laboratory experiments also help address a major constraint in the study of human factors in cybersecurity: the availability of security analysts. Given the scarcity of security analysts, researchers often rely on students from networks and security university programs to conduct their research and collect valuable data regarding experience, situation awareness, and team work. Findings for these populations may not necessarily generalize to experts.

In some cases, researchers use basic experimental paradigms contextualized in security defense but with abstracted stimuli, incentives, and feedback. One example of a daily defense task is security event classification. In complex networks, a preliminary process of analysis and identification is used to decide whether a detected event in the network is a reportable cyber event or not. For example, cyber analysts must assess an event according to the information (attributes) and decide how such event should be categorized before reporting it. A category is a collection of events or incidents that share a common underlying cause for which an incident or event is reported and an event may be associated with one or more categories. Decision-making performance on this task may be studied using a high-fidelity simulation environment or by leveraging an existing experimental paradigm from the cognitive science literature such as CMAB (contextual multi-armed bandit task; Rajivan et al., 2016). Findings from these abstract environments suggest that decision-making using non-descriptive stimuli and unfamiliar contexts lead people to use simple heuristics instead of appropriate analysis processes (Rajivan et al., 2016). Thus, it is important but difficult to find a balance between task abstraction and the preservation of cognitive fidelity. Identifying analogical contexts such as home security and intelligence analysis helps reduce training needs for novices while preserving the cognitive characteristics of a security task.

Modeling and Computational Representations of Behavior

Computational cognitive models represent one or many human cognitive processes such as perception, attention, and decision-making by relying on well-known theories of cognition. Popular modeling techniques include ACT-R (Anderson & Lebiere, 1998), SOAR (Laird, Newell, & Rosenbloom, 1987), and Bayesian approaches (Lee & Wagenmakers, 2014). Modeling the cognitive processes of security analysts is challenging, but many efforts are under development. For example, pattern recognition under uncertainty represents a defender's attempt to find patterns in the attacker's action sequence to predict the attacker's next operation and to provide the best response to it. Cognitive models in ACT-R (Anderson & Lebiere, 1998) and neural networks (West & Lebiere, 2001) account for human processes involved in detecting sequential dependencies, and then use the perceived sequence to predict the possible actions that opponents are most likely to take next. Cognitive models derived from instance-based learning theory (IBLT; Gonzalez, Lerch, & Lebiere, 2003), a theory of decisions from experience in dynamic tasks, can also be used to create cognitive models of the intrusion detection process (Dutt, Ahn, & Gonzalez, 2011).

Game theory has been used to model and capture strategies of defenders and attackers in security situations (Pita et al., 2008). Similarly, game theory has been used to study decision-making in cybersecurity (Alpcan & Basar, 2011; Lye & Wing, 2005; Manshaei et al., 2013; Roy et al., 2010). IBL models have also been used to study the dynamic interaction between attackers and defenders in combination with behavioral game theory experiments (Abbasi et al., 2016), and efforts are underway to scale up current two-person IBL cognitive models to study interactions among multiple decision-makers in social conflicts (Gonzalez, Ben-Asher, & Morrison, 2017).

APPLICATIONS

Given the newness of human factors investigations into cyber defense, there are only a few instances in which results from theoretically grounded basic research have been translated into applied design or technological solutions. We will discuss two broad areas of application with specific examples on how human factors were addressed.

Visualization Design

As aforementioned, excessive information load is detrimental to situational awareness (Taylor, 1990). Hence, before being processed by human analysts, large amounts of security data need to be filtered and visualized. In designing visualization for security analysts, the focus should be more on supporting specific decisions and information requirements and less on the task (e.g., threat detection) in general (D'Amico et al., 2016; Marusich et al., 2017). Cognitive task analyses with security analysts have informed the development of an integrated security visual analytics platform called VIAssist that supports specific threat detection activities such as triage analysis (D'Amico et al., 2005 & 2016). VIAssist integrates disparate yet pertinent security visualizations into a multi-display system that allows analysis at multiple levels, essentially aiding the perception and comprehension phases of situational awareness. VIAssist is intended to assist security analysts in information exploration and processing by enabling analysts to seamlessly move from the holistic view of the network state to the fine details of suspicious events. The holistic visualization of the network's security state allows analysts to quickly perceive deviant behaviors while the integrated visualization of event attributes allows more efficient comprehension of individual events. Furthermore, VIAssist allows analysts to annotate, tag, and highlight parts of the visualization and event details, a feature that is conducive to collaborative threat detection and knowledge transfer between shifts.

Decision Support Systems Design

In dynamic, uncertain environments such as security defense, expert decision-making is imperative. However, in such environments, decision-making is also challenging. Unaided judgments and choices have been shown to systematically deviate from optimal decisions (Kahneman, Slovic, & Tversky, 1982). Designing decision support systems informed by behavioral research would augment security analyst decision-making processes and decision value as demonstrated in other domains such as aviation and healthcare. Decision support systems could aid in structuring the decisions and mitigating cognitive deficiencies. Cognitive models such as instance-based learning theory (IBLT; Gonzalez et al., 2003) and recognition primed decision (RPD; Klein, 1993) models can capture how domain experts make decisions based on similarities between attributes of a given event and individual experience. RPD-based cognitive agents have been used to integrate the three phases of Endsley's (1995) situation awareness model to assist security analysts in developing hypotheses about security situations and refining these hypotheses (Yen et al., 2010). In another study, an IBLT-based cognitive model was developed to analyze network packets to effectively discriminate malicious-intent network behavior from benign network traffic (Ben-Asher, Hutchinson, & Oltramari, 2016). Ben-Asher and colleagues developed an ontology to semantically represent network communications and integrated it with an IBL cognitive model for threat detection. The product, which made decisions from experience as encoded in the memory of the agent, served to detect adversarial network activity.

FUTURE TRENDS

Automation of Security Defense

Security defense is knowledge- and skill-intensive. Nevertheless, the basic independent tasks that a security analyst performs are often simple, manual, and repetitive. To process each security event, analysts follow standard operating procedures and switch between multiple security application screens, which can negatively affect vigilance and cause burnout. A large number of events and a shortage of skilled security professionals also contribute to workload. Hence, the automation of security defense functions and the automation of standard operational procedures are being developed. The objectives of such automation are to support the growing needs of security analysts and to relieve analysts from mundane functions, allowing them to focus on more interesting tasks and increasing defense efficiency in terms of both accuracy and response time. Past human factors research has demonstrated that human operators not only use but also misuse, disuse, and abuse automation (Parasuraman & Riley, 1997). Hence, security automations should be developed using human factors principles to optimize integration with human analysts (Endsley, 2016). For example, presenting security analysts with adequate feedback and making it simpler to switch from automatic analysis mode to manual mode ensures a human remains in the loop. More research is needed on human-automation trust and interaction in cybersecurity, networks, and mobile technologies.

Cognitive Models of Expert Security Analysts

The general approach to security defense is shifting from a compliance-based, reactive method to active, risk-based defense methods that make defense decisions based on projections of the possible set of threats to one's network. However, active methods will not be effective unless the internal mental processes by which attackers and defenders make decisions can be represented computationally, through cognitive decision-making models and/or behavioral game theory. Game-theoretical perspectives must formally capture the space of actions of attacker and defender while pinpointing the likely behaviors of particular attackers. The human factors components discussed earlier, such as expertise, situation awareness, and teamwork, will be essential to the development of effective game-theoretic approaches to security defense. Many research challenges remain in the development of effective security, and we expect that cognitive models of expert decision-makers informed by behavioral research will be needed to achieve human-system integration between analysts and detection tools.

Human Factors of Deception in Security Defense

Security analysts actively employ deception technologies such as honeypots to persuade adversaries to breach into fake computing assets that are closely monitored by security operators. Such deception methods work to divert adversaries from real targets. However, such static methods are often ineffective because adversaries learn quickly and are able to discern when they are operating in a monitored environment. To build adaptive deception technologies, knowledge about effective methods to deceive and persuade attackers in different threat scenarios and with different attack motivations is necessary. Adaptive and personalized deception technologies will be effective only when we know the individual behavioral differences that influence the strategies that an attacker may use against an analyst, as well as the individual differences that influence analysts' abilities to detect and combat deception. We envision that human factors research will inform and advance deception technologies in cybersecurity.

CONCLUSIONS

Cyber security failures can disrupt public infrastructure, influence the outcome of international conflicts, undermine national elections, and even cripple economies. Security analysts are the most potent and agile defense resource available at our disposal. Nevertheless, factors that determine human performance in security defense are largely unknown. In this chapter, we presented a summary of human factors work in the context of security threat detection. We learned from past work that expertise in threat detection is a combination of domain and situated knowledge. Past work on security situation awareness has found that security analysts are inefficient in perceiving security state changes, are susceptible to vigilance decrement from continuous signal detection, and have disparate information requirements for different sub-tasks in the detection process. Information sharing between analysts is often inefficient. To address these issues, there must be a greater focus on the human factor in cybersecurity. For example, leadership in security defense should be made adaptive to operational needs and defense functions. Past efforts to study the human factor in cyber security have predominantly focused on threat detection and largely ignored other critical functions such as human automation interaction, risk management, and deception, to name a few. We envision a thriving research community that makes concerted efforts toward improving the human factors of cybersecurity.

KEY TERMS

Networks and sub-networks. A computer network is a set of computers connected using wired or wireless connections. Computers in a network can share resources, like access to the Internet, printers, file servers, and others. A sub-network is an identifiably separate part of an organization's network. Typically, a subnet may represent all the machines at one geographic location, or in one building.

Network traffic logs. Network traffic monitoring is the process of reviewing, analyzing, and managing network traffic for any abnormality or process that can affect network performance, availability, and/or security. Traffic logs display an entry for the start and end of each session. Each entry includes the following information: date and time; source and destination zones, addresses and ports; application name; security rule applied to the traffic flow; rule action (allow, deny, or drop); ingress and egress interface; number of bytes; and session end reason.

CIA—confidentiality, integrity, and availability. The CIA triad of confidentiality, integrity, and availability is integral to digital information security. Confidentiality is the property that information is not made available or disclosed to unauthorized individuals, entities, or processes. Integrity means maintaining and assuring the accuracy and completeness of data over its entire life cycle. Availability means that the computing systems used to store and process the information, the security controls used to protect it, and the communication channels used to access it must be available and functioning correctly.

Denial-of-service attack. Denial-of-service (DoS) attacks typically flood servers, systems, or networks with traffic in order to overwhelm the victim's resources and make it difficult or impossible for legitimate users to use them. Denial of service typically affects the availability of services.

Threat hunting. The process of proactively and iteratively searching through networks to detect and isolate advanced threats that evade existing security solutions.

Risk management. The process of identifying, assessing, and controlling threats to an organization's digital assets, including proprietary corporate data, a customer's personally identifiable information, and intellectual property.

Forensics analysis. Computer forensics is to examine digital media in a forensically sound manner with the aim of identifying, preserving, recovering, analyzing, and presenting facts and opinions about the digital information.

Honeypot. A computer system that is set up to act as a decoy to lure attackers, and to detect, deflect, or study malicious attempts to gain unauthorized access to information systems. Viewing and logging such activity can provide an insight into the level and types of threat a network infrastructure faces while distracting attackers away from assets of real value.

ACKNOWLEDGMENT

Most of the authors' research reported in this chapter was sponsored by the Army Research Laboratory and was accomplished under Cooperative Agreement Number W911NF-13-2-0045 (ARL Cyber Security CRA). The views and conclusions contained in this document are those of the authors and should not be interpreted as representing the official policies, either expressed or implied, of the Army Research Laboratory or the US Government. The US Government is authorized to reproduce and distribute reprints for government purposes notwithstanding any copyright notation hereon.

REFERENCES

Abbasi, Y. D., Ben-Asher, N., Gonzalez, C., Kar, D., Morrison, D., Sintov, N., and Tambe, M. (2016). Know your adversary: Insights for a better adversarial behavioral model. *38th Annual Meeting of the Cognitive Science Society (CogSci 2016)*. Philadelphia, PA.

Alpcan, T. and Basar, T. (2011). *Network security: A decision and game-theoretic approach*. New York: Cambridge University Press.

Anderson, J. R. and Lebiere, C. (1998). *The atomic components of thought*. Hillsdale, NJ: Lawrence Erlbaum Associates.

Ben-Asher, N. and Gonzalez, C. (2015). Effects of cyber security knowledge on attack detection. *Computers in Human Behavior* 48, 51–61.

Ben-Asher, N., Hutchinson, S., and Oltramari, A. (2016, November). Characterizing network behavior features using a cyber-security ontology. In *Military Communications Conference, MILCOM* 2016-2016 IEEE (pp. 758–763). IEEE.

Buchler, N., Rajivan, P., Marusich, L., Lightner, L., and Gonzalez, C. (in review). Sociometrics and observational assessment of teaming and leadership in a cyber security defense competition. *Journal of Computer and Security*.

Cannon-Bowers, J. and Salas, E. (1993). Shared mental models in expert team decision making. In N. J. Castellan, Jr. (ed.), *Individual and Group Decision Making* (pp. 221–246). Hillsdale, NJ: Lawrence Erlbaum Associates.

Champion, M. A., Rajivan, P., Cooke, N. J., and Jariwala, S. (2012). Team-based cyber defense analysis. In *2012 IEEE International Multi-Disciplinary Conference on Cognitive Methods in Situation Awareness and Decision Support* (pp. 218–221). IEEE.

Chapman, P., Burket, J., and Brumley, D. (2014). PicoCTF: A game-based computer security competition for high school students. Proceedings from *USE-NIX Summit on Gaming, Games, and Gamification in Security Education (3GSE 14)*. San Diego, CA: UNSENIX Association.

Chase, W. G. and Simon, H. A. (1973). Perception in chess. *Cognitive Psychology* 4(1), 55–81.

Chi, M. T., Glaser, R., and Farr, M. J. (2014). *The nature of expertise*. Hillsdale, NJ: Psychology Press.

Childers, N., Boe, B., Cavallaro, L., Cavedon, L., Cova, M., Egele, M., and Vigna, G. (2010, July). Organizing large scale hacking competitions. In *International Conference on Detection of Intrusions and Malware, and Vulnerability Assessment* (pp. 132–152). Springer.

Cooke, N. J. and Shope, S. M. (2004). Designing a synthetic task environment. *Scaled Worlds: Development, Validation, and Application*, 263–278.

Cooke, N. J., Gorman, J. C., Myers, C. W., and Duran, J. L. (2013). Interactive team cognition. *Cognitive Science* 37(2), 255–285. doi:10.1111/cogs.12009.

Crandall, B., Klein, G. A., and Hoffman, R. (2006). *Working minds: A practitioner's guide to cognitive task analysis*. Cambridge, MA: MIT Press.

Crossler, R. E., Johnston, A. C., Lowry, P. B., Hu, Q., Warkentin, M., and Baskerville, R. (2013). Future directions for behavioral information security research. *Computers & Security* 32, 90–101.

D'Amico, A., Whitley, K., Tesone, D., OBrien, B., and Roth, E. (2005). Achieving cyber defense situational awareness: A cognitive task analysis of information assurance analysts. *Human Factors and Ergonomics Society Annual Meeting Proceedings* 49(3), 229–233.

D'Amico, A., Buchanan, L., Kirkpatrick, D., and Walczak, P. (2016). Cyber operator perspectives on security visualization. In *Advances in Human Factors in Cybersecurity* (pp. 69–81). Springer.

DEFCON. (2016). DEFCON CTF Archive. Retrieved October 16, 2017 from https://defcon.org/html/links/dc-ctf.html.

Dishman, L. (2015, June). The New Face of Organized Crime. Retrieved from www.slate.com/articles/technology/ibm/2015/06/the_new_face_of_organized_crime.html.

Dutt, V., Ahn, Y. S., and Gonzalez, C. (2011). Cyber situation awareness: Modeling the security analyst in a cyber-attack scenario through instance-based learning. In Y. Li (ed.), *Lecture Notes in Computer Science* (Vol. 6818, pp. 281–293). Springer.

Endsley, M. R. (1995). Toward a theory of situation awareness in dynamic systems. *Human Factors: The Journal of the Human Factors and Ergonomics Society* 37(1), 32–64.

Endsley, M. R. (2016). From here to autonomy: Lessons learned from human–automation research. *Human Factors* 59(1), 5–27.

Freund, J. and Jones, J. (2014). Measuring and managing information risk: A FAIR approach. Oxford, UK: Elsevier.

GISWC. (2017, May). Global Information Workforce Study: U.S. Federal Government Results. Retrieved from https://iamcybersafe.org/wp-content/uploads/2017/05/2017-US-Govt-GISWS-Report.pdf.

Goodall, J. R, Lutters, W. G., and Komlodi, A. (2004). I know my network: Collaboration and expertise in intrusion detection. In J. Herbsleb and G. Olson (eds), *Proceedings of the 2004 ACM Conference on Computer Supported Cooperative Work* (pp. 342–345). New York: ACM. http://dx.doi.org/10.1145/1031607.1031663.

Goodall, J. R., Ozok, A., Lutters, W. G., Rheingans, P., and Komlodi, A. (2005, April). A user-centered approach to visualizing network traffic for intrusion detection. In *CHI'05 Extended Abstracts on Human Factors in Computing Systems* (pp. 1403–1406). ACM.

Goodall, J. R., Lutters, W. G., and Komlodi, A. (2009). Developing expertise for network intrusion detection. *Information Technology & People* 22(2), 92–108. doi:10.1108/09593840910962186.

Gonzalez, C., Lerch, J. F., and Lebiere, C. (2003). Instance-based learning in dynamic decision making. *Cognitive Science* 27(4), 591–635.

Gonzalez, C. Ben-Asher, N., and Morrison, D. (2017). Dynamics of decision making in cyber defense: Using multi-agent cognitive modeling to establish defense policies. In P. Liu et al. (eds), *Cyber Situation Awareness*, LNCS 10030 (pp. 1–15). doi:10.1007/978-3-319-61152-5_5.

Granasen, M. and Andersson, D. (2016). Measuring team effectiveness in cyber-defense exercises: A cross-disciplinary case study. *Journal of Cognition, Technology, and Work* 18, 121–143. doi:10.1007/s10111-015-0350-2.

Gutzwiller, R. S., Hunt, S. M., and Lange, D. S. (2016, March). A task analysis toward characterizing cyber-cognitive situation awareness (CCSA) in cyber defense analysts. In *Cognitive Methods in Situation Awareness and Decision Support, (CogSIMA), 2016 IEEE International Multi-Disciplinary Conference* (pp. 14–20).

Hackman, J. R. and Katz, N. (2010). Group behavior and performance. In S. T. Fiske, D. T. Gilbert, and G. Lindzey (eds), *Handbook of Social Psychology* (pp. 1208–1251). New York: Wiley.

Henshel, D. S., Deckard, G. M., Lufkin, B., Buchler, N., Hoffman, B., Rajivan, P., and Collman, S. (2016, November). Predicting proficiency in cyber defense team exercises. In *Military Communications Conference, MILCOM 2016-2016 IEEE* (pp. 776–781). IEEE.

Jariwala, S., Champion, M., Rajivan, P., and Cooke, N. J. (2012, September). Influence of team communication and coordination on the performance of teams at the iCTF competition. In *Proceedings of the Human Factors and Ergonomics Society Annual Meeting* 56(1), 458–462. Los Angeles: Sage Publications.

Kahneman, D., Slovic, P., and Tversky, A. (1982). Judgment under uncertainty: Heuristics and biases. New York: Cambridge.

Klein, G. A. (1993). A recognition-primed decision (RPD) model of rapid decision making (pp. 138–147). New York: Ablex Publishing Corporation.

Laird, J. E., Newell, A., and Rosenbloom, P. S. (1987). SOAR: An architecture for general intelligence. *Artificial Intelligence* 33(1), 1–64.
Lee, M. D. and Wagenmakers, E. J. (2014). *Bayesian cognitive modeling: A practical course*. Cambridge, UK: Cambridge University Press.
Liu, S., Chen, Y., and Lin, S. (2013). A novel search engine to uncover potential victims for APT investigations. In C. H. Hsu, X. Li, & X. Shi (eds), *Network and Parallel Computing* (pp. 405–416). Springer.
Lye, K. W. and Wing, J. M. (2005). Game strategies in network security. *International Journal of Information Security* 4(1–2), 71–86. doi:10.1007/s10207-004-0060-x.
Manshaei, M. H., Zhu, Q., Alpcan, T., Bacsar, T. and Hubaux, J. P. (2013). Game theory meets network security and privacy. *ACM Computing Surveys* 45(3). doi:10.1145/2480741.2480742.
Marusich, L. R., Bakdash, J. Z., Onal, E., Yu, M. S., Schaffer, J., O'Donovan, J., and Gonzalez, C. (2016). Effects of information availability on command-and-control decision making: Performance, trust, and situation awareness. *Human Factors* 58(2), 301–321.
McIntire, L., McKinley, R. A., McIntire, J., Goodyear, C., and Nelson, J. (2013). Eye metrics: An alternative vigilance detector for military operators. *Military Psychology* 25, 502–513. IEEE.
McNeese, M., Cooke, N. J., and Champion, M. A. (2011). Situating cyber situation awareness. *Cognitive Technology* 16(2), 5–9.
NCCDC. (2016). The National Collegiate Cyber Defense Competition. Retrieved October 17, 2016 from http://www.nationalccdc.org.
Parasuraman, R. and Riley, V. (1997). Humans and automation: Use, misuse, disuse, abuse. *Human Factors: The Journal of the Human Factors and Ergonomics Society* 39(2), 230–253.
Paul, C. L. and Whitley, K. (2013, July). A taxonomy of cyber awareness questions for the user-centered design of cyber situation awareness. In *International Conference on Human Aspects of Information Security, Privacy, and Trust* (pp. 145–154). Springer.
Rajivan, P., Champion, M., Cooke, N. J., Jariwala, S., Dube, G., and Buchanan, V. (2013, July). Effects of teamwork versus group work on signal detection in cyber defense teams. In *International Conference on Augmented Cognition* (pp. 172–180). Springer.
Rajivan, P. (2014). *Information pooling bias in collaborative cyber forensics* (Doctoral dissertation). Retrieved from ProQuest Dissertations & Theses Global (3666249).
Rajivan, P., Konstantinidis, E., Ben-Asher, N., and Gonzalez, C. (2016, September). Categorization of events in security scenarios: The role of context and heuristics. In *Proceedings of the Human Factors and Ergonomics Society Annual Meeting* 60(1), 274–278. Los Angeles, CA: Sage Publications.
Roy, S., Ellis, C., Shiva, S., Dasgupta, D., Shandilya, V., and Wu, Q. (2010). A survey of game theory as applied to network security. In J. Ralph H. Sprague (ed.), *Proceedings of the 43rd Hawaii International Conference on System Sciences*. Los Alamitos, CA: IEEE.
Saner, L. D., Bolstad, C. A., Gonzalez, C., and Cuevas, H. M. (2009). Measuring and predicting shared situation awareness in teams. *Journal of Cognitive Engineering and Decision Making* 3(3), 280–308.
Salas, E., Dickinson, T. L., Converse, S. A., and Tannenbaum, S. I. (1992). Toward an understanding of team performance and training. In R. W. Swezey and E. Salas (eds), *Teams: Their training and performance* (pp. 3–29). Westport, CT: Ablex Publishing.
SANS (2016). SANS NetWars. Retrieved from https://www.sans.org/netwars.
Sawyer, B. D., Finomore, V. S., Funke, G. J., Mancuso, V. F., Funke, M. E., Matthews, G., and Warm, J. S. (2014, September). Cyber vigilance effects of signal probability and event rate. In *Proceedings of the Human Factors and Ergonomics Society Annual Meeting* 58(1), 1771–1775. Sage Publications.

Webber, S. S. (2002). Leadership and trust facilitating cross-functional team success. *Journal of Management Development* 21(3), 201–214.

Singer, P. W. and Friedman, A. (2014). *Cybersecurity and Cyberwar: What everyone needs to know.* New York: Oxford University Press.

Staheli, D., Mancuso, V., Harnasch, R., Fulcher, C., Chmielinski, M., Kearns, A., Kelly, S., and Vuksani, E. (2016). Collaborative data analysis and discovery for cyber security. Proceedings from *Twelfth Symposium on Usable Privacy and Security (SOUPS 2016).* Denver, CO: USENIX Association.

Taylor, R. (1990). Situational awareness rating technique (SART): The development of a tool for aircrew systems design. In *Situational Awareness in Aerospace Operations (AGARD-CP- 478)* (pp. 3/1–3/17). Neuilly Sur Seine, France: NATO - AGARD.

Terreberry, S. (1968). The evolution of organizational environments. *Administrative Science Quarterly* 12(4), 590–613.

Thompson, R. S., Rantanen, E. M., and Yurcik, W. (2006). Network intrusion detection cognitive task analysis: Textual and visual tool usage and recommendations. In *Proceedings of the Human Factors and Ergonomics Society Annual Meeting* 50(5), 669–673. doi:10.1177/154193120605000511.

West, R. L. and Lebiere, C. (2001). Simple games as dynamic, coupled systems: Randomness and other emergent properties. *Journal of Cognitive Systems Research* 1(4), 221–239. doi:10.1016/S1389-0417(00)00014-0.

Yen, J., McNeese, M., Mullen, T., Hall, D., Fan, X., and Liu, P. (2010). RPD-based hypothesis reasoning for cyber situation awareness. In *Cyber Situational Awareness* (pp. 39–49). Springer.

Case Study 5
Cyber Security Challenges and Opportunities in Kuwait's Oil Sector

Reem F. Al-Shammari

THE STATE OF CYBER SECURITY IN KUWAIT AND THE GCC

Internet users and the use of information systems have grown rapidly in Kuwait and other Gulf Cooperation Council (GCC) states. Although, this has helped to boost prospects for digital economies. At the same time, this digital growth has increased the weakness of the region to cybercrime. While the incidence and effects of cyber crime in the region is difficult to measure precisely (with limitation of availability of official numbers), a number of surveys suggest that cyber attacks are growing rapidly and that the region has become a focal point for such crime. Based on several reports published by security firms, the average malware infection rate is on the rise. Most of the countries in the Middle East have had at least double (in some cases 20 times the global average) the number of infected systems per quarter than the global average.*

The below reasons suggest that the incidence, scale, and impact of cyber attacks are likely to increase further in the future:

1. The number of internet/online users in Kuwait and the GCC has seen fast growth.
2. Significant geopolitical instability in the Middle East has given rise to various hacktivist groups. These hacktivist groups have rained down cyber havoc on governments and public and private institutions in the region on an almost-daily basis since the inception of the turmoil.
3. The GCC countries in the eyes of the global community are perceived to be a region of economic wealth. To cyber criminals who are trying to exploit governments and public and private institutions for financial gains this makes the Middle East a central target for attack.

* Microsoft Security Intelligence Report, Volume 21, January through June, 2016.

Country	Internet users (% of the population)	Mobile broadband subscriptions (per 100 inhabitants)	Fixed broadband penetration (per 100 inhabitants)	Mobile subscriptions (per 100 inhabitants)	Social media penetration (facebook per 100 inhabitants)	International bandwdth per internet user (bit/sec)
Bahrain	93.5	131.8	18.6	185.3	73	42,205
Kuwait	82.0	139.3	1.4	231.8	71	48,619
Oman	74.2	78.3	5.6	159.9	41	59,784
Qatar	92.9	80.0	10.1	153.6	85	71,566
Saudi Arabia	69.6	111.7	12.0	176.6	58	88,669
United Arab Emirates	91.2	92.0	12.8	187.3	94	107,914
GCC Aggregate	76.0	115.0	12.8	184.0	66	84,659

Please check if this was properly captured.

FIGURE CS5.1 GCC internet and mobile penetration rates.

4. With a young population and quick adoption of new technology, the pace of policy and frameworks cannot match that of cyber criminals.
5. Fading boundaries between Operational Technology (OT) and Information Technology (IT) networks and exposure of many digital assets to the Internet of Things (IoT) is creating additional risks through large number of interconnected devices.

As shown in Figure CS5.1 above, GCC states have some of the highest internet and mobile penetration rates in the world Country

Please there is text he

Over the past few years, Cyber attacks on infrastructure in the region have increased rapidly. Major incidents included attacks on key oil and gas companies across the region that impacted their operations for durations lasting from a few hours to months. The impacts of these cyber attacks can be devastating for the organizations as these can result in direct financial losses, the loss of intellectual property/business secrets, a negative impact on a company's competitiveness, an impact on the costs of service/network security/insurance, and, last but not least, a serious impact on an organization's reputation.

CYBER SECURITY IN KUWAIT'S OIL SECTOR

Organizations within the GCC and in Kuwait are taking this challenge head-on and developing a holistic approach towards cyber security. Organizations like Kuwait Oil Company (KOC) have had a leading role in this transformational journey. As part of this journey, KOC initiated a Cyber Security Committee at the KPC Group of Companies level to share capabilities and steer an overall security program. A few years back, KOC started on the journey of cyber security transformation for the organization. Instead of only focusing on the technology aspect of the issue, a three-pronged approach was established, which included:

- people, ensuring skill sets and resources are available to take up this challenge
- processes, through the establishment of organization-wide cyber strategy frameworks, supported by policies, procedures, and standards
- technology, supporting the implementation of controls using best technology solutions

As part of this journey, the KOC has established its cyber security awareness program, which emphasizes cyber security being everyone's responsibility and does not only end with an organization's security/IT teams. As part of the cyber security awareness program, multiple organization-wide awareness initiatives and campaigns have been carried out to enhance user awareness around cyber security. Cyber security steering committees are set up with the participation of KOC top management and other Kuwait oil and gas sector companies. Furthermore, an integrated approach is devised to address risks pertaining to both information technology (IT) and operational technology (OT) infrastructure.

Some key learnings as part of this journey include:

- support from senior management/leadership, key for deriving cyber security initiatives across the organization;
- communicating the value of cyber security to stakeholders and collaborating with business teams in addressing cyber security challenges is of utmost importance in the effective implementation of cyber security initiatives;
- prevention is ideal but it is not always possible with an ever-widening threat landscape and advanced persistent threats. So staying vigilant—to identify cyber attacks on infrastructure—and being resilient and prepared to handle cyber attacks are two key principles that go a long way towards the development of a successful cyber security program in an organization. To support these principles, continuous threat monitoring and cyber incident preparedness are considered key elements within the organization; and
- a holistic and integrated approach to cyber security is essential for its effective implementation. Cyber security in silos of some departments/functions will not lead to a successful and comprehensive implementation of cyber security initiatives.

HUMAN FACTORS AND ERGONOMICS IMPLICATIONS

Although the role of technology and sophisticated cyber security systems is critical to fight against the rising threat of cyber attacks in the increasingly networked world of today. Yet, technology does not exist in isolation and effective implementation, and management of technology assets totally depends on human factors. Some of the most intriguing findings from various security and intelligence reports that have come out over the past few years highlight that the majority of all cyber security incidents involve human error. Many of the cyber attacks over last few years from external attackers have exploited human weakness in order to gain access to the organization's sensitive/critical information assets. Phishing is one such cyber attack technique that has been utilized to send emails with malicious links directly into employee inboxes. Employees accessing or clicking on that link can, in turn, lead to the compromise of the employee system, thereby impacting the overall security posture of the organization.

These are just some examples to highlight the importance of focusing on "people" while planning your organization's cyber security program. Organizations must focus on people, processes, and technology in equal order to face these ever-increasing challenges. Technology provides automated safeguards, and processes ensure consistency. However, even organizations with strong technology and process controls are vulnerable to human error. To address this issue, a cyber security program must attribute equal importance to human factors through employee education and training. This will ensure employee awareness of cyber threats and their responsibility in guarding against these threats. Therefore, keeping organizations safe relies on implementing technology controls, setting up consistent processes, and, most importantly, constantly educating employees about new possible risks and their roles and responsibilities in managing these risks to an acceptable level.

6 Product Service Systems Innovation and Design

A Responsible Creative Design Framework

Girish Prabhu, Beena Prabhu, and Atul Saraf

CONTENTS

INTRODUCTION

Design in the traditional sense has focused on creating practical, experiential, affective, meaningful, valuable, and efficient products, systems, or services, leveraging peoples' emotions and attitudes. As it does today, design greatly contributes to the current consumerist society by crafting experiences that help sell products and services. But, in doing so, it has ignored the significant role it should play in acting as a change agent and an "early warning system" that is truly centered on human values and the matters of people, society, and the world as a whole. Due to deteriorating ecological conditions, changing population dynamics, changing behaviors, and rapid technological advances, we posit that a more responsible, inclusive, and transitionary framework for design is required.

Due to current individualistic consumerist behavior, global debt has increased by $57 trillion since 2007, outpacing world GDP growth.* Global household debt itself has grown from 33 trillion to 40 trillion, at about 5.3% compound annual growth rate (CAGR). This behavior has also led to huge e-waste generated by all developed and developing countries; in 2012, the USA and the EU each generated 10+ million

* Dobbs et al. Debt and (not much) deleveraging. McKinsey Global Institute Report, www.mckinsey.com/global-themes/employment-and-growth/debt-and-not-much-deleveraging.

tons, China a whopping 8 million tons, and India about 3 million tons. This is neither sustainable nor responsible.

At the same time, population dynamics are changing and so are behaviors. The elderly population is expected to grow at a rate of 3.83% annually.[*] In 2016, the percentage of population above 60 years of age was estimated to be 9.3%, or 118 million. In 2026, this number is projected to double and reach 173 million. The aging of the population will affect every environment—private, commercial, and public. Interestingly, there has been a clear and steep increase in the number and percentage of people online who are aged 50 and older. Fifty-three percent of those aged 65 and over, and 77% of those aged 50 to 64, were online as of April 2012.[†] There have been similar increases in "device ownership" by older adults, including their adoption of cell phones, smartphones, notebook computers, tablets, and e-readers.

In contrast, the US millennial generation (born between 1980 and 2000) has shown interesting behavioral trends; some of which are not different from what we have observed with other millennial populations. They are graduating with higher educational debts and lower income-generating possibilities, and hence are willing to stay with their parents for a longer time, pushing out expenses such as marriage, buying a house, and acquiring "luxury" individualistic white goods. The younger generation is also becoming more ecology-considerate and more accepting of "shared" usage models, wherein the focus moves away from owning to paying for a product when used. The advent of Airbnb, Uber, and movements such as the tiny house movement (Ryan, 2009) and local food diet (Martinez, 2010) are examples of products and services that are aligned with this new responsible behavioral trend.

The movement from traditional technology in a controlled environment to weaving technology seamlessly into everyday life started back in the 1990s. Technology and computing has already become part of everyday life, though not as envisioned (Bell & Dourish, 2007). However, the difficulty in realizing the vision of seamlessness lies in the fact that human interaction with things in everyday life is messy, networked, dynamic, and emergent. Technologically driven innovation seems to reduce this world to one person interacting with a single instance of a monolithic system. For example, the iPad, though very capable and appropriate as a shared information and entertainment consumption device, does not enable such use cases. In fact, Apple strategically makes it impossible to share the device with children, seniors, and guests at home. This engineering-based approach has hence led to a series of highly advanced technological prototypes that have not managed to be integrated with everyday life, as had been envisaged (for example, wearable technology, augmented reality, etc.).

Dealing with such transitions due to population dynamics, changing behaviors, technological advances, and deteriorating ecological conditions can produce fear, resistance, and anxiety. As a result, many organizations are retreating from the future. Speaking about urban planning, for example, Isserman (1986) writes that,

* JLL Report 2015, http://d3ajvuw23j7pxv.cloudfront.net/upld_469847659440860642_SeniorLivingSectorinIndia.pdf.

† Older adults and internet use, PewResearch Report, 2012, www.pewinternet.org/files/old-media/Files/Reports/2012/PIP_Older_adults_and_internet_use.pdf.

"We have lost sight of the future … creating increasingly feeble, myopic, degenerative frameworks that are more likely to react to yesterday's events than to prepare the way from here to the future." The effects are habitual blind spots in many modern organizations, making it difficult to discuss or even think about issues of critical change.

Frameworks with an understanding of social problems, marginalized users, and future consequences then become important to help design in such a context. These frameworks will hopefully produce design solutions that focus on human values and create a better present and future for all, in addition to the commercial success of the solutions. Such a framework requires one to move from an engineering- and business-based approach to an inclusive approach to ensure that societal actors work together during the whole research and innovation process. The new framework needs to aim to better align both the process and outcomes, with the values, needs, and expectations of society. It has to foster the creativity and innovativeness of societies to tackle the grand societal challenges that lie before them, while at the same time pro-actively addressing potential side-effects.

FUNDAMENTALS

Inspired and informed by a range of sociological and design research explorations (Suchman, 2007) one needs to take a radically different approach to realize the vision of integrating technological products, services, and spaces with the everyday lives of people. We need an approach that not only considers human experience in everyday life as messy and complex, but also seeks to embrace this messiness to emphasize transition and conceptualize radically innovative solutions.

Through our continuous engagement with contemporary societies, we have come to a fundamental understanding of human behavior: *people inherently encounter, engage, and solve everyday problems (when they are) in transition.* In this approach, the innovator needs to bring the focus of design to the social patterns, user behaviors, roles of technology, and experiences around the emerging societies that are in perpetual transition. With ubiquitous technology, design should focus on human experience as a series of transitions across its various dimensions with porous boundaries: home life, play, work, family, friends, communities, health and well-being, higher meaning, and education (see Figure 6.1).

The framework we have developed is contextual, design-oriented, explorative, collaborative, and human-centered. It is a process that seeks to study and leverage the everyday socio-cultural conditions and practices of the concerned population to envision the solutions. This is achieved by using a deep understanding of user and cultural needs to drive design ideas, business modeling, ecosystem modeling, and technological investigations. Thereby, the research questions are answered through a multidisciplinary approach where we investigate the problem space from the perspectives of the users, business, technology, and design; in other words, desirability, viability, feasibility, and responsibility (Prabhu et al, 2004). Valero and colleagues (2007, 2008) describes responsible design as a higher-level ethical imperative and inclusive of sustainability. Responsible design includes requirements such as environmental sustainability, business sustainability, safety, and respect for people, respect

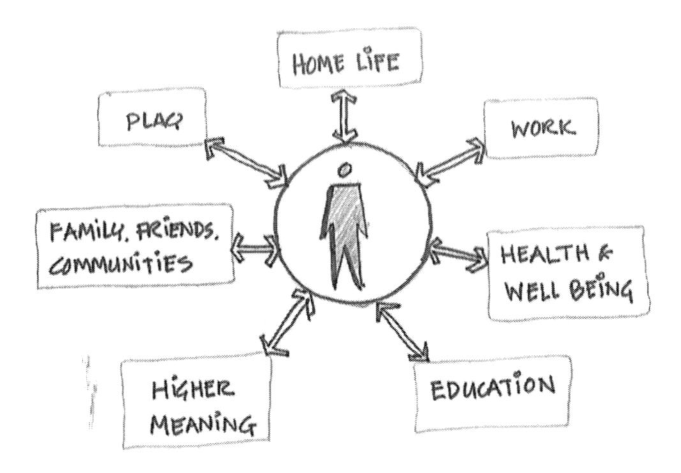

FIGURE 6.1 Transitioning across the various dimensions of human experience.

for cultures, individual professional integrity, and profession-driven responsibilities. This design approach allows designers to look at synergistic means of advancing the safety, health, and welfare of the ecosystem.

In doing so, the framework lays more emphasis on desirability and responsibility but keeps viability (business) and feasibility (technology) in mind. The framework moves away from a pure design thinking approach to include art thinking, which has inspiration, intuition, and imagination as its basic pillars. Art thinking along with design thinking allows designers to be more than just problem solvers.

This new responsible product service system framework also draws from Maxneef's "Barefoot Economics" theories (1992) and Holmgren and Mollison's work (1978) on "Permaculture." As Chilean economist Manfred Maxneef rightly says: "Development needs to be about people and not about objects. And, growth is not the same as development, and development does not necessarily require growth." Inspiration from barefoot economics and permaculture has led the authors to develop a design framework that moves away from the past conventions of designing for desirability-feasibility-viability towards including designing for societal good and bigger impact, as shown in Figure 6.2 below.

We believe that using this framework will help designers discover newer ways of living and working that will help us conserve as we consume. Our framework is based on three core tenets:

1. **Care for context**. Solutions cannot be developed in a vacuum and need to consider the overall ecosystem that they will impact. People and places are part of larger ecosystems that come with their own constraints and strengths. The constraints that existing localities have and the needs of the people living in them are key ecosystem considerations that need to be understood prior to the creation of any solution (Prabhu, 2010).
2. **Care for people**. Systems should be created after understanding the needs, pain points, and aspirations of people. Our human-centered approach goes

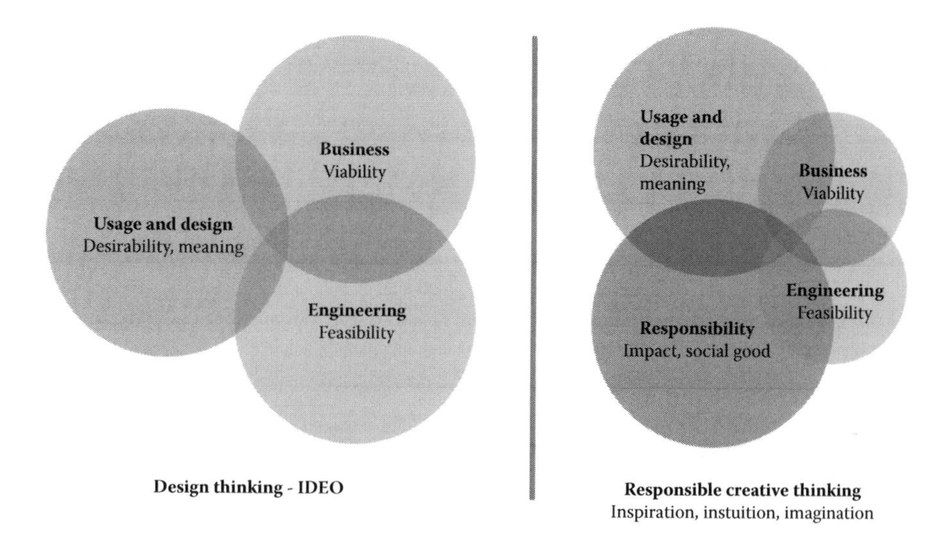

FIGURE 6.2 Including responsible design in product service systems design.

beyond the obvious to uncover and understand the triggers that will help the community move towards economic as well as social and ecological stability.

3. **Fair share**. It is critical to think of economic empowerment of a community rather than of an individual or an organization. Fair share goes beyond the immediate active consumers to passive elements of an ecosystem. For example, when harvesting a crop, how can you leave the soil more enriched rather than devoid of nutrients? This is even more important in countries where the economic gap between people is huge. The ability to yield better economic returns should lie with the individual and the group equally. The social pressure and commitment is expected to prompt individuals to perform better and thereby make them eligible for services that will enhance their lives. For example, in the context of big data, what can we do to give back the benefits of harvested data to individuals and community over time?

METHOD

The methodology for this framework brings together ethnographic, business, design, and technical research in a focused way. This multi-disciplinary method of innovation has helped the authors to develop relevant and contextually appropriate solutions in the past in a range of different initiatives in mobility, entertainment, and the on-demand labor market in India (Prabhu, 2010). The multi-disciplinary methodology applies four different types of lense, namely: conceive, explore, define, and evolve (CEED) in a very non-linear way, as shown in Figure 6.3.* Figure 6.4 also highlights

* Reproduced with permission from Designfold.co.

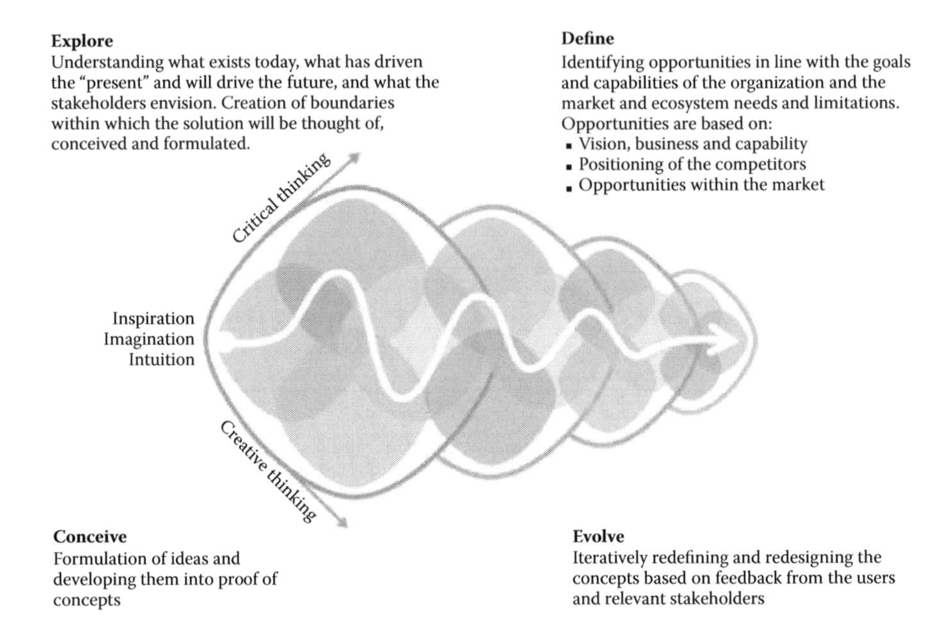

FIGURE 6.3 The four lenses (CEED) of the responsible PSS design methodology.

the objectives of each of these lenses. These lenses of making, research, and analysis weave inspiration, intuition, and imagination of art thinking with creativity and criticality of design thinking.

Instead of the typical user-centered design thinking approach of understand → analyze → conceptualize → build → evolve, this framework starts with the **conceive lens** to promote art thinking and push the designer away from a typical problem-solving approach. Designers operate in two different modes when using this lens. At the early stage, the designer focuses on the formulation of ideas using existing knowledge of the context and through inspiration derived from the surrounding and through one's own intuition. Using various tools such as duct tape, cardboard, crafting supplies, vinyl cutters, laser cutters, foam board, wood work, 3D printers, IoT/sensors, and soldering kits, and technology toolkits such as Makey Makey, Arduino Boards, and Rasberry Pi, designers sketch their "what-if" ideas and scenarios. Seeding possibilities using what-if scenarios, or a feed-forward technique (Prabhu, 2004) allows designers to develop insights that a standard user-centered design approach sometimes hinders. The "what-if" scenarios and ideas can be utilized as possible cultural probes (Gaver et al., 1999) in the explore lens.

At the post-exploration stage, the designer uses the conceive lens to develop proof of concepts through actionable insights. To get actionable insights, it is important to understand the most relevant patterns in the data, which will help in ideation and conceptualization. Patterns are connections in the data or information that is collected; whereas actionable insights are understanding of the data that have implications on the problem being solved. Insight is an interpretation of observation. It requires a point of view on what already exists. They are statements that capture a

Political factors to consider

- When is the country's next local, state, or national election? How could this change government or regional policy?
- Who are the most likely contenders for power? What are their views on business policy, and on other policies that affect your organization?
- Could any pending legislation or taxation changes affect your business, either positively or negatively?

Economic factors to consider

- How stable is the current economy? Is it growing, stagnating, or declining?
- Are customers' levels of disposable income rising or falling? How is this likely to change in the next few years?
- Do consumers and business have easy access to credit? If not, how will this affect your organization?
- How is globalization affecting the economic environment?

Socio-cultural factors to consider

- What social attitudes and social taboos could affect your business? Have there been recent socio-cultural changes that might affect this?
- How do religious beliefs and lifestyle choices affect the population?

Technological factors to consider

- Are there any new technologies that you could be using?
- Are there any new tachnologies on the horizon that could radically affect your work or your industry?
- Do any of your competitors have access to new technologies that could redefine their products?

Demographic factors to consider

- What is the population's growth rate and age profile? How is this likely to change?
- Are generational shifts in attitude likely to affect what you're doing?
- What are your target market's levels of health, education, and social mobility? How are these changing, and what impact does this have?
- What employment patterns, job market trends, and attitudes toward work can you observe? Are these different for different age groups?

Design factors to consider

- What are the new design and UX trends in this area?
- Are there any new design and UX on the horizon that could radically affect your work or your industry?
- Are these designs patented?
- What is the value of design and UX for the particular market?

FIGURE 6.4 Sample questions for PEST-DD trend analysis.

clear and deep understanding of a consumer's attitudes and emotions and are one of the key building blocks that help in the generation of ideas for new products or new services.

Based on the actionable insights, concept ideation using appropriate brainstorming methods and techniques allows formulation of solutions that will deliver the identified value proposition. These concepts are visualized using various techniques, such as storyboards, concept schematics, and concept videos. The main purpose is to articulate the usage models and to narrate the user experience to the audience in an evocative and understandable way. During this lens, concepts are evaluated as part of the iterative design process where feedback from key stakeholders is used to refine the concepts. The output from this lens is the proof of concept; ideas that clearly articulate the user experience to be delivered and the usage models.

With the **explore lens**, designers' role is to understand what exists today, what has driven the present and will drive the future, and what the stakeholders envision. Stakeholder envisioning specifically focuses on understanding the overall goals, vision, and positioning of the ecosystem and the opportunities and challenges perceived by the stakeholders. Also, it is very important for the designer to be well informed about the latest developments in the areas of technology, devices, methods, and level of knowledge or understanding at the present time. Through trend analysis, designers collect and analyze information that will help them in the realization of patterns to predict product/service needs. This lens leads to problem focus—the creation of boundaries within which the solution will be thought of, conceived, and formulated.

PEST-DD, a tool that we developed for trend analysis, helps one to analyze the political, economic, socio-cultural, technological, design, and demographic changes in the business environment. Changes in business environment can create great opportunities—and cause significant threats. Opportunities can come from new technologies that help you reach new customers, from new funding streams that allow you to invest in better equipment, to changed government policies that open up new markets and new design trends. Threats can include deregulation that exposes you to intensified competition, a shrinking market (demographic change), or increases to interest rates, which can cause problems if your company is burdened by debt. This helps designers understand the "big picture" forces of change that they are exposed to, and, from this, take advantage of the opportunities that they present.

Government regulations and legal issues such as tax policies, employment laws, environmental regulations, trade laws, and political stability are some of the variables investigated in order to understand the political trends. Economic trends include buying power of potential users, economic growth, and interest or exchange/inflation rates. A combination of social (and the immediate community or society) and cultural factors covers the influence of religion and traditional beliefs on values and practices, the impact of community on behaviors, and changes due to globalization on behaviors, values, and beliefs. Technological factors such as the state of the art, research and development activities, and government incentives can influence barriers to entry, partnerships required, and other dependencies for the manufacturing/development of product and services. Demographic parameters include gender, age, ethnicity, languages, education, and income. For design trends, one looks at the new ways in which designers are thinking about and implementing design (web design, smart screen design, physical-virtual interaction). Figure 6.4 provides some sample questions that can be used for trend analysis.

The **define lens** allows the designer to identify and scope opportunities in line with the goals and capabilities of the organization, the market, and the ecosystem needs and limitations. This is done using competitive review and contextual ecosystem research. Through competitive review, one collects and analyzes information about current and potential competition, strengths, and weaknesses, and thereby identifies opportunities and threats. User and ecosystem research enables deep understanding of the people and institutions that are affected by a product or service. Contextual user and ecosystem research allows designers to empathize with all stakeholders, including users and ecosystem partners. During this, one has to have the following mindset: observing everything, building empathy, immersing in daily lives, listening openly, and looking for needs, gaps, and problems. The following guidelines help in developing winning insights: (a) listen to what stakeholders **say**; (b) watch what stakeholders **do**; (c) observe what stakeholders **use**; (d) understand what stakeholders **feel**; and (e) appreciate what stakeholders **dream**.

This lens leads to the final value proposition, its delivery mode and business model, users and key stakeholders, and other entities that will play a key role in the success and sustainability of the product/service. Product Ecosystem Map™ (PEM*), is a tool that authors have developed expanding on the business model canvas (BMC) (Osterwalder, 2005). BMC covers business specific parameters such as key partners, key activities, key resources, cost structures, revenue streams, channels, customer relationships, and customer segments, which addresses the viability aspects of the PSS intervention.

Apart from the standard business canvas elements, PEM records trends, behaviors and practices, differentiation, significance, risks and barriers, experience delivery, relevant technology, technology constraints, dependencies, infrastructure needs, and technology partners. PEM hence covers the desirability, feasibility, and responsibility aspects also. Having all these elements in one place helps in the divergent-convergent ideation process to develop responsible product services systems that are viable. These various boxes are populated in a very non-linear fashion based on the insights from the exploration and define stage.

The trend box is populated with the outcome from trend analysis, whereas the differentiation box includes the outcome from competitor analysis. The behaviors and practices box is used to record the relevant insights from contextual ecosystem research. The significance box specifically expects the designer to articulate how the value proposition allows the users to responsibly achieve their overall goals, aspirations, and desires while aligning with their current behavior. The risks and barriers box lists the socio-cultural, technology adoption, infrastructure, technical feasibility, and economic barriers and risk. The emotions and thinking of the users towards the role of people, organizations, devices/platforms, and technology in delivering the value proposition (in other words, channels and customer relationships) is recorded in the experience delivery box. As the designer populates each of the boxes, she continuously tweaks the value proposition.

The **evolve lens** employs a human-centered iterative design process and redefines and redesigns the concepts based on feedback from the users and relevant stakeholders. Designers should use different prototyping tools to create a model or mockup of a product, service, or process for the purpose of articulating its form factor/design language and/or the user experience it delivers. Exposing stakeholders to concepts and collecting their feedback and reactions helps the designer to enhance and refine the concepts. The output of this lens is final concepts—concepts that incorporate feedback from users and stakeholders, and deliver a solution that is not only desirable, feasible, and viable, but also responsible.

* http://designfold.co/consulting/.

APPLICATION

The AutoRaja service, conceptualized by the authors for auto-rickshaw drivers in India, is an example of usage of this framework. An auto-rickshaw is a three-wheeled indigenously manufactured vehicle and is a common form of urban transport in many countries around the world for private and hire uses. With our own understanding of this ecosystem (see Figure 6.5) and inspirations from other cooperative business models (Sapovadia and Patel, 2013), the team conceptualized multiple ideas and what-if scenarios, as schematized in Figure 6.6, which could possibly create a win-win solution for both consumers and auto-rickshaw drivers.

During the explore and define lenses, the team undertook multi-pronged research to build a better understanding of overall ecosystem of the driver community. We conducted onsite research in different parts of Bengaluru, connecting with individual drivers and informal self-help groups to develop a deeper understanding of their key concerns. The what-if scenarios were introduced to the participants as cultural probes. The research included competitive review, one-on-one interviews, and unstructured conversations with smaller groups and individual drivers using commutes for informal interaction. Using secondary research, the team explored existing models (of NGOs) that are prevalent in India that worked with auto-rickshaw communities in providing more jobs, financial aid, vehicle ownership, or education and health awareness.

The contextual user and ecosystem research helped in uncovering complex facets of the drivers' ecosystem (as shown in Figure 6.7) with insights related to their lifestyle and economy, aspirations, social behaviors, role of language, law and political context, influence of government policies, infrastructure implications, rising new age competition, and, ultimately, their relationships with end consumers.

FIGURE 6.5 Early user research.

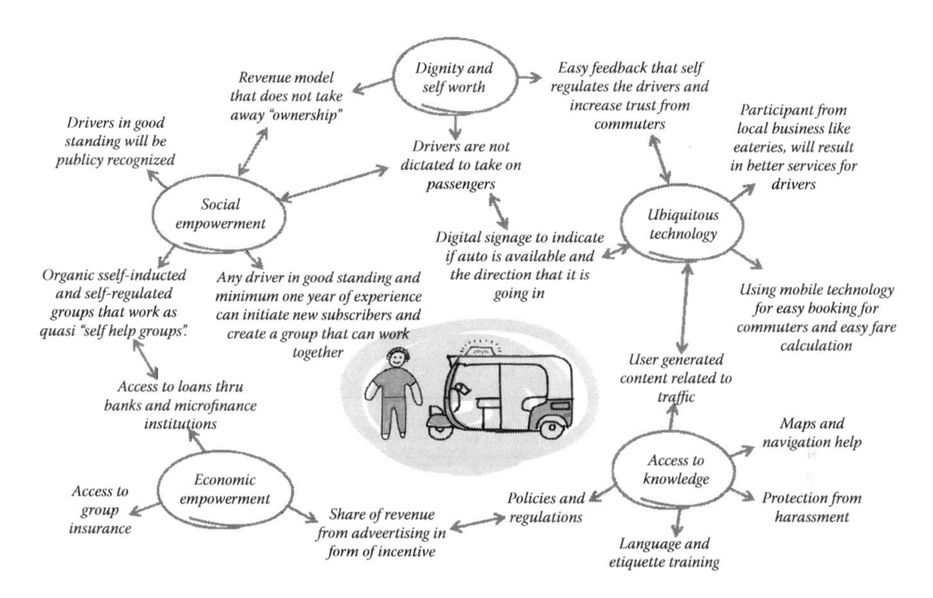

FIGURE 6.6 Auto-Raja concept.

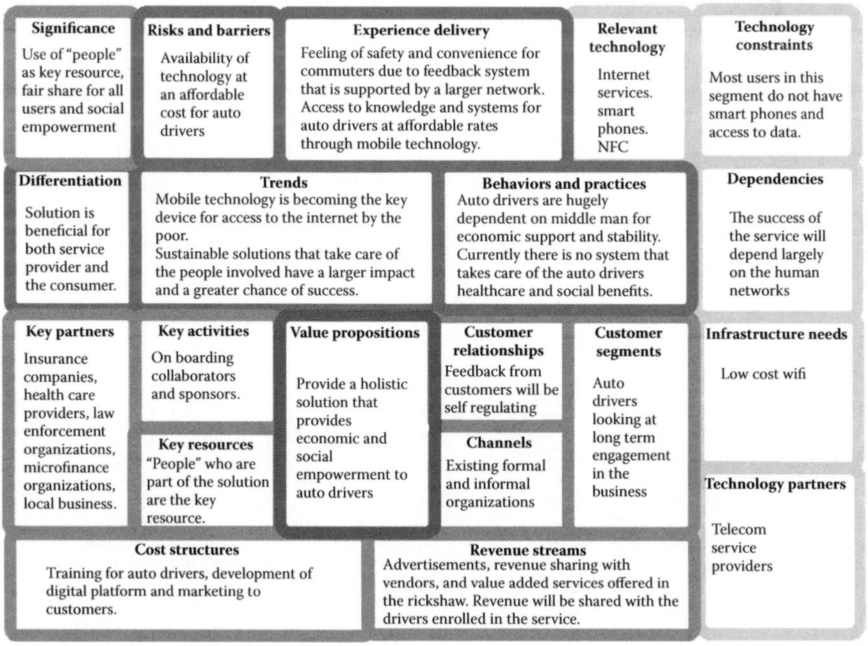

Significance
Use of "people" as key resource, fair share for all users and social empowerment

Risks and barriers
Availability of technology at an affordable cost for auto drivers

Experience delivery
Feeling of safety and convenience for commuters due to feedback system that is supported by a larger network. Access to knowledge and systems for auto drivers at affordable rates through mobile technology.

Relevant technology
Internet services. smart phones. NFC

Technology constraints
Most users in this segment do not have smart phones and access to data.

Differentiation
Solution is beneficial for both service provider and the consumer.

Trends
Mobile technology is becoming the key device for access to the internet by the poor.
Sustainable solutions that take care of the people involved have a larger impact and a greater chance of success.

Behaviors and practices
Auto drivers are hugely dependent on middle man for economic support and stability. Currently there is no system that takes care of the auto drivers healthcare and social benefits.

Dependencies
The success of the service will depend largely on the human networks

Key partners
Insurance companies, health care providers, law enforcement organizations, microfinance organizations, local business.

Key activities
On boarding collaborators and sponsors.

Key resources
"People" who are part of the solution are the key resource.

Value propositions
Provide a holistic solution that provides economic and social empowerment to auto drivers

Customer relationships
Feedback from customers will be self regulating

Channels
Existing formal and informal organizations

Customer segments
Auto drivers looking at long term engagement in the business

Infrastructure needs
Low cost wifi

Technology partners
Telecom service providers

Cost structures
Training for auto drivers, development of digital platform and marketing to customers.

Revenue streams
Advertisements, revenue sharing with vendors, and value added services offered in the rickshaw. Revenue will be shared with the drivers enrolled in the service.

FIGURE 6.7 Auto-Raja product ecosystem map.™

The key features of this modular product service system were to provide a holistic solution using ubiquitous technology that provides economic and social empowerment to the drivers. The services included are: financial participation services, access to regulatory policies for drivers, and safe consumer relationship management. The overall solution works as follows:

1. Drivers subscribe to the service in groups and they can avail the service as a community. This creates a need for compliance and self-regulation that is reinforced with ratings and feedback for an individual driver that in turn decide the rating for the group.
2. Each group has access to behavioral, financial, and soft skills training, both off-line and online.
3. As the overall system is non-hierarchical and inclusive, the additional revenue generated from advertising is shared with the community, which encourages them to work together as a team.
4. By using the smart communities framework, this solution empowers communities socio-economically by integrating the existing driver communities and using people as key resources to make the system work, and, in doing so, creates a fair-share model. The revenue model does not take away their "ownership," which is important for the drivers.
5. The system provides incentives such as healthcare, life insurance, bank loans, and other social benefits to the participants.
6. Through offline training and online tracking through mobile apps, the system provides safety, convenience, and feedback to make the drivers successful individually and as a group.

In the second round of the conceive lens process, insights from early explorations were used to tweak the existing scenarios and to create multiple concepts. Based on the research, the team identified "professional growth" and "socio-economic empowerment" as critical areas for developing the solution, as shown in Figure 6.8.

Some of the key concepts focused on creating a service that increase the operational efficiency, managing finance and monitoring their daily performance.

Concepts were organized in the service blueprint that provided the overall service roadmap connecting stakeholders, artefacts, and activities within relevant scenarios to demonstrate value. Concepts for specific touch points were developed and mocked up. Figures 6.9 and 6.10 below show a scenario around initiation and some concepts related to self-regulation, motivation for safe driving, navigational help, and relevant information

In the evolve phase, mockups of concepts were tested with drivers to validate the understanding and test the usefulness, usability, and desirability of the proposed solution. The learning from the validation was documented and used for informing subsequent development of the overall solution.

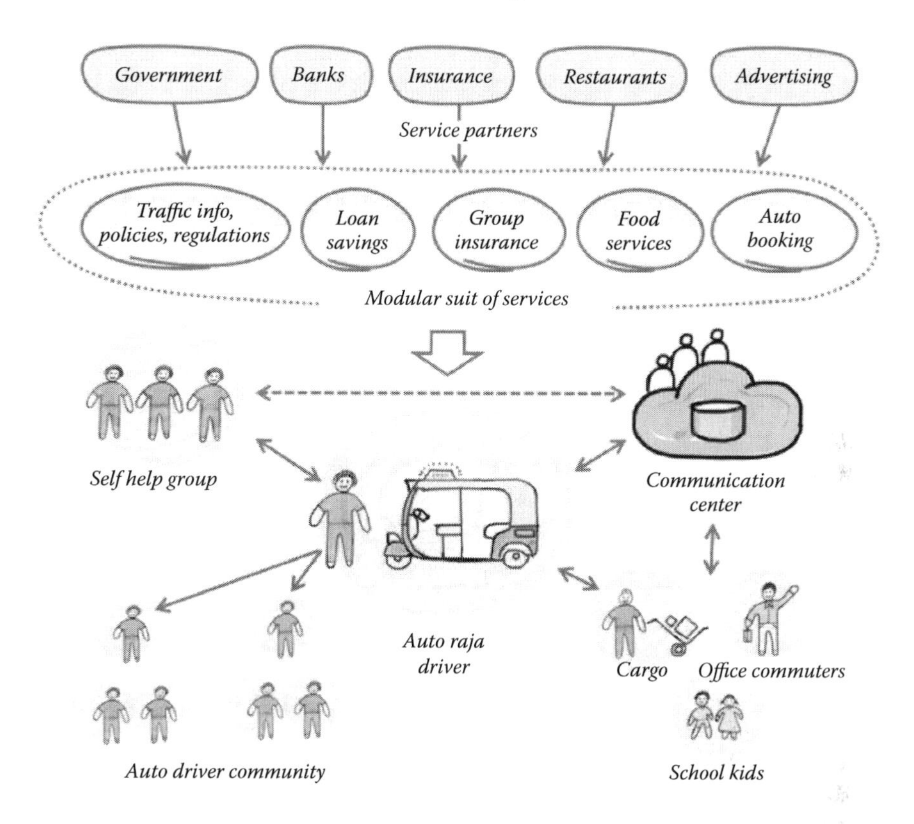

FIGURE 6.8 Auto-Raja product service system.

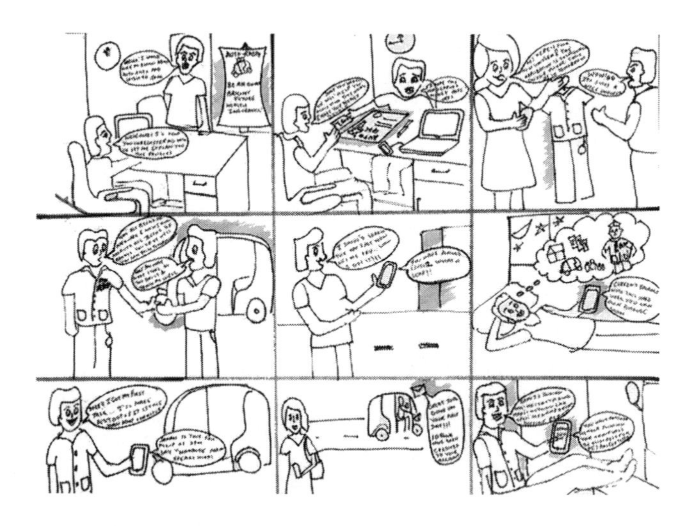

FIGURE 6.9 Initiation scenario.

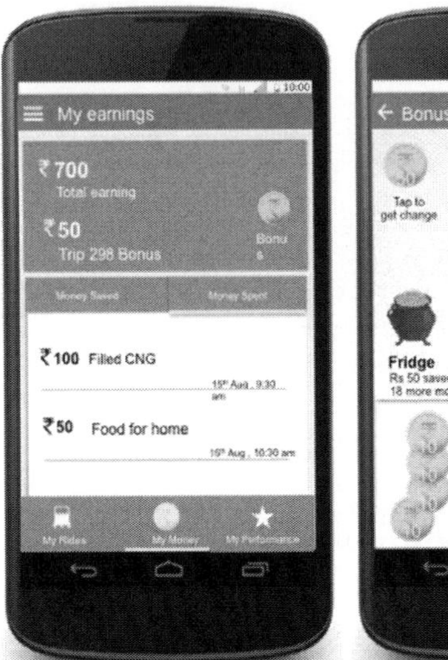

Encouraging self-regulation

Using "persuasion" for driving saving habits

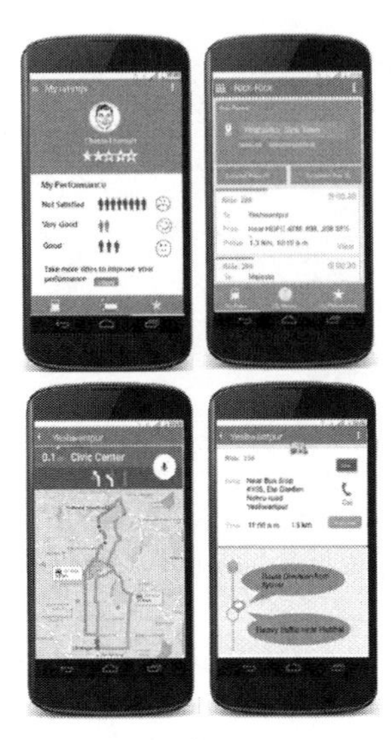

Providing traffic navigation support, information and feedback

FIGURE 6.10 Concept mockups.

FUTURE TRENDS

Product service systems design draws from various fields such as participatory design, transition design, and sustainable and responsible design.

1. Participatory design: apart from the field and center-based research and ideation agenda involving the partners and members, it is important to create living labs consisting of user panels, not only as an arena for co-creation but also as an arena for understanding future user behaviors. The aim of the living labs is to create a setting for continuous engagement in co-creation, exploration, and evaluation of novel ways of designing for transition, by providing a consistent group of lead users as stakeholders in the process (see Figure 6.11). Living labs allow the design, building, and utilization of novel technology and services as cultural probes, very close to actual-use settings. Living labs can also utilize prototyped new technologies and services to understand changes in user behavior.*
2. Transition design: conceived at the School of Design at Carnegie Mellon University in 2012, transition design (Irwin et al., 2015) argues that design has a key role to play in societal transitions to more sustainable

* Contribution from Dr. Naveen Bagalkot.

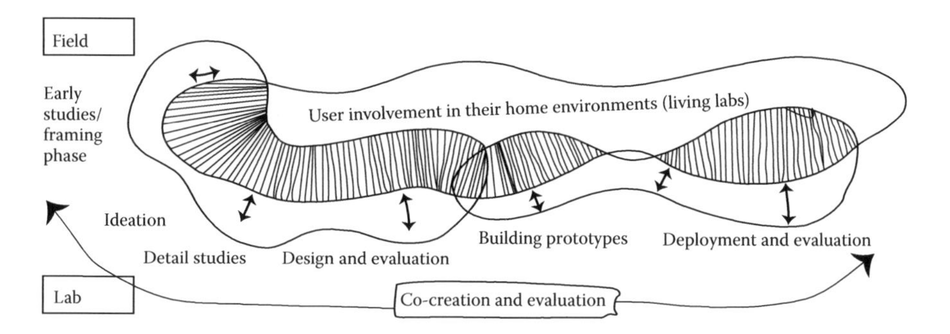

FIGURE 6.11 Living lab concept.

futures. It applies an understanding of the interconnectedness of social, economic, political, and natural systems to address problems to improve quality of life and advocates the reconceptualization of entire lifestyles, with the aim of harmonizing them with the natural environment. Transition design focuses on the need for a lifestyle that is place-based and regional, yet global in its awareness and exchange of information and technology. This is in contrast to the dominant economic paradigm that is predicated upon unbridled growth and an imperative to maximize profit.

3. Sustainable and responsible design: Melles and colleagues (2011) identified critical features of socially sustainable product design, suggesting that Papanek's (1985) agenda for socially responsible and sustainable design must be developed further through the changed role of the designer as facilitator of flexible design solutions that meet local needs and resources. The focus of design has shifted from merely commercial to sustainability and social concerns. Many of the systemic social, economic, and environmental concerns require the designer to argue for change through eco-design and inclusive design. In business and corporate contexts, design strategists are looking at a triple bottom line of social, environmental, and economic factors.

CONCLUSION

As shown through the Auto-Raja examples, human-centered innovation and design is not just "user-driven" in its traditional sense, but more about "designing for emerging usages and technologies." A new approach that combines creative art and design thinking, grounded in the critical thinking of the humanities to push the boundaries of both the technological and business aspects of innovation, along with appropriate tools, was presented in this chapter. This approach is shown to be successful in developing a product-service system (PSS) that not only is desirable, viable, and feasible, but is also responsible.

KEY TERMS

Multi-disciplinary innovation. This approach involves drawing appropriately from multiple academic disciplines to redefine problems outside normal boundaries and reach solutions based on a new understanding of complex situations.

Product-service system design. The design of a marketable set of products and services capable of jointly fulfilling a user's needs.

Contextual ecosystem research. Research of core elements (end users, stakeholders, suppliers, partners, service providers, etc.) and the connections between these elements in their actual context.

Product ecosystem map. A diagram showcasing connections within a product-service system innovation ecosystem formed by the interaction of the participating elements.

Human-centered iterative design. A design methodology based on a cyclic process of exploring, prototyping, testing, analyzing, and refining a product or process in which human needs are always at the center.

Responsible design. A design that is ethical, honest, and socially and environmentally sustainable.

REFERENCES

Bell, G. and Dourish, P. (2007). Yesterday's tomorrows: notes on ubiquitous computing's dominant vision. *Personal Ubiquitous Computing* 11, 2 (Jan. 2007), 133–143.

Gaver, B., Dunne, T., and Pacenti, E. (1999). Design: Cultural probes. *Interactions Magazine*, Volume 6, Issue 1, Jan/Feb 1999, 21–29, ACM New York, USA.

Irwin, T., Kossoff, G., Tonkinwise, C., and Scupelli, P. (2015). Transition Design 2015: A new area of design research, practice and study that proposes design-led societal transition toward more sustainable futures. http://transitiondesign.net/wp-content/uploads/2015/10/Transition_Design_Monograph_final.pdf.

Martinez, S. (2010). Local Food Systems Concepts, Impacts, and Issues (PDF). Economic Research Service. Retrieved October 12, 2016.

Max-Neef, M. (1992). *From the Outside Looking In: Experiences in Barefoot Economics*. Zed Books Ltd; New edition

Melles, G., Vere, I., and Misic, V. (2011). Socially responsible design: thinking beyond the triple bottom line to socially responsive and sustainable product design. *CoDesign* Vol. 7, Issue. 3–4.

Mollison, B. and Holmgren, D. (1978). *Permaculture One: A Perennial Agriculture for Human Settlements*. Ealing: Corgi.

Osterwalder, A., Yves P., and Tucci. C. L. (2005). Clarifying business models: Origins, present, and future of the concept. *Communications of the Association for Information Systems* 16.1.

Papanek, V. (1985). *Design for the Real World: Human Ecology and Social Change*. Chicago: Academy.

Prabhu, G., Frohlich, D., and Greving, W. (2004). Contextual Invention – A multi-disciplinary approach to develop business opportunities and design solutions. FutureGrounds conference, Melbourne, Australia.

Prabhu, G., (2010). Usage Ecosystems: Dynamics of Emerging Markets. In Chavan, A. L. and Prabhu, G. V. (eds), *Innovation Solutions: What Designers Need to Know for Today's Emerging Markets*. Boca Raton, FL: CRC Press.

Ryan, M. (2009). What is the tiny house movement. *The Tiny Life*. Retrieved October 12, 2016.

Sapovadia, V. K. and Patel, A. V. (2013). What Works for Workers' Cooperatives? An Empirical Research on Success & Failure of Indian Workers' Cooperatives. Available at SSRN: https://ssrn.com/abstract=2214563.

Suchman, L. (2007). *Human-Machine Reconfigurations: Plans and Situated Actions*. New York: Cambridge University Press.

Vallero, D. A. and Vesilind, P. A. (2007) *Socially Responsible Engineering: Justice in Risk Management*. Wiley & Sons. ISBN: 0-471-78707-8.

Vallero, D. A. and Brasier, C. (2008) *Sustainable Design: The Science of Sustainability and Green Engineering*. Wiley & Sons. ISBN: 978-0-470-13062-9.

Weiser, M. (1999). The computer for the 21st century. *SIGMOBILE Mobile Computing and Communications Review* 3, 3 (July), 3–11.

Case Study 6
The Challenges of Simplification

Waleed A. Alsaleh

BADIR PROGRAM FOR TECHNOLOGY INCUBATORS AND ACCELERATORS

The Badir Program for Technology Incubators and Accelerators is one of King Abdulaziz's initiatives for the City for Science and Technology program. It is a comprehensive national program that seeks to activate and develop technical business incubators. It provides a wide range of services to business entrepreneurs, such as the provision of office space and assistance in developing work plans for the incubated projects, preparing workshops of different subjects to develop the individual skills of the incubated party, offering legal, administrative, and marketing advice, assisting in obtaining the necessary funding by facilitating the access to financial support sources, along with ongoing follow-ups and support to develop the project and make it a success.

PRODUCT DESIGN IN THE KINGDOM OF SAUDI ARABIA

The design process of a product starts with an idea that is illustrated, validated technically, and proved feasible to manufacture; then a 3D model of the product is created using modern software and 3D printers to test and verify the practicality of the product; finally, an industrial design of the product is developed for mass production. The KSA currently has the most advanced prototyping centre in the GCC, if not the Middle East, which follows the product development process stated above in the design of any product.

Product design is expected to increase over the next few years in the KSA, given that many large companies have begun supporting and adopting new products developed by many start-up initiatives. Among the many products developed by start-ups in the Badir program, two will be discussed below: Yatooq and Controllex.

YATOOQ ARABIC COFFEE MACHINE

Yatooq is a company founded by Lateefa Alwaalan, a Saudi entrepreneur, who recognized a need to come up with ready-made coffee blends and an easier, time-saving brewing method for Arabic coffee preparation. Lateefa set out to design a coffee maker that looked like a traditional Arabic coffee pot, one that can take care of all the brewing

FIGURE CS6.1 Yatooq Arabic Coffee Machine.

steps with as little human input as possible, so that even the least experienced coffee drinker can prepare a great-tasting pot of Arabic coffee (see Figure CS6.1).

Among the problems that Lateefa and her team faced in the original design was the electrical circuit. Arabic coffee requires steps in brewing; therefore, they needed to introduce sensors, relays, and feedback loops for temperature and time control. This circuit was originally placed in the bottom-most part of the pot, and upon user-testing the team found that washing the pot would cause a short in the circuit. This was addressed by introducing proper water insulation across the bottom part of the pot, which proved challenging at first and took a few iterations to perfect. Another problem was the blooming nature of coffee when brewing, so the team came up with the idea to add window-like slots in the top of the internal filter to allow the coffee to circulate within the pot rather than flowing out of it when blooming.

CONTROLLEX: A HOME AUTOMATION SOLUTION

Controllex is currently part of the Badir incubation program, which manufactures and sells home automation solutions. What Controllex did differently with home automation solutions was making them both affordable and easy to install. The company designed individual units for each room that allow users to remotely control home appliances in specific rooms using mobile phone applications (see Figure CS6.2), thus achieving a very flexible home automation system with minimal capital cost to the end user.

FIGURE CS6.2 Controllex: A Home Automation Solution.

When the software was being designed, attention was given to making it simple and easy to use and compatible with all iOS, Android, and web applications. Design features allowed users to give orders to any specific Controllex device, with just two taps from the main screen for faster order execution, especially when users need to control appliances in more than one room. Controllex was also made compatible with smart watches to allow an even simpler and faster control mechanism.

The CEO of Controllex clarified that for identifying the characteristics of their brand they had to highlight the reliability and scalability of their product. They invested in well-designed hardware units that blend in with the room and are aesthetically pleasing, as well as creating branded boxes with clear and easy packaging and installation instructions to show clients their high-quality product. These are just some small design decisions that were made to make the customer experience as easy and enjoyable as possible.

HUMAN FACTORS AND ERGONOMICS DESIGN IMPLICATIONS

The two products share one thing in common: a goal for automation. Both products were designed to simplify human input in certain tasks and revolved around designing for the end-user during the design process. Yatooq wanted to design a product that resonates with its Arabic coffee lovers and for them to enjoy the traditional and modern elements of a coffee pot. Controllex added smart-watch compatibility for customers to quickly access their device for more convenience and control while ensuring ease-of-use features with minimal interaction.

Also, both companies invested in automating their sales transactions to promote an enjoyable user experience. This proved to have a substantial impact on customer satisfaction due to the ease, convenience, and efficiency of purchasing using online payment gateways. The CEO of Controllex stated that "Handling 200 orders a day is much easier using computers and automation than handling them using people, so we are investing in the experience of the purchase to make it simple, fast, easy and user-friendly."

7 Environmental Design

Erminia Attaianese

CONTENTS

INTRODUCTION

Human well-being is a systemic and holistic condition that integrates physical, cognitive, and psychological needs in expressing overall satisfaction, happiness, and quality of life (WHO, 2006). Comfort is a basic component of well-being, commonly expressing a state or situation in which one is relaxed without any unpleasant feelings caused by external factors. It is conditioned by how human activities are performed, as comfort level may foster or hamper people performing their tasks due to the effects of human physiological and psychological perceptions and reactions to different situations and environmental conditions.

The history of architecture is full of examples showing the ambition of architects to shape buildings to the proportions of the human body, with the belief that a beautiful form is one in which people can live better. For instance, the ancient Roman architect Vitruvius' architectural orders were dimensioned in relation to human size; in Renaissance buildings, a complex dimensional system represented the idea of an anthropomorphic space; and in the case of the Modern Movement the principles of the "Existenzminimum" or minimum requirements for living inspired architects rationalists to theorize a fixed size suitable for main living functions, and applied this to an optimum housing dimension (Attaianese, 2016). This anthropocentric model gradually began to change from the 1960s, thanks to the bioclimatic approach and regionalism in architecture, in which the human body was interpreted as a complex physiological and psychological system that reacts to environmental stimuli in built environments (Olgyay, 1963). Concurrently during this time the growing field of environmental psychology proposed models of the physical space surrounding humans (Proshansky, 1970), supported by studies on the welfare of confined spaces, particularly focusing on the relationship between the physical parameters of an environment, the physiological parameters of people, and the perception of well-being expressed by peoples themselves, which successively was transformed into technical standards for thermal comfort in the built environment (Fabbri, 2015).

Please check sense of this sentence – it could probably be reworded (or shortened).

Environmental design conditions based on ergonomics and human factors approaches ensure that the physical factors of space design are used to support users' health and well-being to enhance performance and improve comfort. Theoretical principles, methods, and data are used to understand interactions among humans and system elements to identify how environmental conditions fit users' needs and expectancies. Environmental design conditions that are important elements in designing a healthy and productive life space may be described through a range of physical factors, including temperature settings, lighting, sound, and spatial layout of buildings.

FUNDAMENTALS

Thermal Environments

Thermal comfort is defined as a condition of the mind (ISO, 2005) and is influenced by a host of individual differences (i.e., age, gender, cultural differences; see Table 7.1 for a summary). Together with clothing insulation and your metabolic rate, along with your anatomy and activity level, these factors shape human-environment thermal interaction and are the basic variables of thermal comfort (Fanger, 1980). Research demonstrates that comfort is influenced by exposure time to weather conditions. For instance, evidence findings demonstrated that while Pakistani office workers perceived as comfortable indoor temperatures of up to 31°C, people living in Antarctic winter conditions are comfortable in an indoor environment of around 6°C (Goldsmith, 1960). Thus, geographic backgrounds influence an individual's behavioral adaptation. Many cultures have rules of proper social behavior that are manifested in their clothing to reflect their climate (Jendritzky and Tinz, 2009).

The perception of thermal comfort is also affected by indoor lighting, color, and texture such that higher comfort and feeling warmer is associated with a low color temperature of illumination (Huebnera et al., 2016). Without any changes to thermal parameters, feeling warm is also experienced by adding wood panels and carpets in the workplace (Rohles, 2007). The role of air movement has also been found to have a significant impact on thermal comfort. For instance, people want to control air movement even in cold weather to improve their comfort and thermal perception (de Dear et al., 2013), and to increase fresh air rates as natural ventilation is perceived as a priority for occupant's satisfaction in both home and workplace settings (Clements-Croome, 2014).

TABLE 7.1
Personal Factors and Perception of Comfort

Personal Factors	Thermal Comfort	Auditory Comfort
	Age	
Elderly	Show a smaller range of comfort indoor temperature than younger (Roelofsen, 2014) and less ability to heat adaptation, due not only to physiology and an age-related decline in health, but mainly to their fears and anxieties about extreme heat (Yun et al., 2014)	Are less tolerant to building noise (Sakellaris et al., 2016)
Young and children	Prefer temperature lower than the standardized one (Yun et al., 2014)	Are more tolerant to building noise (Sakellaris et al., 2016)
	Gender	
Female	Particularly sensitive to environmental conditions; they express more dissatisfaction than males in the same thermal environment, and their heat demand is higher (Karjalainen, 2011)	
Male	Less sensitive to environmental conditions, their heat demand is lower (Karjalainen, 2011)	
	Geographic Background, Cultural Habits and Costumes	
Inhabitants from temperate climatic regions	Weak tolerance to the heat stress (Kenawy and Elkadi, 2013)	
Inhabitants from tropical and dry countries	More tolerant to the heat stress (Kenawy and Elkadi, 2013)	
People from Japan	High formality of employees restricts their ability to adjust clothing in workplaces, obstructing their comfort (Mishra et al., 2013)	
People from North America	The lack of formal restrictions encourages people to wear comfortable clothes in the workplace, promoting comfort (Mishra et al., 2013)	
People from Southern Europe		Less tolerant to building noise (Sakellaris et al., 2016)
People from Central and North Europe region		More tolerant to building noise (Sakellaris et al., 2016)

Environmental characteristics that enable adaptation are crucial for comfort of occupants in variable conditions. For instance, adaptive tools that improve thermal perception may include personal control of mechanical conditioning (thermostats, fans, heaters, coolers), operating windows and solar shades, building layouts that allow free movement within a space/between spaces with different temperature to find more comfortable conditions, and rooms and workstations that allow posture adjustments and changes of activity (Clements-Croome, 2014).

Auditory Environments

Auditory environments influence hearing of occupants and their overall comfort, including their physical and mental health, performance, and overall satisfaction. Human sound responses may be pleasant or unpleasant. For instance, ambient long-term background sound becomes normal to listeners, while transient or sporadic sounds generally are more annoying and distracting (Evans and Lorrane, 2005). Additionally, unexpected sounds generated by others tend to be more annoying than sounds that are predictable, unavoidable, and under the individual's control. Also, distraction effects are high if sound has high information content. Specifically, hearing meaningful sounds decreases cognitive performances, especially in open spatial layout plans where speech has been found to be the most annoying and distracting type of noise (independently from its loudness), as the human ear is most sensitive to frequency ranges used in the human voice (Sakellaris et al., 2016).

Excessive silence can also be as stressful and distracting as excessive noise. It may provide a sense of isolation and disorientation, especially when people move in an unfamiliar space. Blocking out noise completely in a space means blocking out opportunities for occupants to connect and be conscious of their surroundings. Hence, environmental physical factors may be designed to provide acoustical conditions for sound to not be elevated to be dangerous/annoying or intrusive, but also to not become so low as to be undetectable (Vischer, 2005). Beyond the physical effects of sound, noise also has significant psychological impacts. For example, the feeling of displeasure associated with a disturbing sound is expressed by annoyance that may increase stress and decrease human performance. These reactions are not necessarily associated with perceived sound intensity, but rather with a noise tolerance level that depends on personal and contextual factors (Oseland & Hodsman, 2017).

While there is no single acoustic solution that can be universally applied, spatial planning, room shaping, and architectural materials are fundamental elements in the design of acoustic performance. Since sound moves through building spaces in a variety of ways, the shape of rooms and indoor materials used may be selected and assembled to control sound and influence reverberation time (i.e., the way a space "sounds"). Specifically, concave surfaces tend to concentrate or focus reflected sound in one area, whereas convex surfaces tend to disperse sound in multiple directions. Additionally, the use of absorptive and diffusing materials can control the amount of reflected sound within a room, diminishing the strength of the sound and diffusing its effects. In small square rooms, for example, low frequencies seem to be predominant, so that materials placed on the ceiling surface may lessen any annoying speech effect; in round rooms, sound diffusing elements should be positioned on curved surfaces to mitigate and disperse sound in many directions (GSA, 2011; Knauf, 2013). Furthermore, panels, doors, and partitions can be used as physical barriers to block sound between spaces. Electronic background sound may be introduced to mask sounds from air conditioning systems and other equipment, or indirect speech within a space so that it is either not intelligible or even audible to unintended listeners (ASA, 2016). See Table 7.1 for the influence of individual differences (i.e., age, cultural background) on sound comfort.

Environmental design considerations are essential in multifunctional environments. Numerous types of tasks are performed simultaneously, and spatial zoning allows the association of specific acoustic requirements for each activity with specific layout areas, in order to support their optimal location and contiguity. For example, interaction areas for loud activities should be grouped and isolated from focus areas that support individual concentrative tasks, and from privacy areas where confidential or personal work and conversations need to be protected (Roy, 2015).

Luminous Environments

Indoor space is lit by a combination of daylight entering through openings to the outdoor environment and via electric-light sources. Environmental space is a function of brightness, contrast, and light color rendering so that specific luminous patterns may have a consistent effect on an occupant's subjective impression of an indoor environment and impact their emotional well-being in terms of spaciousness/confinement, visual clarity/haziness, relaxation/activation, and private/public surrounding. For instance, excessive bright light may occur with windows, luminaries, or reflecting materials causing eye tiredness and consequent errors and injuries. Glass surfaces and glossy finishing of walls, flooring, and furniture may also cause similar effects and uncomfortable veiling reflections on objects (CIBSE, 2015). Thus, room surface reflectance influences the uniformity of lighting and the sensation of brightness of a room, and consequently need to be adequately selected in interior design to avoid glare.

Uniformity of lighting can be desirable or less desirable depending on the function of the space and type of activities. A completely uniform space is usually undesirable because light contrasts can improve visibility without increasing light level, whereas non-uniform lighting may cause distraction and discomfort. For instance, spatial wideness and perceptual clarity of a room are evoked by uniform bright light from the ceiling and higher horizontal luminance in a central location and preferably with cold-colored lighting. Conversely, non-uniform environmental lighting is induced with high brightness on walls and low brightness in areas to reinforce the impression of spatial privacy with preferably warm colors. Human behavior is also influenced by lighting as relaxation is cued by the non-uniformity of wall lighting and lower light levels, especially with warm colors, whereas higher levels of bright light improve levels of arousal. Moreover, lighting may be used to direct behavior by aiding orientation to a space to move them in a direction to the bright light. Thus, the brightness of paths and walls may focus attention and support way-finding.

Research demonstrates that luminous environments impact humans both visually (light quality affects visual comfort and visual task performance) and non-visually (it also evokes mental states and the emotional impression of space and produces long-term biological effects on human health) (Boyce, 2014). Visually, higher illuminance levels induce greater arousal (Gifford, 1988), and lead to more communication and louder conversations (Veitch and Galasiu, 2012; Boyce, 2014). Light intensity plays a role also in alertness regulation, since bright light may improve wakefulness, or, the opposite, dim light may decrease it (Górnicka, 2007). Non-visual effects of light are numerous, where variations of light parameters and colors can affect mood and trigger emotions.

Daylight and outside views have also been found to be relevant factors in visual performance and comfort, particularly when the outside opening focuses on natural surroundings, which reduces the sensitivity to discomfort glare (Boyce, 2014). The balanced spectrum of colors and wavelengths of natural light, varying over the day with latitude and seasons, helps to control not only vision, but a wide range of non-visual functions, including nervous and endocrine system activation and circadian rhythm balancing (National Institute of Building Sciences, 2013). Even though from a subjective perspective people's moods are influenced by daylight factors such that lower levels of daylight in a room spread a sense of gloom, higher levels of sunlight can improve a state of forcefulness (Steffy, 2002). Thus, when direct daylight is missing, lighting techniques that are close to daylight are used to enhance visual task performance, health, and positive well-being (Edwards and Torcellini, 2002).

Table 7.2 provides recommended guidelines for temperature, noise, and lighting factors to improve comfort and performance of occupants.

Spatial Environments

According to the International Standards Organization, the dimensional minimum unit of operation space in a defined productive building is by the workstation (ISO 6385, 2001). These defined spatial dimensions will impact the occupant's accessibility and postural/moving conditions to move freely and reach all points of their workspace (Vischer, 2005). The interior spatial layout and the integration of thermal, auditory, and luminous factors provide ambient conditions that affect perception, mood, thoughts, behavior, and health of occupants (Clements-Croome, 2014). For instance, shapes, colors, textures, proportions, relations, and sequence of spaces may exert a range of different psychological impacts that, depending on the task, may produce positive or adverse effects.

Affordance theories suggest that human interaction with the environment is largely conditioned by the affordances of the building elements and spaces as we recognize interior space according to our understanding of the functions that they provide us (Koutamanis, 2006). Building coherence, which is expressed as the level of comprehensibility of building elements and forms, influences the effectiveness of task performance. For instance, purposive actions require legible interiors, where users can simply deduce the identity, meaning, and locations of objects inside the buildings. This is enhanced by regularity of geometric shapes, distinctive building markers, and multiple repetitive features. Specifically, building design may influence people behaviors, such that circular-shaped settings with complex visual scenes are considered appropriate for creativity (Kristensen, 2004) and high ceilings encourage

TABLE 7.2
Guidelines of Quantitative Parameters of Environmental Factors Associated with Occupant Comfort, Health, and Performance

Temperature

- Ideal temperature for the average employee is approximately 21°C (Sundstrom, 1986).
- Concentration decreases if the office temperature is higher than the norm, but when it is too hot, employees take up to 25% longer than usual to complete a task (Seppanen et al., 2006).
- Human performance increases with temperatures up to 21–22°C and decreases with temperatures above 23–24°C (Seppanen et al., 2006).
- Ventilation rates up to about 25 l/s per person tend to lead to reductions in sick-building syndrome symptoms, absenteeism due to illness, and respiratory diseases (Sundelletal, 2011).
- An increase of 1 l/s per person in fresh air ventilation rate over the range 2–20 l/s per person is associated with a 1.0–1.5% decrease in illness absenteeism (Mendell et al., 2013).

Noise

- Elevated exposures to high-intensity noise affect human health, since strong sound pressure gradually damages the auditory system.
- A habitual noise level higher than 85 dB (sound of clock alarm) cause gradual hearing loss in a significant number of individuals.
- Loss of auditory sensitivity occurs initially around a frequency of 4 kHz.
- With high-pitched sounds, auditory sensitivity gradually increases to both lower and higher frequencies, if exposure is continuous.
- Exposure to higher noise intensity, starting from 120–130 dB (the loudness of a rock concert or a jackhammer) may cause dizziness, nausea, or balance disorders; these are usually reversible after the end of the sound stimulus, but may lead to deafness if a high level of sound persists for a long time (Roy and Johnson, 2015).

Light

- Performance of visual tasks is affected by quantitative parameters, since a minimum level of illuminance is necessary for a clear vision without tiredness (NIBS, 2013).
- Excessively abundant illumination can be perceived as uncomfortable.
- A lower illuminance level is preferred for office VDT and paper-based horizontal tasks, while higher levels are associated with visually demanding tasks such as reading and writing (NIBS, 2013).
- Increasing levels of illumination in the environment and in the working area improves visual performance, but a visually comfortable luminous environment depends also on other lighting-related factors, as the distribution of the light sources in the space, light color characteristics, and the limitation of glare (CIBSE, 2015).
- Recommended average illumination is generally fixed according to the room typology and activity, and light level ratios between the task area and its surroundings are given (DiLaura et al., 2011).

abstract creative thought (Anthes, 2009). Furthermore, the extent to which the spaces are distant or interconnected influences social regulation capabilities. Social interaction is promoted by the size, location, and permeability of interior rooms, their proximity, and by creating focal points. Conversely, privacy is induced by spatial hierarchy within buildings as size and location stimulate the sense of intimacy in the environment.

METHODS

Holistic Methodology for the Human-Centered Design of Buildings

Human-centered design of built environments is a holistic approach that is structured in three macro-activities. First, it aims to identify the context of user specification by identifying the users in physical and organizational terms, and the tasks they need to perform using the artefact. Functional goals and environmental context, particularly elicited by different occupant expectations, are gathered and finalized. Attention is reserved to occupant analysis to define users' profiles and cluster settings that consider personal, cognitive, and behavioural factors. For example, human variability is assessed with variations in anthropometric body size and shape that are due to inter-individual differences (i.e., age) and intra-individual variations (i.e., gender, genetic diversities, ethnicity/culture). Thus, anthropometric data design may be developed to identify body size to shape environmental design. This aspect is relevant for designing accessible and inclusive buildings (City of Calgary, 2010).

To determine occupant interaction with the building, a task analysis should be conducted by observing directly and/or indirectly how occupants can achieve their goals using the buildings, and consequently how buildings' technical features (forms, volumes, dimensions, layout, ambient conditions) need to be designed to support and satisfy occupants' needs and expectations. General and technical requirements become a strategic step at this stage for the design process to ensure that all actual users' demands and needs will be translated into technical requirements and will be in the building's features. Then, a creative activity follows where architectural details are depicted on users' requirements and conceptual diagrams and mock-ups are used to integrate technical drawings/renderings to support the designer's development of both the designer and occupants' mental models in the assessment stage. In fact, the iterative design process should involve occupants at several stages of design to gradually control coherence of solutions against settled requirements and correct previous problems before the design becomes concrete.

The validation of human-related design solutions may involve several techniques, such as checklists for incompliance, surveys and questionnaires, participatory sessions and focus groups with stakeholder/experts, and heuristic evaluations. Following improved design solutions (with minor or major revisions), the iterative cyclic process is reinitiated until the optimal version of the architectural project is released (Attaianese, 2016). Evaluating the building systematically and comprehensively after user occupancy may also be employed using the post-occupancy evaluation (POE) survey. This process differs significantly from other conventional building surveys because it uses direct, unmediated experiences of building occupants as the basis for evaluating how a building works for its intended use. Focusing mainly on occupants' needs allows them to participate in the evaluation process and gives them more ownership in the building (Charkas et al., 2016). POE should operate throughout the life cycle of a building with continuous feedback as building use is often changing and evolving. Major areas for evaluation include the original purpose for which the building was designed, the process by which the building was built, the building's physical performance and impact on the occupant, and the operation and maintenance of the building.

APPLICATION

WORKPLACE

Workplace environmental design is continuously evolving with the global changes that are modifying lifestyles and work systems. Diversity is a leading design concern with more diverse occupants shaping the social systems of work and life to improve productivity, accessibility, sustainability, and inclusion. In addition to diversity, the aging population is on the rise in the workplace. Particularly in Western countries, large variations among the aging population require an environment that addresses their physical needs in terms of tolerance and acceptable range of motion. Moreover, information and communication technologies (ICT) are having a profound environmental impact, producing complex work models that are flexible and unpredictable, allowing workers to be outside of enclosed spaces during fixed times. Thus, workplace features have changed from a backdrop-passive setting, to actively supporting work and productivity based on various space typologies to provide opportunities to perform different and often conflicting tasks (Vischer, 2005). Occupants' survey-based guidelines of office workspace recommendations for environmental conditions are provided in Table 7.3.

SMART HOMES

Smart homes are interactive spaces that assist occupants to live independently and comfortably with the help of technology. The concept was originally developed with the focus on quality of life and improving security and energy savings. Users' energy consciousness and personal control have a strong impact not only on perceived comfort, but also on operation costs. System-state monitoring captures all indications, as far as feasible, about decay conditions and gives the possibility to adjust their working to assure comfort and well-being in a wide range of use conditions, so that end-users and/or tenants play a key role in maintenance operations (Attaianese, 2011). Over time, smart home technology has increasingly targeted users with restricted abilities due to age or disability. Thus, traditional goals centered on control and energy-resource monitoring have been implemented with more pervasive aims. Identification and prevention of emergency situations, monitoring medical conditions, and gathering information on changes in health status are functions that have been provided to assist occupants in daily activities through home systems and devices that are context-aware, and invisible in use, to interact with people in a natural way, mainly involving usability issues and inclusive design (Andreoni and Pizzagli, 2006). The new capabilities of smart technology and embedded computing raise considerations of how the future house looks regarding its structure, interior design, materials, operating modes, and activities, and its position and interrelationship with the built environment (Bitterman and Shach-Pinsly, 2015).

GREEN ERGONOMICS

Green ergonomics focuses on the bi-directional connections between human systems and nature to improve health, productivity, and effectiveness in engineered

TABLE 7.3
Recommendations for Workplace Environmental Design

Layout

- Organize the floor plate to maximize natural light penetration onto the floor. For example, place enclosed spaces around the core of the building and open spaces at the perimeter where windows are located.
- Minimize the appearance of long corridors or paths by introducing color, pattern, or texture changes.
- Provide a variety of work settings in the right proportion to support a variety of work functions:
  - Focus work: provide quiet zones or spaces for concentrated work.
  - Collaboration: emphasize small group collaboration and provide diverse settings (formal and informal).
  - Learning: consider the workplace to be an educational environment that supports learning and mentoring by providing e-learning and in-person, one-on-one learning.
  - Socializing: provide a variety of informal spaces that accommodate work and casual communication while fostering informal collaboration and innovation.
- Provide zoned temperature controls or, if possible, individual controls in each enclosed space.

Rooms

- Integrate strategies for achieving a non-intrusive level of speech privacy. They include absorption (through acoustical ceiling, fabrics, and carpet), blocking (through furniture system, panels, walls, partitions, and screens), and covering (sound masking).
- In open-plan environments, ensure that people are sitting near those with similar work patterns or subjects of study.
- Provide a quiet space and a collaboration space.
- Reduce the impact of "dense" space and the impact of seeing a significant number of people at once by orienting individual workspace openings such as workstation openings or desk positions in an office to minimize views into others' workstations while seated.
- When possible, provide views to windows to reduce perception of crowding. This could be accomplished by moving circulation to the perimeter of the space and relocating fixed elements such as offices or conference rooms to the interior of the space.

Windows

- Provide operable windows or operable window coverings to maximize sunlight, airflow, and temperature control.
- Favoring cross-ventilation, if possible.
- Use glass where visual privacy is not required.

Textures and Colors

- Consider introducing "texture" into a space by using natural materials such as wood, cork, plants, natural fibers.
- Use color strategically to promote desired behaviors and feelings based on psychological reactions.
- Vary color use through the workplace. Use it as a design technique to identify circulation or the changing character of space.
- Use lighter colors to help reflect light through the space and increase the amount of natural light.

(*Continued*)

TABLE 7.3 (CONTINUED)
Recommendations for Workplace Environmental Design

Lighting

- Provide adjustable lighting levels; changes in lighting levels should be gradual and employees should have local control of lighting levels

Views and Surroundings

- Provide outdoor areas for use by employees. Encourage employees to go outside for breaks, even if just for a few minutes.
- Give preference of "nice views" to shared spaces.
- Provide outside views on natural elements.

Personalization and Participation

- Provide work spaces that enable visibility, openness, and greater employee mobility to foster engagement. When workers are more likely to see each other, they are more likely to connect and collaborate.
- Include workers' opinions and preferences to shape design solutions.

environments (Thatcher, 2012). Since human quality of life is difficult to maintain in an environment that itself is not healthy, the goals of ergonomics (effectiveness, efficiency, health, safety, and usability) may be closely aligned with the goals of design for environmental sustainability. Supported by a larger system-design perspective, the green ergonomics approach to building design is focused on reducing the human impact on ecosystems through ergonomic design to diminish or avoid natural crises. The concept of biophilic describes the extent which humans are hard-wired to connect with nature. Thus, well-being and productivity are improved by pro-nature aspects of indoor environmental quality, particularly in work settings where the introduction of natural light, fresh air, external windows, and aesthetic views, and indoor plants to rise indoor air quality and remove pollutants. Biophilic design is an extension of the concept of biophilic and incorporates natural light and materials with vegetation and nature views of the natural world into the built environment. Ryan and colleagues proposed biophilic design principles that inspire green design solutions by promoting innovative building performance to improve human health and effectiveness (Ryan et al., 2015). Integrated research findings from neurosciences, endocrinology, and other science disciplines found design parameters of architecture to enhance cognitive performance and positive well-being. The authors recommended several design patterns, including a visual connection with nature (seeing nature), non-visual connection with nature (using auditory, haptic, olfactory, or gustatory stimuli), access to thermal and airflow variability (thermal comfort feelings similar to those experienced in nature), the presence of water (seeing, hearing, or touching water), complexity and order (the presence of rich sensory information in a coherent spatial hierarchy similar to nature), prospect (the inclusion of unobstructed views similar to nature), and mystery (providing sensory stimuli that encourage understanding and exploration similar to nature). Through deep understanding of the connections between nature and health while keeping eco-efficiency awareness levels high, green ergonomics of designed environments may foster an occupant's behavior change toward sustainability.

FUTURE TRENDS

Sensory Design Architecture

The concept of sensory design architecture recently emerged to acknowledge the role of total multi-modal sensory experience in influencing our attitudes and behaviors. Taking an occupant-centered approach, sensory environmental design is aimed at optimizing "health," in terms of the health of an individual occupant, the health of a building's effectiveness, and the health of its ability to harmonize with surrounding environments. It focuses on the effects of architecture on occupants to understand how they can be better attuned through sensory design for a healthier mind and body connection physiologically, cognitively, emotionally, behaviorally, and spiritually (Leheman, 2011; Keelin et al., 2012).

Sustainability and Building Design

Sustainability is imperative for all developmental elements, including humans. Looking at the main goals of sustainable design, a broader reflection on the role of human factors/ergonomics must be considered to enhance design process of sustainable buildings. It is crucial that solutions are selected based on user's needs and their related tasks, considering human capabilities and limitations, diversities, and uniformities. Thus, more ergonomics issues are needed in whole-building design, particularly to prevent unexpected disadvantages for occupants in terms of safety, perceived comfort, and well-being that design solutions for energy savings frequently provoke (Attaianese, 2014). Despite several studies demonstrating the rising interest of socio-technical aspects of energy efficiency in buildings, the explicit reference to the human side of building sustainability is only partially assigned to the occupants' role. Moreover, it is far from being balanced, biasing towards buildings assessment methods rather than occupant needs. Human-related factors in current rating systems are largely undervalued, confirming that social dimension of building sustainability in practices are overlooked and need to be increased through ergonomics (Hedge, 2013; Thatcher and Milner, 2014).

Urban Planning

Ergonomics science and socio-technical systems applied to urbanism may provide new ways forward for the design and development of our cities and towns, since it has been noticed that land use planning and urban design have not had the tools to manage the complexity and often competing priorities that are presented in urban development (Smith et al., 1994). With this understanding, ergonomic approaches may allow for multidisciplinary perspectives and empirical evaluation of the complex interactions between humans, technology, and artifacts within modern urban environments (Masztalski and Michalski, 2014; Stevens, 2016).

CONCLUSIONS

Environmental design is concerned with the relationship between human performance and well-being within the designed environment. The ergonomic criteria of environmental design are to optimally arrange the setting to aid people in achieving their objectives, to create effective and efficient workstations that promote comfort and productivity, and to provide ambient conditions that promote health and well-being.

The literature of environmental design and their respective evidence-based design principles are characterized by three key concepts: variability, adaptability, and greenery. Variability affects occupants with a range of differences in personal capabilities and behaviors (i.e., gender, age, ethnicity/cultural differences), time variability, and body and mental modifications occurring at different stages of life. Work demands and tasks are changeable, requiring increased creativity and decision-making abilities. Technology's pervasiveness is removing all barriers between work and personal life, such that physical and social factors are blending with almost intangible spaces that are pervading our quality of life. Adaptability in environmental design is expressed in different ways: via behavioral and cognitive personal controls; achieved through design solutions allowing occupants direct action over the environment with easy interpretation of spatial and temporal situations and events; by physical settings that are adjusted and personalized; and by offering opportunities for users to be empowered by active participation in the design process. Lastly, environmental design needs to be green-oriented to improve the health and restorative effects of nature to occupants. Spaces should have an open surrounding with pleasant outside views of green elements that allow access of natural light and natural ventilation, and with the placement of operable windows.

KEY TERMS

Building and architecture. The term building refers to any human-made structure used or intended for supporting or sheltering any use or continuous occupancy. In a broader meaning it indicates not only a structure that has a roof and walls and stands more or less permanently in one place, but also its inside and outside designed spaces. It is one of the main elements of the built environment. Architecture is the art or practice of designing and building edifices and other elements of built environment for human use, taking both aesthetic and practical factors into account. It is the process of creating buildings and spaces that motivates occupants that facilitate them to do their jobs, that bring people together, and that become, at their best, works of art that people can move through and live in.

Occupant well-being. Well-being may be defined as a systemic and holistic measure that integrates many facets including the physical, cognitive, and psychological needs of people. When referring to building occupants it is connected with happiness and overall satisfaction in relation to the quality of living conditions in built environment; it is consistent with World Health Organization statement about health, as a state of complete well-being, by a physical, social, and mental perspective.

Comfort. A basic component of wellbeing, as it commonly expresses a state or a situation in which one is relaxed and does not have any unpleasant feelings caused by external factors. It affects, and is conditioned by, how human activities are performed, since comfort level may foster or hamper people performing their tasks, due to the effects of human physiological and psychological perceptions and reactions to environmental conditions, in different situations that may occur.

Performance. The accomplishment of a given task measured against preset known standards of accuracy, completeness, cost, and speed. Organizations need highly performing individuals in order to meet their goals, to deliver the products and services they specialized in, and to achieve competitive advantage. Performance is also important for the individual. Accomplishing tasks and performing at a high level can be a source of satisfaction, with feelings of mastery and pride. Low performance and not achieving the goals might be experienced as dissatisfying or even as a personal failure.

Built environment supportivity. The term built environment refers to aspects of our surroundings that are built by humans, that is, distinguished from the natural environment. It includes not only buildings, but the human-made spaces between buildings, such as parks, and infrastructures for human activity. From an ergonomic perspective, built environment may be considered a facility able to support people acting in and around it during their everyday life. Thus, built environment supportivity expresses the degree to which physical characteristics of the built environment are able to support human and organization performance.

Human-centered design. An approach to systems development that aims to make systems usable and useful by focusing on the users, their needs and requirements, and by applying human factors/ergonomics, usability knowledge, and techniques. This approach enhances effectiveness and efficiency, improves human well-being, user satisfaction, accessibility, and sustainability; and counteracts possible adverse effects of use on human health, safety, and performance.

Human variability and diversity. The multiplicity of human characteristics, including not only the variety of personal attributes such as age, gender, body size, physical capacity, and limitations, but also cognitive attributes such as intellectual abilities, attitudes, motivation, lifestyle, education level, or culture of origin. Particularly, inter-individual human variations cover all the changes that individuals undergo during their life; intra-individual human variations concern the differences among individuals, in relation to the great geographic and ethnic social groups and the variations between genders.

REFERENCES

Al horr, J., Kaushik, A., Arif, M., and Elsarrag, E. (2016). Occupant productivity and office indoor environment quality: A review of the literature. *Building and Environment*, 105.

Andreoni, G. and Pizzagli, M. (2006). Relevance of crgonomic in domotics and ambience intelligence. In Karwowski, W. (ed.), *International Encyclopedia of Ergonomics and Human Factor.* Second Edition. Vol. 1, Boca Raton, London, New York: CRC Press.

Anthes, E. (2009). How Room Designs Affect Your Work and Mood Brain research can help us craft spaces that relax, inspire, awaken, comfort and heal. *Scientific American Mind,* April 22, pp. 1–5.

Attaianese, E. (2011). Human factors in maintenance for a sustainable management of built environment, in Wellbeing and innovation through ergonomics. In Lindfors, J., Savolainen, M., and Väyrynen, S. (eds), *Proceedings of NES2011* (pp. 129–134).

Attaianese, E. (2014). Human factors in design of sustainable buildings. In Soares, M. and Rebelo, F. (eds), *Advances in Ergonomics is Design and, Usability and Special Population Part III* (pp. 392–403). AHFE Conference.

Attaianese, E. (2016). Ergonomic design of built environment, In *Anais do VI Encontro Nacional de Ergonomia do Ambiente Construído & VII Seminário Brasileiro de Acessibilidade Integral.* Blucher Design Proceedings, v.2 n.7. São Paulo: Blucher.

Bitterman, N. and Shach-Pinsly, D. (2015). Smart home—A challenge for architects and designers. *Architectural Science Review* 58(3), 266–274.

Boyce, P. R. (2014). *Human factors in lighting.* Third Edition. Boca Raton, London, New York: CRC Press.

Canter, D. V. (1983). The Purposive Evaluation of Places. A Facet Approach. *Environment and Behavior* 15(6), 659–698.

Canter, D. V. (1970). The place of architectural psychology: A consideration of some findings. London: Proceedings of the Architectural Psychology Conference.

Charkas, M. N., Ibrahim, M. A., and Farghaly, T. A. (2016). Towards Environmentally Responsive Educational Buildings: A Framework for User-Centered Post Occupancy Evaluation (POE). *International Journal of New Technology and Research (IJNTR)* 2(2), 34–42.

City of Calgary (2010). *Universal Design Handbook. Building Accessible and Inclusive Environment.*

Clements-Croome, D. (2014). Sustainable Intelligent Buildings for Better Health, Comfort and Well-Being. Report for Denzero Project.

de Dear, R. (2004). RP-884 Project. Available from: http://aws.mq.edu.au/.

de Dear, R. J., Akimoto, T., Arens, E.A., Brager, G., Candido, C., Cheong, K. W. D., Li, B., Nishihara, N., Sekhar, S. C., Tanabe, S., Toftum, J., Zhang, H., and Zhu, Y. (2013). Progress in thermal comfort research over the last twenty years. *Indoor Air* 23(6), 442–461.

DiLaura, D. L., Houser, K. W., Mistrick, R. G., and Steffy, G. R. (eds) (2011). *The Lighting Handbook* (10th ed.). New York: Illuminating Engineering Society of North America.

Edwards, L. and Torcellini, P. (2002). A Literature Review of the Effects of Natural Light on Building Occupants. *National Renewable Energy Laboratory.* Golden, CO.

Evans, G. and Lorrane, E. N. (2005). Noise and human behavior. In Stanton et al. (eds), *A. Handbook of Human Factors and ergonomic Methods.* Boca Raton, London, New York: CRC Press.

Evans, G. W. (2003). The Built Environment and Mental Health. *Journal of Urban Health: Bulletin of the New York Academy of Medicine* 80(4), 236–255.

Evans, G. W. and McCoy, J. M. (1998). When buildings don't work: the role of architecture in human health. *Journal of Environmental Psychology* 18, 35–94.

Fabbri, K. (2015). A Brief History of Thermal Comfort. In Fabbri, K. (ed.), *Indoor Thermal Comfort Perception.* Springer. On line.

Fanger, P. O. (1980). *Thermal Comfort, Analysis and Application in Environmental Engineering.* Copenhagen: Danish Technical Press.

Gifford, R. (1988). Light, decor, arousal, comfort and communication. *Journal of Environmental Psychology* 8(3), 177–189.

Górnicka, G. B. (2007). Effect of lighting level and colour temperature on performance and visual comfort. *Sleep-Wake Research in The Netherlands* 18, 53–57.

GSA Public Buildings Service (2011). *Sound Matters: how to achieve acoustic comfort in the contemporary office.*

Hedge, A. (2013). The Importance of Ergonomics in Green Design. Proceedings of the Human Factors and Ergonomics Society 57th annual meeting.

Heerwagen, J. (1998). Design, productivity and well being: What are the Links? Paper presented at: The American Institute of Architects Conference on Highly Effective Facilities Cincinnati, Ohio March 12–14, 1998.

Helander, M. (2006). *A Guide to Ergonomics of Manufacturing.* Second Edition. Boca Raton, London, New York: CRC Press.

Huebnera, G. M., Shipwortha, D. T., Gauthierb, S., Witzelc, C., Raynhamd, P., and Chana, W. (2016). Saving energy with light? Experimental studies assessing the impact of colour temperature on thermal comfort. *Energy Research & Social Science* 15, 45–57.

ISO 6385 (2001). Ergonomic principles in the design of work systems.

ISO 7730 (2005). Ergonomics of the thermal environment.

Jendritzky, G. and Tinz, B. (2009). The thermal environment of the human being on the global scale. *Global Health Action* 2. On line.

Karjalainen, S. (2012). Thermal comfort and gender: a literature review. *Indoor Air* 22, 96–109.

Kenawy, I. and Elkadi, H. (2013). The impact of cultural and climatic background on thermal sensation votes. In PLEA 2013: Proceedings of the 29th Sustainable Architecture for a Renewable Future Conference, Munich, Germany, pp. 1–6.

Knauf (2013). *Acoustic Design According to Room Shape.* Danoline. On line.

Koutamanis, A. (2006). Buildings and affordances. In: Gero, J. S. (eds), *Design Computing and Cognition.* Dordrecht: Springer.

Kristensen, T. (2004). The Physical Context of Creativity. *Creativity and Innovation Management* 13(2), 73–141.

Leheman, M. V. (2011). How sensory design brings value to buildings and their occupants. *Intelligent Buildings International* 3(1), 46–54.

Masztalski, R. and Michalski, M. (2014). The human factor in urbanism of medium-sized cities in Poland. In Charytonowicz, J. (ed.), *Advances in human factors and sustainable infrastructures.* Springer International Publishing Switzerland.

Mishra, A. K. and Ramgopal, M. (2013). Field studies on human thermal comfort. An overview. *Building and Environment* 64, 94–106.

National Institute of Building Sciences (2013). Design Guideline for the Visual Environment: Version 4.3.

Olgyay, V. (1963). *Design with Climate.* Princeton, NJ: Princeton University Press.

Oseland, N. and Hodsman, P. (2017). Psychoacoustics: Resolving Noise Distractions in the Workplace. In Hedge, A. (ed.), *Ergonomic Workplace Design for Health, Wellness, and Productivity.* Boca Raton, London, New York: CRC Press.

Proshansky, H. M. (1970). *Environmental Psychology: Men and His Physical Setting.* Holt, Rinehart & Winston of Canada Ltd.

Rodemann, P. A. (2009). Psychology and perception of patterns in architecture. *Architectural Design* 79(6), 100–107.

Roelofsen, P. (2014). Healthy ageing and the built environment Journal. *Intelligent Buildings International* 6(1), 3–1.

Rohles, F. H. (2007). *Temperature and Temperament. A psychologist looks at comfort.* ASHRAE, February, www.healthyheating.com/Thermal_Comfort_Working_Copy/downloads/Rohles_view_only.pdf.

Please check page range for corrections. Should this be “1–3”?

Roy, K. P. and Johnson, J. (2015). *Acoustic Comfort in the Workplace: Getting Back to the Basics*. INSIGHT to the point. Allsteel.

Ryan, C. O., Browning, W. D., Clancy, J. O., Andrews, S. L., and Kallianpurkar, N. B. (2015). Biophilic design patterns. Emerging Nature-Based Parameters for Health and Well-Being in the Built Environment. *International Journal of Architectural Research* 8(2), 62–76.

Sakellaris, I. A., Saraga, D. E., Mandin, C., Roda, C., Fossati, S., de Kluizenaar, Y., Carrer, P. Dimitroulopoulou, S., Mihucz, V. G., Szigeti, T., Hänninen, O., de Oliveira Fernandes, E., Bartzis, J. G., and Bluyssen, P. M. (2016). Perceived Indoor Environment and Occupants' Comfort in European "Modern" Office Buildings: The OFFICAIR Study. *International Journal of Environmental Research Public Health* 25(13), 5.

Smith, M. J., Carayon, P., Smith, J., Cohen, W., and Upton, J. (1994). Community ergonomics: a theoretical model for rebuilding the inner city. Proceedings of the Human Factors and Ergonomics Society 38th annual meeting.

Steffy, G. R. (2002). *Architectural Lighting Design*. John Wiley & Sons.

Stevens, N. J. (2016). Sociotechnical urbanism: New systems ergonomics perspectives on land use planning and urban design. *Theoretical Issues in Ergonomics Science* 17.

Thatcher, A. (2012). Green ergonomics: Definition and scope. *Ergonomics* 56(3), 443–451.

Thatcher, A. and Milner, K. (2014). Changes in productivity, psychological wellbeing and physical wellbeing from working in a "green" building. *Work* 49, 381–393.

Veitch, J. A. and Galasiu, A. D. (2012). The Physiological and Psychological Effects of Windows, Daylight, and View at Home: Review and Research Agenda. NRC-IRC Research Report RR-325.

Vischer, J. (2005). *Spaces Meets Status: Designing Workplace Performance*. Oxford: Routledge.

WHO (2006). *Constitution of the World Health Organization*. Basic Documents, Forty-fifth edition, Supplement, October.

Yun, H., Nama, I., Kima, J., Yanga, J., Leeb, K., and Sohna, J. (2014). A field study of thermal comfort for kindergarten children in Korea: An assessment of existing models and preferences of children. *Building and Environment* 75, 182–189.

Case Study 7

Bahrain World Trade Center: An Environmental Experimental Design

Fay Abdulla Al Khalifa

SUSTAINABILITY CONCERNS IN THE GCC

Member states of the Gulf Cooperation Council (GCC) share the world's concerns about the mitigation and management of environmental degradation, climate change, youth booms, and demographic demands. They, like the rest of the world, are also facing the general security challenges of safeguarding food, water, and energy. Nevertheless, today, countries of the council are also fronting security challenges of their own; new realities within the changing political economy of the oil-dependent region (Ulrichsen, 2011).

Since the end of the twentieth century, the GCC states initiated strategies that are aimed at increasing the local gross domestic product (GDP) through industrialization and the production of goods and services to uphold economic sustainability and reduce the region's overall dependency on depleting oil revenues (Al-Khalifa, 2012). Ever since, researchers have been racing to come up with solutions to maintain healthy economic growths for the member states while also safeguarding the environment (Dalkmann et al., 2004; Ulrichsen, 2011). Research suggested that sustainable development plans, agendas, and policies must be supported by an informed, educated general population to stand a better chance at succeeding in establishing its goals (Al-Khalifa, 2016). However, current environmental management systems of the GCC are centralized, mostly understaffed, and within inexperienced authorities (Al-Saqri & Sulaiman, 2014). To address such problems, research suggested the urgent need to integrate environmental education into the curricula's of undergraduate engineering in the GCC (Abdulwahab & Abdulraheem, 2003).

THE CASE OF BAHRAIN

Bahrain is the smallest of the GCC member states; approximately 15 times smaller than Qatar and 2809 times smaller than Saudi Arabia, Bahrain faces pressing environmental challenges, some of which could be a real threat to the sustainability of life on its small islands (Al-Madany et al., 1991). Because of its small size, high density, and limited resources, the argument of sustainability in Bahrain is the most crucial in the region (Al-Khalifa, 2015). While most of the state members can still enjoy experimenting with environmental ideas and design solutions, Bahrain has the utmost need for ideas that can be implemented today, with immediate positive results for both the economy and environment. This could assist in reducing the public debt to GDP that reached an overwhelming 82.1% in 2016, compared to 9.2% in Oman, 13.1% in KSA, 18.6% in Kuwait, 19.1% in UAE, and 47.6% in Qatar (*Trading Economics*, 2017). Also, effective design solutions could help in solving environmental issues like those relating to the continued reclamation of the local waters in attempts to expand the land area to accommodate the increasing population of the small island country (Al-Madany et al., 1991; Naser, 2011).

A growing number of applauded buildings exist today in the GCC, most of which are very experimental, designed and constructed by foreign companies that lack mature knowledge about the culture and the context in which their designs are to be situated (Al-Khalifa, 2016). Those experiments do not have any noticeable positive impact on the environment or any substantial empirical evidence to support their efficiency. Most of the buildings with claims to sustainability in the region include photovoltaic panels, turbines, or a mixture of both. Some won't introduce any new technology but are argued to be sustainable because of their passive design and unique cooling intervention (Alnaser, 2008).

BAHRAIN WORLD TRADE CENTER (BWTC): AN EXPERIMENTAL DESIGN

In Bahrain, the most famous and distinct environmentally designed building is the BWTC, an office building designed and built by Atkins in 2008 on the waterfront of Manama with three wind turbines connecting its twin towers. The BWTC is the first commercial building in the world to integrate three large wind turbines into its design (BWTC, 2015), another example of experimental design in the GCC. The BWTC is the most referred-to building in any conversation about sustainable design in Bahrain. The three very noticeable wind turbines connecting the elegantly sleeked triangular twin towers make a landmark out of the building, which stands out from its surroundings in the Bahraini landscape.

Atkins claimed that their design was technically validated and that the turbines are capable of generating a considerable percentage of the buildings' electricity demand each year (Killa & Smith, 2008). Not long after the completion of the building, research showed that, to begin with, the average annual wind speed and direction in Bahrain (wind speed at 10 m height is 4.8 m/s, mostly north to north-west) is not economically viable for the production of wind energy (Bachellerie, 2012). Furthermore, recent studies on wind tunnel testing and computational fluid dynamics simulations showed that the building was literally designed in the wrong direction; if the building was flipped 180 degrees, the turbines would be 14% more efficient and the building would yield greater wind energy output annually. Moreover, moving the wind turbines further back would have increased the turbines' competence to 31% (Stathopoulos, 2017). This supports the thesis that while it is in the best interests of decision-makers in the GCC to benefit from foreign consultancy and expertise, unfortunately, the resulting designs are mostly very experimental and yield minimal benefits to the environment.

Despite the proven letdown of the wind turbines, the BWTC integrated a number of potentially more successful passive environmental solutions into the design, including the use of the car park as a buffer zone to the south between the exterior exposed walls and interior air-conditioned spaces, using deep gravel for installation, the use of overhangs to provide shading, connecting the building to the district cooling system, and the use of low pressure-loss distribution for the central air and water circulation systems that would theoretically reduce the power requirements for fans and pumps (BWTC, 2015). There is, however, little empirical evidence that proves the efficiency of those solutions and future research in environmental design in Bahrain needs to investigate the competence of the many existing experimental solutions. The building also integrates fluorescent lighting as opposed to more sustainable LED lighting solutions. The architects claimed that the fluorescents lights used are efficient with high frequency and zonal control, but that still does not justify the decision when better solutions that are more environmentally sustainable are readily available.

FUTURE POTENTIALS THROUGH HUMAN FACTORS AND ERGONOMICS

In the future, the BWTC could benefit from interventions to investigate human factors and ergonomics (HFE) in relation to post-occupancy evaluations of the current facilities. HFE were not explicitly influential during the design or engineering processes of the building, but are probably a good starting point for any further investigations of the efficiency of its operations. Further studies could evaluate the extent to which the building design and environmental solutions are affecting the daily lives of the occupants. Data obtained from such studies could then be used to inform decisions concerning other environmental buildings that are to be designed in the region to optimize both the well-being of their occupants and their performances, as opposed to depending on foreign consultancies and experts who have limited knowledge about the culture and the context in which future environmental designs are to be situated.

REFERENCES

Abdulwahab, S. and Abdulraheem, M. (2003). The need for inclusion of environmental education in undergraduate engineering curricula. *International Journal of Sustainability in Higher Education* 4(2), 126–137.

Al-Khalifa, F. (2016). Achieving Urban Sustainability in the Island City of Bahrain: University Education, Skilled Labor, and Dependence on Expatriates. *Urban Island Studies* 2, 95–120.

Al-Khalifa, F. (2012). *An Urban Healing Agenda for Reform in Bahrain: Where the Dweller Falls into the Urban Gap and the Sailing Boat Hits the Skyscraper.* University of Sheffield.

Al-Khalifa, F. (2015). *Urban Sustainability in the Transforming Culture of the Arabian Gulf: The Case of Bahrain.* University of Sheffield.

Al-Madany, I. M., Abdalla, M. A., and Abdu, A. S. E. (1991). Coastal zone management in Bahrain: An analysis of social, economic and environmental impacts of dredging and reclamation. *Journal of Environmental Management* 32(4), pp. 335–348. Available at: http://dx.doi.org/10.1016/S0301-4797(05)80070-2, accessed November 8, 2012.

Al-Saqri, S. and Sulaiman, H. (2014). Comparative Study of Environmental Institutional Framework and Setup in the GCC States. *Journal of Environmental Protection* 5, 745–750.

Alnaser, N. (2008). Towards Sustainable Buildings in Bahrain, Kuwait and United Arab Emirates. *The Open Construction and Building Technology Journal* 2, 30–45.

Bachellerie, I. J. (2012). *Renewable energy in GCC countries: Resources, potential, and prospects*, Jeddah: Gulf Research Centre.

BWTC (2015). Architecture, Design and Awards.

Dalkmann, H., Herrera, R. J., and Bongardt, D. (2004). Analytical Strategic Environmental Assessment. Developing a New Approach to SEA. *Environmental Impact Assessment Review* 24, 385–402.

Killa, S. and Smith, R. (2008). Harnessing Energy in Tall Buildings: Bahrain World Trade Center and Beyond. In *CTBUH 2008 8th World Congress*. Dubai: Council of Tall Buildings and Urban Habitat.

Naser, H. A. (2011). Effects of reclamation on macrobenthic assemblages in the coastline of the Arabian Gulf: A microcosm experimental approach. *Marine Pollution Bulletin* 62(3), pp. 520–524. Available at: http://www.sciencedirect.com/science/article/pii/S0025326X10005217#!, accessed October 14, 2017.

Stathopoulos, T. (2017). Urban Wind Energy: Potential and Challenges. In B. C. et al., eds. *The International Conference on Wind Energy Harvesting*. Coimbra, Portugal: University of Coimbra.

Trading Economics (2017). Government Debt to GDP.

Ulrichsen, K. C. (2011). *Insecure Gulf: The End of Certainty and the Transition to the Post-Oil Era*. London: Columbia University Press.

8 Human Factors and Ergonomics in Health Care

Participation, Systems, Medical Devices, and Increasing Capacity

Sue Hignett, Alexandra Lang, and Gyuchan Thomas Jun

CONTENTS

INTRODUCTION

The core business of health care for both public and private sector organizations is fundamentally different to most other industries. In contrast to technology-centered industries where the human role is to monitor the equipment or supervise small numbers of other staff, the health care sector is both "people-centred" and "people-driven" (Van Cott, 1994). Additionally, the environment and tasks involved in caring for people can involve contaminated, physically demanding, and emotionally challenging work in situations where the patient can be both physically and mentally vulnerable.

HFE input in any industrial sector often follows a major incident or change in legislation (Hignett, 2016). In the United Kingdom (UK), for instance, one of the HFE opportunities in the health care industry was due to a change in legislation in 1986. The removal of Crown Immunity from prosecution under the Health and Safety Act 1974 (Seccombe, 1995) meant that the UK National Health Service (NHS) had to comply with safety legislation as hospitals and other care locations were considered to be places of work. From the 1980s to 2000s, HFE input focused on occupational health (Straker, 1990), building design (Hilliar, 1981), and systems approaches to embed HFE as part of the health care organizational culture (Hignett, 2001).

The interest in safety moved from staff to patients after the Bristol heart scandal (Department of Health, 2002) and seminal publications on the higher level of iatrogenic harm (Kohn, Corrigan, and Donaldson, 2000; Department of Health, 2000) where it was reported that at least 10% of patient admissions could result in some form of harm. Furthermore, in the United States of America (USA), "never" events were defined in 2007 as identifiable and preventable medical errors that result in serious consequences. As an incentive to reduce such occurrences, health care providers are penalized with no reimbursement for associated costs (investigation, treatment, and additional duration of stay), so there is a considerable motivation to reduce and eliminate both the total number and associated injuries (National Quality Forum, 2007).

FUNDAMENTALS

HFE provides an understanding of physical, cognitive, and social human capabilities and limitations with respect to interactions between physical, social, and informational aspects of the environment at a micro-level (e.g., using tools), a meso-level (e.g., local sociotechnical teams), and a macro-level in networks of health care organizations (Dul et al., 2012).

Physical HFE

The physical environment has a significant impact on the well-being, safety, and performance to support caregivers and patient care at both a micro- and macro-level. This includes design challenges of patient confidentiality, infection risks, and staff travel time. There have been many recommendations for the spatial requirements in health care facility design for treating and caring for patients since 1866 (Hignett and Lu, 2010). Some guidance is based on empirical evidence (Nuffield Provincial

Hospitals Trust, 1955; Villeneuve, 2004; Hignett and Lu, 2007, 2008) and is included in national policy (AIA, 2010; Department of Health, 2008). The integration of HFE into the building design process has been encouraged since the 1980s (Hillier, 1981) and most recently with an evidenced-based resource (Safety Risk Assessment (SRA) tool) with agreement across six safety domains; infection control, patient handling, medication safety, falls, behavioural health, and security from the USA (Center for Health Design, 2015). They concluded that the "the SRA process offered a value proposition to not only improve the design, but foster a culture of safety within the organization and build consensus" (Taylor et al., 2015).

Cognitive HFE

Within complex health care systems, the significance and role of cognitive HFE should not be underestimated. Specifically, research suggests that the cognitive well-being of both caregivers and patients is closely linked to their interactions, performance, and quality of care (Robertson and Cooper, 2010; Black, 2012). For patients, cognitive and mental well-being is an important factor in their recovery from injury and illness (Lamers et al., 2012). For staff, cognitive HFE factors provide an essential component in understanding perceived and actual work, how the health care systems operate and where improvements can be sought.

Workload, as a concept, has been used to understand, predict, assess, and avoid cognitive overload or underload, for individual, team, and system performance (Wickens, 1984; HSE, 2016). Workload was defined by Wickens (1979) as the "human's limited processing resources." It is a multifaceted cognitive concept involving cognitive load, processing (perceptual, central, and sensory), decision-making, and output. An integrated approach using the measurement of cognitive workload in addition to physical parameters, alongside an understanding of the human interactions within a system, provides the most comprehensive assessment of a sociotechnical system (STS). There are extrinsic influencers (e.g., fatigue, time pressure, resources) and intrinsic cognitive factors (e.g., emotion, stress, anxiety) that can compromise human reliability leading to use errors, slips, lapses, and mistakes (HSE, 2016); so STS should be designed to avoid and minimize these effects. The aim is to ensure that the cognitive workload within a new design or change management process will offer the best opportunity to achieve inclusive implementation for all the people in the system (at micro- and macro-levels) and provide benefits for both recipients and providers of health care.

HFE typically addresses empirical design where the "fit [to] the human is seen as the priority," and only considers training when the former is not possible (Dul et al., 2012). However, training approaches have been used to address workload and associated cognitive stress, particularly to support explicit (correct) knowledge becoming implicit and eventually tacit. Increasingly, health information technology (HIT) solutions (including eHealth and mHealth) are being developed to support human behavior and performance, to reduce workload, and ultimately to improve patient safety (Bates et al., 2001). There is still a need for improved HFE integration in the HIT design life cycle and evaluation to ensure consideration of the person and system requirements (Patel & Kannampallil, 2014).

Systems HFE

In order to support the study and improvement of overall health care system performance, various health care system-specific HFE frameworks were developed such as Systems Engineering Initiative for Patient Safety (SEIPS) by Carayon and colleagues (2006) and DIAL-F by Hignett (2013; Figure 8.1). Much of the background and motivation for these frameworks derives from systems theory with a focus on interrelationships between organizations, teams, and technology. Geels (2004) suggests that STS could be described as "linkages between elements necessary to fulfil societal functions (e.g. transport, communication)," where the "production, distribution and use of technologies [is] a sub-function" are related to the activities of human actors, who are "embedded in social groups which share certain characteristics (e.g. certain roles, responsibilities, norms, perceptions)."

The DIAL-F model (Figure 8.1) includes staff (care givers) and patients (care recipients) in an open, dynamic system (Hignett et al., 2013). The rotary telephone DIAL (used for telephone design from the 1920s to the 1980s; F for falls) represents the active elements with the most transience (dynamic change and motion) in the outer rings and the most stability in the inner rings.

An approach to unpacking system complexity is to follow the progress of a system user. For example, Wilson (2014) described the "systems of systems" in healthcare as nested and overlapping (parent/sibling) systems; "bed in a hospital is a system, the

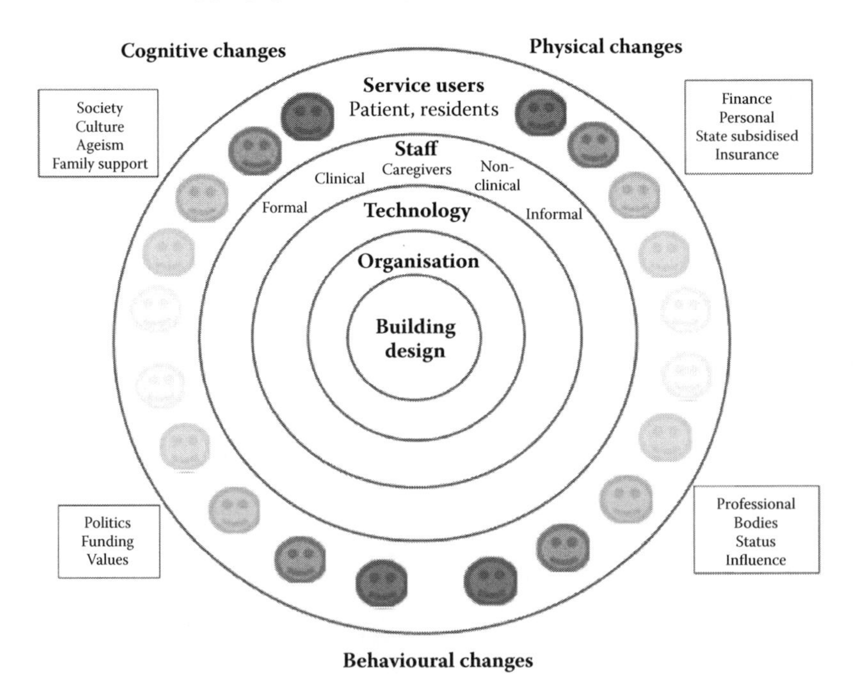

FIGURE 8.1 DIAL-F model of Falls risk management. (From Hignett, S., Why Design Starts with People, The Health Foundation: Patient Safety Resource Centre, http://patientsafety.health.org.uk/sites/default/files/resources/why_design_starts_with_people.pdf, 2013.)

patient monitoring equipment is a sibling system, the two together plus the patient's room comprise another system … whereas the radiology or scanning equipment, the drugs dispensary, the beds, the ambulances are all systems, but together can be seen as a system of systems when looking at maintenance and replacement regimes." An alternative approach is to categorize physical and cognitive HFE as micro-level human-machine system interactions, drawing a distinction with macro- or organizational systems-level approaches when applying socio-technical systems theory to the design of complex work systems (Karsh et al., 2014).

METHODS

Participatory Ergonomics for HFE in Healthcare

Participatory ergonomics (PE) can be very simply described as a concept involving the use of participative techniques and various forms of participation in the workplace (Vink and Wilson, 2003). Wilson (1995) defined participation in HFE projects as "the involvement of people in planning and controlling a significant amount of their own work activities, with sufficient knowledge and power to influence both processes and outcomes in order to achieve desirable goals." The definition of participatory approaches includes interventions at macro- (organizational, STS), meso- (teams), and micro- (individual) levels with the opportunity and power to use personal knowledge to address HFE problems relating to individual working activities. PE provides a flexible approach to health care design and process change, meeting the specific needs of people (both workers and patients) in the context of interventions (van Eerd, 2010).

There are differences in the understanding and application of PE projects between the USA and Europe. In the USA, PE tends to be used at a macro-ergonomics level, for the development and implementation of technology (Imada, 1991; Brown, 2000). In Europe, PE approaches have been applied at all levels of HFE interventions, with the key factor being the involvement of all stakeholders in the project (Kuorinka, 1997). Projects in the UK and USA using participatory HFE have tended to be at a micro-level, with mixed results; however, macro-level interventions also have much to offer (Hignett et al., 2005).

There is evidence that PE interventions have a positive impact on musculoskeletal symptoms, reducing injuries and workers' compensation claims, and a reduction in lost days from work or sickness absence (Rivilis et al., 2008). One example of a PE intervention program in a UK hospital was evaluated by the Health and Safety Executive. The intervention used a range of HFE methods to tackle musculoskeletal disorders (MSDs) in all staff groups (Hignett, 2001) and was reported to have saved over £3.6 million in three years.

System Mapping Methods

There are also various systems mapping/modeling methods derived from (cognitive) systems engineering that have been applied to health care to represent/analyze the structure and behavior of health care systems. These diagrammatic representations are used to capture system requirements, assess system-wide risk and performance variability, and generate system improvement recommendations that take into account the contextual information of the wider system. The methods include, but are not limited to, Systems Modelling Language (SysML; Jun et al., 2010), Work Domain Analysis (WDA; Lim et al., 2016), Functional Resonance Analysis Method (FRAM; Laugaland et al., 2014) and Causal Analysis based on Systems Theory (CAST; Leveson et al., 2016). They have been applied to address various system safety issues in health care that involve multiple organizations, teams, and individuals to different degrees, e.g., medication safety in care homes (Lim et al., 2016); preoperative medication errors (Leveson et al., 2016); hospital discharge processes of the elderly (Jun et al., 2009); prostate cancer diagnosis processes (Jun et al., 2010).

When applied appropriately, these methods offer potential for system safety improvement in health care, but there are some practical difficulties in implementing these methods into existing quality and safety improvement practices (Hignett et al., 2015). Unlike some quality improvement (QI) methods, which are relatively easily used by healthcare professionals after minimal training, these HFE methods require more in-depth understanding of humans and systems and should be applied by or with systems HFE experts.

APPLICATION

Participatory System Design: Integrated Pathway for Safer Medicines Management

Healthcare processes are highly pressured and complex with an ever-present potential for error and accidents (Clarkson et al., 2004). The scope for error in all parts of the system is high, and although most research studies have tended to focus on only limited components of the complex system, HFE professionals have long recognized that enhancing performance and reducing risk requires an emphasis on design at the system level. Jun and colleagues (2014) demonstrated how to carry out system design by applying system mapping and risk assessment methods in a participatory framework to create integrated pathways for safer medicines management with the participation of healthcare workers in primary and secondary care systems and service users in the community. There is growing recognition of the need to consider the healthcare work delivered by non-professionals especially in the community setting (Amalberti and Vincent, 2016).

With an aging population and increasing number of people with chronic conditions living in the community, more health care work is delivered away from traditional clinical settings by patients (self-care) and their families (informal and formal carers). This creates new challenges for health service design and previous patient safety strategies developed for the acute hospital setting. The involvement of all stakeholders and the inclusion of their input in system design is typical of the participatory systems approach (e.g., Waterson, 2015; Hettinger et al., 2015).

In the safer medicines management project, a stakeholder diagram (Figure 8.2) and flow diagram (Figure 8.3) were created to support participatory group discussion to represent, firstly, the structure of stakeholder communication for medicine

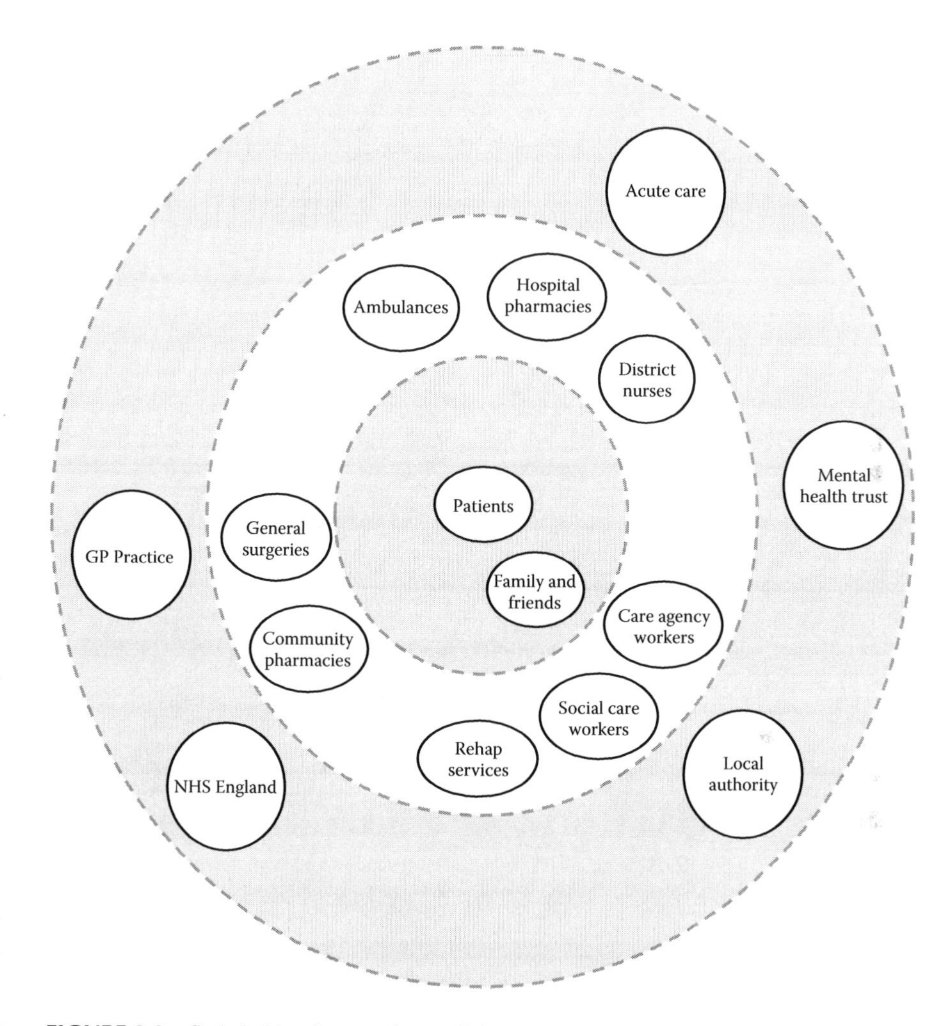

FIGURE 8.2 Stakeholder diagram for medicine management. (Adapted from Jun, G. et al., *Ergonomics*, published first online, DOI:10.1080/00140139.2017.1329939, 2017.)

management and, secondly, the behavior of the repeat prescription processes (patient states, transition actions, and conditions).

After developing a shared understanding of the whole system, a risk assessment method was applied with stakeholders to identify, prioritize, and mitigate the risks in the system. Various risk assessment methods are available including, but not limited to, HAZOP, FMEA, what-if analysis, barrier analysis, etc. (Ward et al., 2010). Due to time constraints in health care stakeholder participation, a relatively simple risk assessment method, "what-if analysis" was chosen, with results shown in Table 8.1.

Further group discussion was carried out on risk prioritization and risk mitigation ideas, with two specific challenges identified (Jun et al., 2014). Firstly, the appropriate selection and application of the HFE methods and tools for the participatory systems approach are not straightforward and require careful consideration and

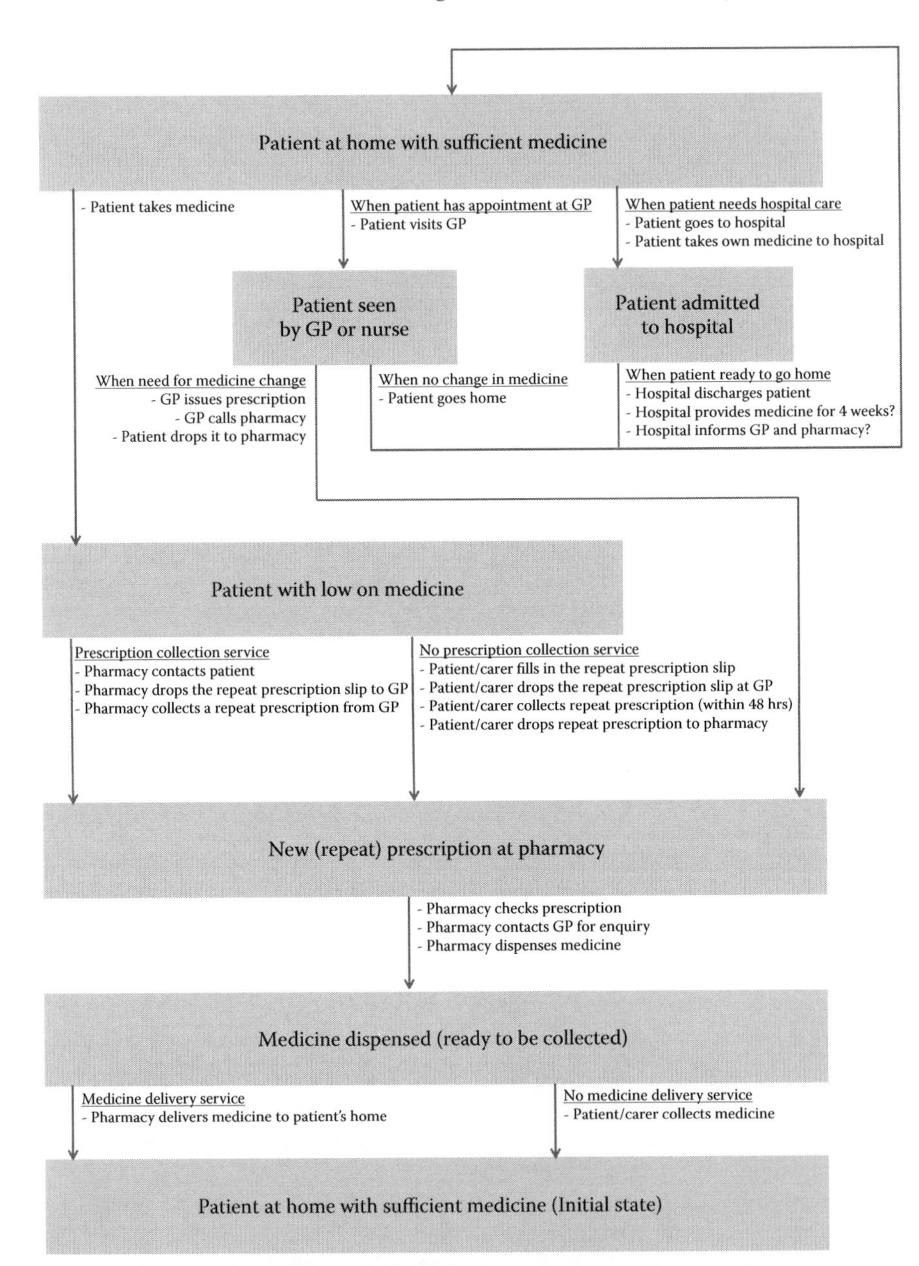

FIGURE 8.3 Flow diagram of repeat prescription processes. (Adapted from Jun, G. et al., *Ergonomics*, published first online, DOI:10.1080/00140139.2017.1329939, 2017.)

TABLE 8.1
"What-if Analysis" for Safer Medicines Management

Process Steps	Potential Failures	Causes	Consequences
1. Prescribing medicines	Prescribe too many medicines	– Poor coordination between prescribers	Hard to take all the medicines
2. Reordering medicines	Reorder medicine too late	– Patient forgets	Running out of medicine
3. Getting medicines	Fail to get medicines on time	– Limited pharmacy opening times	Running out of medicine
4. Taking medicines	Take wrong medicines	– Poorly informed patient	Patient harm
	Intentionally do not take	– Side effects	Worsening condition
5. Managing changes	Delayed changes	– Poor communication between health care professionals	No medicine
6. Reviewing medication	No (infrequent) review Poor review	– No turn-up	Over-prescription

Source: Adapted from Jun, G. et al., *Ergonomics*, published first online, DOI:10.1080/00140139.2017.1329939, 2017.

balance between various factors such as problem type, design stage, level of stakeholder engagement, and availability of resources (time, money, data, and expertise). Secondly, the highly distributed nature of the target service requires the participation of many different key stakeholders for the success of design projects (Smith & Fischbacher, 2005).

Patient Handling

The presence of musculoskeletal disorders (MSD) and illness in the health and social care workforce has been reported in many epidemiological studies (Fanello et al., 2002; Smedley et al., 1995) with reports of high prevalence rates in nursing and related personnel. The NEXT study (conducted in seven countries) found that high levels of musculoskeletal and psychosocial risk factors are still prevalent in healthcare workers in the twenty-first century (Simon et al., 2008). They reported a correlation between physical lifting and bending and the prevalence of back and neck pain, with psychosocial factors showing a stronger link with disability from MSDs; staff in hospitals reported the lowest availability of lifting equipment compared to nursing homes and home care.

Since 1992, the management of manual handling risks to workers in the European Union (EU) has been directed by the Manual Handling Directive (Council Directive, 1990). A range of HFE and other approaches have been used to reduce the risks, e.g., assessment and management, training, equipment provision, and culture change

(Hignett, 2003a). Five systematic reviews (Amick et al., 2006; Bos et al., 2006; Dawson et al., 2007; Hignett et al., 2003a & b; Martimo et al., 2008) reported very little high-quality evidence and a limited quantity of moderate evidence to show reductions in the rate of MSDs, especially from multi-factorial interventions (Amick et al., 2006). The complexity of evaluation has been addressed in a Tool for Risk Outstanding in Patient Handling Interventions (TROPHI; Fray and Hignett, 2013) to include metrics for safety culture, musculoskeletal health, quality of care, workload and patient condition, and staff and patient injuries.

Medical Device Design

In the UK, design has been accepted as an important component in patient safety since the 2000s and there have been design projects on electronic infusion devices, medication labeling, computer interfaces, and ambulances (NHS, 2015). This work continues with national initiatives to embed HFE in health care across both product and systems led by Health Education England "to ensure that the practices and principles of human factors are integrated into all training and education … to develop the future healthcare workforce by ensuring it contains individuals with the right skills, attitudes, behaviours and training, to enable the delivery of excellent healthcare and drive improvements for the quality of care provided and the safety of our patients" (Health Education England, 2015).

In the USA, the Food and Drug Administration (FDA) followed the publication of the 1993 European Medical Devices Directive 93/42/EU (MDD 93/42/EEC, 2007) with novel guidance resources (Sawyer, 1996) and proposals for "Reporting Use Errors with Medical Devices" (FDA, 2003), and "Incorporating HFE into Risk Management" (Kaye and Crowley, 2000), the combination of which provided precursor documents for future HFE/usability engineering design standards.

Accessing equipment and consumables has been a perennial challenge. Cox and colleagues (2005) commented on a procurement dichotomy of "clinical effectiveness versus cost effectiveness" in terms of the "rights of clinicians to design and specify requirements individually without any real consideration of the commercial consequences of their actions." This describes a tension that HFE is well-placed to support through product design and evaluation and user trials (Hignett, 1998; 2001). A usability and HFE standard for medical device development was established in 2007 (IEC/ISO 62366) and has been updated as regulatory bodies modify it for their purposes and improve clarity of requirements for industry. The standard makes explicit links to the associated risk management standard (ISO 14971, 2007) to ensure that manufacturers consider potential risks of system use and integration and, also, where relevant, the life cycle of medical device software (IEC 62304, 2006). In the UK, the Medicines and Healthcare products Regulatory Agency (MHRA) is developing HFE guidance for medical devices and drug delivery devices, to replicate the progress made in the USA by the FDA and the Association for the Advancement of Medical Instrumentation (AAMI; MHRA, 2016).

The standards and related guidance introduced by IEC/ISO (2007), FDA (2016), and AAMI (ANSI/AAMI 2013) are trailblazing. This regulation has formalized HFE integration in medical device development and was developed due to the quantity of evidence supporting the need to reduce use errors, ensure correct use, and improve performance, efficiency, and adherence (Martin et al., 2008). Additionally, the FDA Home Use Devices Initiative (FDA, 2010, 2014) provides an important resource to support the medical device industry in the consideration of the unique risks created by interactions among the user, device, and home or non-clinical use environment.

Despite a wealth of evidence for the inclusion of HFE practices in medical device design, challenges still exist to rigorous HFE practice being undertaken in this sector (Privitera et al., 2017; Vincent et al., 2014). Whilst progress has been made to ensure the role of HFE in the design of medical technologies, it is important that HFE is also considered a key part in technology deployment strategies and medical technology integration (Pelayo and Ong, 2014). This includes feedback mechanisms to ensure that these interventions remain fit for purpose, that they can be deployed harmoniously alongside existing technologies and meet the needs of the complex healthcare systems in which they are operating.

Lang et al. (2016) looked at the deployment and integration of a mobile electronic patient observation tool (eObs). The results demonstrated not only the need for eliciting views of users on design and usability of the device and interface but also on the health information technology (HIT) intervention and use in context over time. The evaluation of the new HIT system discovered a range of benefits to staff despite challenges during the initial deployment with both intended and unanticipated benefits of use; including real-time accessible information, distributed working, and decision-making, in addition to organizational learning points for future HIT deployments.

FUTURE TRENDS

Increasing HFE Knowledge in Health Care

There is considerable interest in increasing HFE healthcare capability. However, the limited availability of HFE education may be a barrier when, for example, in the UK only five universities offer accredited qualifying courses in HFE (www.ergonomics.org.uk/degree-courses/), compared with 99 universities providing accredited education in nursing (www.nmc.org.uk/education/approved-programmes/). There are two paths to increase the capacity of HFE knowledge and experience in health care in the short- to medium-term. Firstly, education and training for health care staff on postgraduate courses with at least 600 hours study on accredited short courses (postgraduate certificate). Secondly, by encouraging HFE professionals from other industrial sectors to apply their knowledge in health care offers substantial opportunities for health care organizations to capitalize on their knowledge and experience.

Increasing Resilience in Health Care Systems

Safety II coined by Hollnagel and colleagues (2015) is a new way of managing safety. The basic concept of Safety II is to ensure that as many things as possible go right instead of ensuring that as few things as possible go wrong (Safety I). The safer medicines management example is based on a typical Safety I "find-and-fix" approach to (i) look for (potential) failures, (ii) try to find the causes, and (iii) eliminate those causes or introduce barriers (Hollnagel et al., 2015). The Safety II approach, as complimentary to Safety I, aims to understand the condition where performance variability, instead of failures, becomes difficult or impossible to monitor and control (Hollnagel et al., 2015).

Medical Device Regulation and Medical Informatics

For medical informatics, there are still many gray areas with regards to the ethics and governance associated with healthcare data. For instance, the UK Health and Social Care Information Centre (HSCIC; http://systems.hscic.gov.uk/infogov/security) provides three cornerstone principles of information security; confidentiality, integrity, and availability. Any development of HIT or medical devices with information and communication technology (ICT) capabilities must no longer just consider the risks associated with use in a physical sense but also must consider the context of a connected world, with integrated health ICT services and commercially available health and well-being technology products, to capitalize on the benefits associated with these capabilities and mitigate the risks that they simultaneously pose.

CONCLUSION

The fundamentals and applications for HFE discussed in this chapter demonstrate the benefits of HFE design, integration, and participation of stakeholders. The three future trends offer opportunities for health care to benefit from human factors integration (MOD, 2015a, b) by increasing awareness and capacity (through education), achieving a better theoretical understanding of the complexity of health care systems (Safety II), and by embedding HFE in the infrastructure with changes in regulations for design and procurement of medical devices and informatics. These will all help to establish HFE in health care as an integrating approach/discipline for work design and systems analysis with device usability, human error, teamwork, and safety culture (Norris, 2012). There are opportunities for HFE knowledge and experience in other sectors (e.g., defence, rail) to share motivators, barriers, and experience from HFE progress over the last 30 years. As time progresses and the health care industry "catches up" with those HFE experienced domains there is then opportunity for this uniquely complex domain to proffer its own methods and solutions.

KEY TERMS

"Never" (*patient safety*) events. Identifiable and preventable medical errors that result in serious consequences.

Cognitive workload. A multifaceted cognitive concept involving cognitive load, processing (perceptual, central, and sensory), decision-making, and output.

Sociotechnical system (STS). Defined by Geels (2004) as the "linkages between elements necessary to fulfil societal functions (e.g., transport, communication)" where the "production, distribution and use of technologies [is] a sub-function" are related to the activities of human actors, who are "embedded in social groups which share certain characteristics (e.g., certain roles, responsibilities, norms, perceptions)."

Participatory ergonomics (PE). A concept involving the use of participative techniques and various forms of participation in the workplace (Vink and Wilson, 2003). Includes interventions at macro- (organizational, STS), meso- (team), and micro- (individual) levels.

Quality improvement (QI). The QI-HFE relationship presents some practical difficulties in implementing HFE methods into existing QI practices (Hignett et al., 2015). Some QI methods are relatively easily used by healthcare professionals after minimal training; HFE methods require more in-depth understanding of humans and systems and should be applied by or with systems HFE experts.

Safety II. A new way of managing safety. The basic concept of Safety II is to ensure that as many things as possible go right instead of ensuring that as few things as possible go wrong (Safety I; Hollnagel et al. 2015).

REFERENCES

AIA (2010). *Guidelines for the Design and Construction of Health Care Facilities* (5th ed.). The American Institute of Architects. Washington, DC, www.fgiguidelines.org/2010guidelines.html.

Amalberti, R. and Vincent, C. (2016). *Safer Healthcare – Strategies for the Real World.* Cham: Springer.

Amick, B., Tullar, J., Brewer, S., Irvine, E., Mahood, Q., Pompeii, L., Wang, A., Van Eerd, D., Gimeno, D., and Evanoff, B. (2006). *Interventions in Health-Care Settings to Protect Musculoskeletal Health: A Systematic Review.* Toronto, ON: Institute for Work and Health.

ANSI/AAMI HE75:2009(R) (2013). Human factors engineering—Design of medical devices, www.aami.org/productspublications/ProductDetail.aspx?ItemNumber=926#sthash.z3tHlcCS.dpuf.

Bates, D. W., Cohen, M., Leape, L., Overhage, M., Shabot, M. M, and Sheridan, T. (2001). Reducing the Frequency of Errors in Medicine Using Information Technology. *Journal of the American Medical Informatics Association* 8(4), 299–308.

Black, C. D. (2012). Why healthcare organisations must look after their staff. *Nursing Management* 19(6), 27–30.

Bos, E. H., Krol, B., Van Der Star, A., and Groothof, J. W. (2006). The effect of occupational interventions on reduction of musculoskeletal symptoms in the nursing profession. *Ergonomics* 49(7), 706–723.

Brown, O. Jr. (2000) Participatory approaches to work systems and organisational design. In: *Proceedings of the 14th Triennial Congress of the International Ergonomics Association and the 44th Annual Meeting of the Human Factors and Ergonomics Society*, San Diego, California. Santa Monica, CA: The Human Factors and Ergonomics Society, July 29–August 4, 535–538.

Carayon, P., Hundt, A. S., Karsh, B. T., Gurses, A. P., Alvarado, C. J., Smith, M., and Flatley Brennan, P. (2006). Work system design for patient safety: The SEIPS model. *Quality and Safety in Health Care* 15(suppl 1), i50–i58.

Center for Health Design (2015). Safety Risk Assessment. https://www.healthdesign.org/insights-solutions/safety-risk-assessment-toolkit-pdf-version.

Please provide author names.

Clarkson, P. J. et al. (2004). *Design for patient safety: A scoping study to identify how the effective use of design could help to reduce medical accidents*, No. 0954524306, University of Cambridge.

Council Directive (1990). *Council Directive of the 29 May 90/269/EEC on the minimum health and safety requirements for the manual handling of loads where there is a risk particular of back injury to workers "the Manual Handling of Loads Directive,"* www.europe.osha.int/legislations/directives/A/1/2/05.

Cox, A., Chicksand, D., and Ireland, P. (2005). Sub-Optimality in NHS Sourcing in the UK: Demand-Side Constraints on Supply-Side Improvement. *Public Administration* 83(2), 367–392.

Dawson, A. P., McLennan, S. N., Schiller, S. D., Jull, G. A., Hodges, P. W., and Stewart, S. (2007). Interventions to prevent back pain and back injury in nurses: A systematic review. *Occupational and Environmental Medicine* 64, 642–650.

Department of Health (2000). *An Organisation with a Memory: Report of an Expert Group on Learning from Adverse Events in the NHS.* London: Department of Health.

Department of Health (2002). *Learning from Bristol. The Department of Health's response to the report of the Public Inquiry into children's heart surgery at the Bristol Royal Infirmary 1984-1995.* London: The Stationary Office.

Department of Health (2008). *Health Building Note 09-02 Maternity Care Facilities.* The Stationary Office, London.

Dul, J., Bruder, R., Buckle, P., Carayon, P., Falzon, P., Marras, W. S., Wilson, J. R., and van der Doelen, B. (2012). A Strategy for Human Factors/Ergonomics: Developing the Discipline and Profession. *Ergonomics* 55(4), 377–395.

Fanello, S., Jousset, N., Roquelaure, Y., Chotard-Frampas, V., and Delbos, V. (2002). Evaluation of a training program for the prevention of lower back pain among hospital employees. *Nursing and Health Science* 4(1–2), 51–54.

FDA (2003). GHTF/SG2/N31R8 *Medical Device Post Market vigilance and Surveillance: Proposal for Reporting of Use Errors with Medical Devices by their Manufacturer or Authorised Representative*, www.fda.gov/ohrms/dockets/98fr/04d-0001-bkg0001-08-sg2_n31r8.pdf.

FDA (2010). *Medical Device Home Use Initiative.* Centre for Devices and Radiological Health. US Food and Drug Administration, www.fda.gov/downloads/MedicalDevices/ProductsandMedicalProcedures/HomeHealthandConsumer/HomeUseDevices/UCM209056.pdf.

FDA (2014). *Design Considerations for Devices Intended for Home Use.* Guidance for Industry and Food and Drug Administration Staff. Centre for Devices and Radiological Health. US Food and Drug Administration, www.fda.gov/downloads/MedicalDevices/DeviceRegulationandGuidance/GuidanceDocuments/UCM331681.pdf.

FDA (2016). *Applying Human Factors and Usability Engineering to Medical Devices.* Food and Drug Administration, www.fda.gov/downloads/MedicalDevices/.../UCM259760.pdf.

Fray, M. and Hignett, S. (2013). TROPHI: Development of a tool to measure complex, multi-factorial patient handling interventions. *Ergonomics* 56(8), 1280–1294.

Geels, F. W. (2004). From sectoral systems of innovation to socio-technical systems: Insights about dynamics and change from sociology and institutional theory. *Research Policy* 33(6–7), 897–920.

Health Education England (2015). *Learning to be Safer*, https://hee.nhs.uk/work-programmes/human-factors-and-patient-safety/.

Hettinger, L. J., Kirlok, A., Goh, Y. M., and Buckle, P. (2015). Modelling and simulation of complex sociotechnical systems: Envisioning and analysing work environments. *Ergonomics* 58(4), 600–614.

Hignett, S. (1998). Ergonomic Evaluation of Electric Mobile Hoists. *British Journal of Occupational Therapy* 61(1), 509–516.

Hignett S. (2001). Embedding ergonomics in hospital culture: Top-down and bottom-up strategies. *Applied Ergonomics* 32, 61–69.

Hignett, S. (2003a). Intervention strategies to reduce musculoskeletal injuries associated with handling patients: A systematic review. *Occupational and Environmental Medicine* 60(9), e6 (electronic paper), www.occenvmed.com/cgi/content/full/60/9/e6.

Hignett, S. (2003b). Systematic review of patient handling activities starting in lying, sitting and standing positions. *Journal of Advanced Nursing* 41(6), 545–552.

Hignett, S. (2013). Why Design Starts with People. The Health Foundation: Patient Safety Resource Centre, http://patientsafety.health.org.uk/sites/default/files/resources/why_design_starts_with_people.pdf.

Hignett, S. (2016). Human Factors training in Healthcare. *The Ergonomist*. March 11, 2016.

Hignett, S. and Crumpton, E. (2005) Development of a patient handling assessment tool. *International Journal of Therapy and Rehabilitation* 12(4), 178–181.

Hignett, S., Wilson, J. R., and Morris, W. (2005). Finding Ergonomic Solutions – Participatory Approaches. *Occupational Medicine* 55, 200–207.

Hignett, S., Fray, M., Rossi, M. A., Tamminen-Peter, L., Hermann, S., Lomi, C., Dockrell, S., Cotrim, T., Cantineau, J. B., and Johnsson, C. (2007). Implementation of the Manual Handing Directive in the Healthcare Industry in the European Union for Patient Handling tasks. *International Journal of Industrial Ergonomics* 37, 415–423.

Hignett, S. and Lu, J. (2007). Evaluation of Critical Care Space Requirements for Three Frequent and High-Risk Tasks. *Critical Care Nursing Clinics of North America* 19, 167–173.

Hignett, S. and Lu, J. (2008) Ensuring bed space is right first time. *Health Estate Journal*. February www.healthestatejournal.com/Story.aspx?Story=3395.

Hignett, S. and Lu, J. (2010). Space to care and treat safely in acute hospitals: Recommendations from 1866–2008. *Applied Ergonomics* 41, 666–673.

Hignett, S., Griffiths, P., Sands, G., Wolf, L., and Costantinou, E. (2013). Patient Falls: Focusing on Human Factors rather than Clinical Conditions. In *Proceedings of the HFES 2013 International Symposium on Human Factors and Ergonomics in Health Care*. Baltimore, USA, 11–13 March 2013.

Hignett, S., Jones, E., Miller, D., Wolf, L., Modi, C., Shahzad, MW, Banerjee, J., Buckle, P., and Catchpole, K. (2015). Human Factors & Ergonomics and Quality Improvement Science: Integrating Approaches for Safety in Healthcare. *BMJ Quality & Safety* 24(4), 250–254.

Hilliar, P. (1981). The DHSS Ergonomics Data bank and the Design of Spaces in Hospitals. *Applied Ergonomics* 12209–12216.

Hollnagel, E., Wears, R., and Braithwaite, J. (2015). *From Safety I to Safety II: A white paper, The Resilient Health Care Net:* Published simultaneously by the University of Southern Denmark, University of Florida, USA and Macquarie University, Australia.

HSE (2016). *Human Factors: Workload. Why is workload important?* Health and Safety Executive, www.hse.gov.uk/humanfactors/topics/workload.htm.

Imada, A. S. (1991). The rationale and tools of participatory ergonomics. In: Noro, K., Imada, A. S., (eds), *Participatory Ergonomics*. London: Taylor and Francis, pp. 30–49.

IEC (2006). IEC 62304:2006. *Medical device software – Software life cycle processes*. Geneva: International Organization for Standardization.

IEC (2007). ISO 62366-1. *Medical devices – Part 1: Application of usability engineering to medical devices*. Geneva: International Organization for Standardization.

ISO (2007). ISO 14971: *Medical devices — Application of risk management to medical devices as a methodology for assessing and documenting product safety and effectiveness*. Geneva: International Organization for Standardization.

Jun, G., Ward, J., Morris, Z., and Clarkson, J. (2009) Health care process modelling: Which method, when? *International Journal for Quality in Health Care* 21(3), 214–224.

Jun, G., Ward, J. R., and Clarkson, P. J. (2010). Systems modelling approaches to the design of safe healthcare delivery: Ease of use and usefulness perceived by healthcare worker. *Ergonomics* 53(7), 829–847.

Jun, G., Canham, A., Altuna-Palacios, A., Ward, J. R., Bhamra, R., Rogers, S., Dutt, A., and Shah, P. (2017) A participatory systems approach to design for safer integrated medicine management. *Ergonomics*. Published first online. DOI:10.1080/00140139.2017.1329939.

Karsh, B.-T., Waterson, P. E., and Holden, R. (2014), Crossing levels in systems ergonomics: A framework to support "mesoergonomic" inquiry. *Applied Ergonomics* 45, 45–54.

Kaye, R. and Crowley, J. (2000) *Medical device use-safety: Incorporating human factors engineering into risk management*. US Department of Health and Human Services, Food and Drug Administration.

Kohn, L. T., Corrigan, J. M., and Donaldson, M. S. (eds) (2000). *To Err Is Human: Building a Safer Health System*. National Academy Press, Washington, DC.

Kuorinka, I. (1997). Tools and means of implementing participatory ergonomics. *International Journal of Industrial Ergonomics* 19, 267–270.

Lamers, S. M. A., Bolier, L., Westerhof, G. J., Smit, F., and Bohlmeijer, E. T. (2012). The impact of emotional well-being on long-term recovery and survival in physical illness: A meta-analysis. *Journal of Behavioral Medicine* 35(5), 538–547.

Lang, A., Pinchin, J., Brown, M., and Sharples, S. (2016). *Handheld Technologies in the Ward. A Human Factors Evaluation of eObs deployment and uptake in Nottingham Universities Hospital Trust,* www.nuh.nhs.uk/media/2239905/unott_hfrg_report_handheld_technologies_in_the_ward_mar2016.pdf.

Laugaland, K., Aase, K., and Waring, J. (2014). Hospital discharge of the elderly—an observational case study of functions, variability and performance-shaping factors. *BMC health services research* 14(1), 1–15.

Leveson, N., Samost, A., Dekker, S., Finkelstein, S., and Raman, J. (2016). A Systems Approach to Analyzing and Preventing Hospital Adverse Events. *Journal of Patient Safety*, January 11 (E-pub ahead of print).

Lim, R., Anderson, J. E., and Buckle, P. W. (2016). Work Domain Analysis for understanding medication safety in care homes in England: An exploratory study. *Ergonomics* 59(1), 15–26.

Martimo, K. P., Verbeek, J., Karppinen, J., Furlan, A. D., Takala, E. P., Kuijer, P., Jauhianen, M., and Viikari-Juntura, E. (2008). Effect of training and lifting equipment for preventing back pain in lifting and handling: Systematic review. *BMJ* 336, 429–431.

Martin, J. L., Norris, B. J., Murphy, E., and Crowe, J. A. (2008). Medical device development: The challenge for ergonomics. *Applied Ergonomics* 39(3), 271–283.

MDD 93/42/EEC (2007). Council of the European Communities. *Council Directive of 14th June 1993 concerning medical devices*. 93/42/EEC.

MHRA (2016). Draft for comment Human Factors and Usability Engineering – Guidance for Medical Devices Including Drug-device Combination Products. www.gov.uk/government/news/human-factors-and-usability-engineering-guidance-for-medical-devices-including-drug-device-combination-products

MOD (2015a). JSP 912. Human Factors Integration for Defence Systems. Part 1: Directive, www.gov.uk/government/uploads/system/uploads/attachment_data/file/483176/20150717-JSP_912_Part1_DRU_version_Final-U.pdf.

MOD (2015b). JSP 912. Human Factors Integration for Defence Systems. Part 2: Guidance, www.gov.uk/government/uploads/system/uploads/attachment_data/file/483177/20151030-JSP_912_Part_2_DRU_version_Final-U.pdf.

National Quality Forum (2007). *Serious Reportable Events in Healthcare 2006 Update. A Consensus Report.* Washington, DC: National Quality Forum.

NHS (2015). *Design for Patient Safety*, www.nrls.npsa.nhs.uk/resources/collections/design-for-patient-safety/ (accessed April 17, 2015).

Norris, B. J. (2012). Systems human factors: How far have we come? *BMJ Quality &Safety* 21(9), 713–714.

Nuffield Provincial Hospitals Trust (1955). *Studies in the functions and design of hospitals.* Oxford University Press, London.

Patel, V. L. and Kannampallil, T. G. (2014). Human factors and health information technology: Current challenges and future directions. *Yearb Med Inform* 9, 58–66.

Pelayo, S., Ong, M., and Section Editors for the IMIA Yearbook Section on Human Factors and Organizational Issues. (2015). Human Factors and Ergonomics in the Design of Health Information Technology: Trends and Progress in 2014. *Yearb Med Inform* 10(1), 75–78.

Privitera, M. B., Evans, M., and Southee, D. (2017). Human factors in the design of medical devices – Approaches to meeting international standards in the European Union and USA. *Applied Ergonomics* 59, Part A: 251–263.

Rivilis, I., Van Eerd, D., Cullen, K., Cole, D. C., Tyson, E. J., and Mahood, Q. (2008). Effectiveness of participatory ergonomic interventions on health outcomes: A systematic review. *Applied Ergonomics* 39, 342–358.

Robertson, I. and Cooper, C. (2010). *The Boorman Report on the Health and Well-Being of NHS Staff: Practical advice for implementing its recommendations*, www.robertsoncooper.com/files/boorman_download.pdf (accessed July 4, 2016).

Sawyer, D. (1996). *Do it by design: An introduction to human factors in medical devices.* US Department of Health and Human Services, Food and Drug Administration.

Seccombe, I. (1995). Sickness absence and health at work in the NHS. *Health Manpower Management* 21(5), 6–11.

Simon, M., Tackenberg, P., Nienhaus, A., Estryn-Behar, M., Conway, P. M., and Hasselhorn, H.-M. (2008). Back or neck pain related disability of nursing staff in hospitals, nursing homes and home care in seven countries – results from the European NEXT study. *International Journal of Nursing Studies* 45, 24–34.

Smedley, J. Egger, P., Cooper, C., and Coggon, D. (1995). Manual handling activities and the risk of low back pain in nurses. *Occupational and Environmental Medicine* 52, 160–163.

Smith, A. M. and Fischbacher, M. (2005). New service development: A stakeholder perspective. *European Journal of Marketing* 39(9–10), 1025–1048.

Straker, L. M. (1990). Work-Associated Back Problems: Collaborative Solutions. *Occupational Medicine* 40, 75–79.

Taylor, E., Quan, X., and Joseph, A. (2015). Testing a tool to support safety in healthcare facility design. *Procedia Manufacturing* 3, 136–143.

Van Cott, H. (1994), Human Errors: Their Causes and Reduction. In M. S. Bogner (ed.), *Human Error in Medicine* (pp. 53–65). Hillsdale, NJ: Erlbaum.

van Eerd, D., Cole, D., Irvin, E., Mahood, Q., Keown, K., Therberge, N., Village, J., St Vincent, M., and Cullen, K. (2010). Process and implementation of participatory ergonomic interventions: A systematic review. *Ergonomics* 53(10), 1153–1166.

Villeneuve, J., (2004). Participatory Ergonomic Design in Healthcare Facilities. In Charney W. and Hudson A. (eds), *Back Injury among Healthcare workers* (pp. 161–178). Boca Raton, FL: Lewis Publishers.

Vincent, C. J., Yunqui, L., and Blandford, A. (2014). Integration of human factors and ergonomics during medical device design and development: It's all about communication. *Applied Ergonomics* 45(3), 413–419.

Vink, P. and Wilson, J. R. (2003). Participatory ergonomics. In *Proceedings of the XVth Triennial Congress of the International Ergonomics Association and The 7th Joint conference of the Ergonomics Society of Korea/Japan Ergonomics Society, "Ergonomics in the Digital Age,"* Seoul, Korea: August 24–29.

Ward, J., Clarkson, J., Buckle, P., Berman, J., Lim, R., and Jun, T. (2010). *Prospective Hazard Analysis: Tailoring Prospective Methods to a Healthcare Context.* Research project report, Patient Safety Research Programme of the Department of Health, www.birmingham.ac.uk/Documents/college-mds/haps/projects/cfhep/psrp/finalreports/PS035-RevisedPHAFinalReportv11withToolkitJuly2010.pdf.

Waterson, P. (2015). Sociotechnical design of work systems. In Wilson, J. R. and Sharples, S. *Evaluation of human work* (4th ed) (pp. 753–769). Boca Raton, FL: CRC Press.

Wickens, C. D. (1979). *Measures of Workload, Stress and Secondary Tasks. Mental Workload: Its Theory and Measurement.* N. Moray. Boston, MA, Springer US: 79–99.

Wickens, C. D. (1984). Processing resources in attention. In R. Parasuraman and D. R. Davies (eds), *Varieties of Attention* (pp. 63–102). New York: Academic Press.

Wilson, J. R. (1995) Ergonomics and participation. In Wilson, J. R., Corlett, E. N. (eds), *Evaluation of Human Work: A Practical Ergonomics Methodology* (2nd ed) (pp. 1071–1096). London: Taylor and Francis.

Wilson, J. R. (2014). Fundamentals of Systems Ergonomics/Human Factors. *Applied Ergonomics* 45, 5–13.

Case Study 8
Health Informatics at Dubai Health Authority

Mazin Gadir

DUBAI HEALTH AUTHORITY

Dubai Health Authority (DHA) is a government organization overseeing the health system of Dubai, United Arab Emirates. It was formed in 2007 under the directive of H. H. Sheikh Mohammed bin Rashid Al Maktoum, the Vice President, Prime Minister, and Ruler of Dubai. Beyond general oversight of Dubai's health care sector, DHA provides health care services through hospitals and other health care facilities that fall under its direct jurisdiction. These include Latifa Hospital, Dubai Hospital, Rashid Hospital, and Hatta Hospital, in addition to other specialty centers and DHA primary health centers throughout Dubai. Other DHA services include creating and ensuring the execution of policies and strategies for healthcare in Dubai's public and private health care sectors; enabling partnerships between health care providers; licensing and regulating health care professionals and facilities; and increasing the transparency and accountability of the health care system. With the Mandatory Health Insurance Law 11 in November 2013, DHA was required to transition to a fully transparent and sustainable health information system with international practices and expertise. This allowed the implementation of performance measures that enabled patients and their families to make informative decisions in choosing their health care providers.

DUBAI HEALTH STRATEGY 2021

In January 2016, Sheikh Mohammed launched Dubai Health Strategy 2021. The aim is to provide the highest quality of medical care to patients with chronic diseases, to promote a culture of early detection, and to raise awareness about the importance of regular check-ups. The strategy features four main approaches, six objectives, 15 programs, and 93 initiatives, all of which are set to improve patient satisfaction and empower clinical staff to provide equitable and equal access to care. The four approaches are health and lifestyle, excellence in providing service, smart health care, and governance. The six objectives include:

- ensuring a healthy and safe environment for the people of Dubai
- ensuring the provision of a high-quality, comprehensive, and integrated health service system

- improving efficiency in providing health care
- creating an integrated database to be used for smart government policy in making decisions
- creating an effective ecological system for the health care sector in Dubai in collaboration with the private and public sectors
- promoting innovation

DHA EHEALTH INFORMATION NETWORK

Some of DHA's selected projects include:

1. Electronic medical record (EMR). DHA is in the middle of implementing an EMR across public hospitals and primary health centers (PHC), with the first hospital operating in 2017. The aim of automating clinical processes and workflow is to reduce human errors as data is stored centrally and can be accessed remotely. Also, medical devices are integrated to ensure a continuous stream of data flow drives efficiency for clinical staff to monitor the patient's journey with minimal avoidable errors. The human-system interface in this project is critical to the success of care continuum and patient journey.
2. Health information exchange. A Dubai-wide health information exchange project was also planned to standardize and enable the exchanging of patient data across all health information systems deployed throughout Dubai in 2017. It is intended to integrate seamlessly with the National Unified Medical Record, which will be developed by United Arab Emirates Ministry of Health via an overarching health information exchange infrastructure that allows the sharing of health information at the national level. This will have great impact on UAE residents and citizens as well as enabling health care professionals to reduce duplicate data entry and diagnostic tests.
3. Health Information Interoperability Standards. DHA published the Health Information Interoperability Standards for Dubai in January 2011. The objectives are:
    a. to serve and establish a cooperative partnership between the public and private sectors to achieve a widely accepted and useful set of standards that will enable and support widespread interoperability among health care software applications in a city-wide e-health information network for Dubai
    b. to harmonize relevant standards in the health care industry to enable and advance interoperability of healthcare applications, and the interchange of health care data, to assure accurate use, access, privacy, and security, both for supporting the delivery of care and public health

HUMAN FACTORS AND ERGONOMICS IMPLICATIONS

The focus of Dubai Health Strategy 2021 is to improve quality, safety, efficiency, and effectiveness in patient care. In the implementation of the SALAMA project in DHA facilities, hardware and equipment were selected carefully to be suitable for the clinical staff job roles. This required the positioning of computers on wheels (COWs) to facilitate the movement of patients while clinicians are recording patient data. To reduce human error, the clinical workflow automation was a major milestone that was applied before implementation. Furthermore, the clinical decision support mechanism was tested and evaluated to ensure patient and staff safety when collecting accurate data. Hence, the implications of HFE with the implementation of these health care information systems existed, along with developing job role definitions and responsibilities to determine the mandated skills and competencies of clinical staff.

The Health Information Exchange and Interoperability Standards projects discussed above are planning the future of health data-sharing across all the health care facilities in Dubai. This will constitute building the base for a Dubai unified medical record for every patient in Dubai. Therefore, these vital projects will have a significant impact to reducing human errors. One example is the establishment of a single repository for medication management. The DHA pharmacy department is fully automating pharmacy-related processes in government facilities by introducing smart pumps for inpatient medications to be connected to the Dubai medication repository. This repository is critical to reduce human errors and ensures patient safety on the intravenous administration of drugs in hospitals. In conclusion, the continued automation and implementation of smart solutions and digital connectivity would enhance the benefits of HFE on reducing human errors.

In the area of medical training and simulation, DHA is exploring the implementation of various methods that can be used in health care education to replace or amplify real patient experiences with scenarios designed to replicate real health encounters, using lifelike mannequins, physical models, or computers. Hence, the DHA is developing a fully accredited state-of-the-art medical training facility that incorporates simulation, research, and development to address the following areas:

- human patient simulation
- augmented reality/virtual reality/mixed reality
- task trainers through an AI-based platform
- computerized simulation
- room scale simulation models

The implementation of these technologies will enhance the productivity of the clinical staff and safeguard the patients in the delivery of high-quality services.

9 Human Factors and Ergonomics in Energy Industries

Alexandra Fernandes and Kine Reegård

CONTENTS

INTRODUCTION

Safety is one of the main key words when referring to HFE in the energy context. Most of the contributions of HFE standards, guidelines, analysis, and research work have the goal of improving safe operation by minimizing the probability of errors and optimizing the allocation of tasks between humans and machines. Another important dimension relates to safety culture, a common term that relates to an attitude of a certain organization that intends to minimize its exposure to situations that can be dangerous or harmful (e.g., Cooper, 2000). It is visible in the beliefs, norms, roles, and practices that are implemented and reinforced in everyday work processes. Although safety culture is not simply "engineered" or implemented in an organization, there are a set of practices related with safety management that can guide the organization's growth into a culture of safety, shared and maintained by all employees (e.g., Cooper, 2000).

The birth of HFE in the energy industry can be tied to the assessment and analysis of severe accidents. In 1979, the Three Mile Island (TMI) Power Plant suffered a partial reactor meltdown. This was the largest incident in a commercial power plant in the USA, and the emphasis on the role that poorly designed interfaces and insufficient training can have in the cause and/or progression of accident scenarios became clear for the nuclear industry (Kemeny, 1979). After the TMI accident the role of HFE became vital for the industry, especially regarding control room operation, with a set of guidelines being presented regarding the design of the systems and its support of human performance. More recently, the Fukushima Daichi accident in Japan in 2011 marked a new era for HFE with a focus on resilient operations, crew adaptive responses, and training for unexpected events in safety critical industries. The accident was initiated by a tsunami after a large earthquake that introduced a series of faults that resulted in nuclear meltdown (IAEA, 2015).

In the petroleum industry another large accident that signaled the need for HF was the explosion and fire in the Piper Alpha oil platform in the North Sea in 1988, resulting in the deaths of 167 workers, leaving only 61 survivors. In the aftermath of the accident, a set of factors were pointed out as contributing to its severity, such as the design of the platform; the processes and practices regarding work orders and shift handling; the restricted emergency procedures; and poor overall communication (Cullen, 1990). More recently, the Deepwater Horizon accident (Macondo) that started on April 20, 2010 marked the petroleum industry. An explosion and subsequent fire in the offshore drilling rig in the Gulf of Mexico resulted in 11 deaths and the evacuation of 115 workers with several injured. The accident analysis report indicates that a series of events led to the explosion, covering rig design issues, failures in equipment, wrong interpretation of signals, and inadequate monitoring of conditions.

The analysis of these events shows that accident causes are complex and multifaceted—it is not one specific event or error that leads to an accident, but rather a(n unlikely) combination of factors (Reason, 1995). In the presented cases, the initial technical faults could have been of no concern if other factors had not exacerbated its consequences or failed to mitigate them. A number of crucial lessons learned can be drawn from these accidents and used to improve the industry's capacity of response to future events, increasing the safety standards and preventing similar accidents or incidents to occur in the future.

FUNDAMENTALS

In this section we present a quick overview of some key concepts and perspectives that dominate the HFE field in the energy industries.

Process Control

The concept of process control is important in relation to the joint effort of human and machine to achieve a task (Kantowitz & Buck, 1983). It allows consistent production in the industry and is used particularly in continuous processes such as power plants, oil and gas production and refining, and chemical industries. As such, the control room has been a major setting for HFE work in the energy industries. The interface between the process elements and the human operators monitoring and controlling them is one of the key topics in HFE for energy. In process control, the literature often differentiates between task interface and interaction interface (e.g., Nachreiner, Nickel, & Meyer, 2006). While task interface refers to distribution of tasks between the human and machine (classically defined by Fitts, 1951, and challenged by Dekker and Woods, 2002) the interaction interface relates to the design of the controls the human will need to operate, and can be linked to design requirements. The task interface is tightly related with the amount of automation that is included in the system and the optimization of allocation between human and machine. For the interaction interface, information-processing models have determined the design approach (Wickens & Hollands, 2000), but other approaches, such as ecological design, have proven to be relevant (e.g., Burns & Hajdukiewicz, 2004). Many incident and accident reports in the past have shown that a large amount of process control systems are not optimized according to HFE standards (e.g., Nachreiner, Nickel, & Meyer, 2006), thereby being a relevant topic for further work. In energy, process control is highly linked with safety issues, where an overview of the systems and a good understanding of the status of the plant are essential for an acceptable performance.

Human-Computer Interaction

Human-computer interaction encompasses knowledge from various disciplines, including cognitive science, behavioral science, design, media studies, and computer science. A central aspect in human-computer interaction has been the usability of the systems, or "the extent to which a product can be used by specified users to achieve specified goals with effectiveness, efficiency and satisfaction in a specified context of use" (ISO-DIS 9241-11, 1998). Aspects such as the way controls and alarms are presented to the operators can be crucial in incident and accident scenarios. For instance, the number and lack of prioritization of alarms can result in mental overload of the operators and might result in a delayed or faulty response (e.g., Nachreiner, Nickel, & Meyer, 2006). Another relevant aspect relates to the overall structure of the displays in digital interfaces. Traditionally, in the analog control rooms the control panels would follow the structure of the piping and instrumentation diagrams (P&IDs); however, in their everyday work the operators will perform specific tasks that might imply monitoring and controlling different elements belonging to different systems. Consequently, the concept of task-based displays was introduced (e.g., Burns & Hajdukiewicz, 2004), adjusting the interfaces to the needs of the human operator when using the interface, instead of focusing on a more traditional engineering view of how the process works. This enables simpler interfaces that require less navigation, minimizing the errors and optimizing the time in the task (e.g., Braseth, Veland, & Welch, 2004).

Socio-Technical Systems and Integrated Operations in Petroleum

A socio-technical perspective considers that "a system's performance, including its safety performance, emerges from the pattern of dynamic activities within and between its social and technical components" (Kleiner, Hettinger, DeJoy, Huang, & Love, 2015). A number of explicitly socio-technical approaches to systems design have emerged, for example, cognitive systems engineering (Hollnagel & Woods, 1983), resilience engineering (Hollnagel, Woods, & Leveson, 2006), and macro-ergonomics (Hendrick & Kleiner, 2001; Kleiner, 2006). Taking on a socio-technical perspective when understanding work systems means that one is trying to understand both the main effect of each individual component as well as their interaction effects on performance.

In the petroleum industry, the need for socio-technical thinking has been particularly recognized in recent years in conjunction with developments of new concepts for operation. For instance, the concept of integrated operations (IO) is linked to oil and gas operations in the North Sea and is used to describe the new ways of performing oil exploration and operations by use of information and communication technologies. Technological developments have allowed for new types and increased availability of sensory data, more remote monitoring and control options, and closer cooperation between people across distances both disciplinary, organizationally, and geographically (Rosendahl & Hepsø, 2013). The concept soon broadened to include not only technology aspects, but also focusing on how work practices enable the offshore oil and gas industry in the North Sea to improve decision-making processes and operational efficiency (e.g., Besnard & Albrechtsen, 2013).

The industry's early experience with IO implementation projects led to a realization that they needed to address the multifaceted issues of integrated operations through an integrated approach (Edwards, Mydland, & Henriquez, 2010). Often, the issues pertaining to the softer part of the system did not receive enough attention in implementation projects due to large technical challenges that needed to be solved to operate the field (Drøivoldsmo & Nystad, 2016). The result would then be a technologically advanced field that can be run more efficiently, yet operated in much the same way as the less technologically advanced (ibid.). The capabilities approach to integrated operations (Drøivoldsmo, Reegård, & Farbrot, 2014; Henderson, Hepsø, & Mydland, 2013; Reegård, Drøivoldsmo, Rindahl, & Fernandes, 2014) was generated within the Center for Integrated Operations in the petroleum industry (IO Center, 2008) to help the industry in identifying the critical elements that are required to obtain the benefits of technology implementation already in the early project phases (Drøivoldsmo & Nystad, 2016). In this approach, the combination of human, technology, process, and governance resources is what constitutes the basis for performance, and not any individual resource alone (Reegård, Rindahl, & Drøivoldsmo, 2015). For example, several oil and gas companies have invested in collaboration technologies that can be used in several settings for getting hold of non-collocated expertise for problem solving and decision-making for safe and efficient production. However, it is the combination of resources that enable performance delivery. This means that the companies will reap the benefits of the collaboration technologies only if it enables and empowers its people to use it and design its work processes to include its appropriate use. Through the work in the IO Center, a method for development of capabilities was created and piloted focusing on the translation of targeted business goals into resource combinations and development needs on a practical level (see the Methods section).

METHODS

In this section, we will describe three main methods representative of the literature of HFE in energy. These include complex approaches to human performance and safety aspects, and often involve a systemic approach where the human operator is taken as a part of a larger context involving the machines s/he interacts with, the context where the actions are performed, the physical states of the controlled processes, and the overall organizational aspects and global setting of the activities.

Human Performance and Reliability Analysis

Human performance is a central aspect for HFE in all disciplines. Within the energy sector, in relation with the emphasis on safety, the field of human reliability analysis (HRA) is particularly relevant since it enables error identification and quantification, and might guide measures towards the reduction of errors (e.g., Gertman, Blackman, Marble, Byers, & Smith, 2004). To reach this, the reliability analyst needs to decompose the operator chore into specific tasks through a procedure known as task analysis. Task analysis includes a description of task/sub-task durations, frequency, allocation, complexity, context, tools, and other factors impacting on task completion (e.g., Kirwan & Ainsworth, 1992). It can be of several types, of which two are most commonly referred to: cognitive task analysis (identifying tasks that require memory, attention, reasoning, and decision skills) and hierarchical task analysis (identifying subtasks for every main task of the operator). HRA can present a quantitative and/or a qualitative focus (e.g., Oxstrand, Kelly, Shen, Mosleh, & Groth, 2012; Bye et al., 2017). The quantitative approach focuses on attributing a probability to non-fulfillment of a human failure event, which is an event or safety barrier in which human action is involved. This then feeds into probabilistic risk/safety assessment. Methods such as the Technique for Human Error Rate Prediction (THERP) by Swain and Guttmann (1983) or the Standardized Plant Analysis Risk – Human Reliability Analysis (SPAR-H) by Gertman and collaborators (2004) are often used in nuclear and petroleum contexts. Other approaches allow retrospective as well as prospective analysis of human performance such as the Cognitive Reliability and Error Analysis Method (CREAM) (Hollnagel, 1998). The qualitative focus (Davies, Ross, Wallace, & Wright, 2003) uses similar methods, but does not include the quantification of the probabilities of human error, being more focused on the detailed analysis of how tasks are performed and how that can be improved to enable success paths for the operating crews.

Verification and Validation

Verification and validation (V&V) assessments are intended to determine whether the system designs correspond to requirements and to comprehensively evaluate whether the systems support the operators in fulfilling their tasks and operate the plant, enabling their operational and safety goals (O'Hara, Higgins, Fleger, & Pieringer, 2012). V&V assessments are critical in the nuclear industry and correspond to a detailed plan, done before building a new power plant or when initiating modernization processes in existing plants. They include a description of all the expected activities, involved staff and responsibilities, from the assessment of needs to the implementation of the new designs in a control room. One of the most discussed topics within V&V is integrated system validation (ISV). This relates to an assessment of whether the control room allows safe operation of the plant or not (O'Hara et al., 2012). The goals of ISV are to validate the acceptability of staffing levels, assignment of tasks, coordination between individuals and groups of individuals (e.g., control room crew and support centers); validate the design features for alerting, informing, monitoring, and controlling the process, as well as its feedback to the operators; validate that the interfaces minimize personnel error and allow for error detection and recovery; validate that specific tasks are completed within reasonable time and with acceptable levels of workload and effective situational awareness (O'Hara et al., 2012). This evaluation is achieved by combining a set of methods that include observations of simulations, self-rated questionnaires, and expert assessment to identify whether the overall control room supports safe operations.

Methods for Integrated Operations in Petroleum

Through the work in the IO Center, a method for capability development was created and piloted focusing on the translation of targeted business goals into resource combinations and development needs that are required for establishing a new way of working that takes advantage of the technology implementation (Henderson, Hepsø, & Mydland, 2012; Reegård, Drøivoldsmo, & Rindahl, 2015; Reegård et al., 2014). The method consists of four main steps: (1) define the operational context; (2) identify and prioritize target capabilities; (3) define target capabilities; and (4) define resource development for the targeted capabilities. The method consolidates the top-down business requirements and bottom-up operational requirements, and the first output of the method (step 1 to 3) is a consensus on the definition of what is to be developed. The second output (step 4) is a resource development plan that is meant to function as an overall development plan for the capability, thereby pinpointing other activities and methods that needs to be used and when. A new addition to the method was inspired by Curtis, Hefley, and Miller's (2001) people capability maturity model (PCMM) for improved management of the human assets of a workforce, focusing on continuous improvements. In the capability development method, the PCMM concept is adjusted and expanded to include the concept of scalability, referring to the organizations' ability to successfully scale a pilot project into a common practice (Reegård et al., 2014). This new development allows for more detailed resource development plans at the company level that encompass the different project phases from the first piloting to company-wide scaling and continuous improvement efforts.

While the capability method is focused on laying out a development plan that spans across the project from concept phase to operations, there is also a need for more detailed analysis in the early project phases in order to establish an organization for the operations. The Integrated Operations – Man Technology Organization (IO – MTO) function allocation method has been developed specifically in the IO context and is used to adapt the organization to the opportunities inherent in IO (Drøivoldsmo & Nystad, 2016). It is a structured method for analyzing and (re)allocating tasks between physical locations, organizational units, and human and machine agents (Nystad, Drøivoldsmo, Lunde-Hanssen, & Heimdal, 2012). The method consists of defining the project goal, mapping of tasks for the relevant roles, testing of hypotheses for reallocation of tasks including an assessment of potential consequences for health, safety, and environment (HSE), and modeling of the proposed changes. In addition, a third-party evaluation of the combined consequences is proposed. An example of the use of the method is provided in *applications* section.

APPLICATION

There is a vast set of applications of HFE to the energy industries. Knowledge and competency from HFE has been used to create new concepts of design for control room settings by integrating principles from human cognition; developing new tools for the operators to increase efficiency and safety; analyzing safety; improving the type and style of training for operators and broadening the scope and timings of training; enhancing the organizations' capabilities to adapt and evolve in the markets; changing consumers behaviors towards a more sustainable energy use.

A Tool for Designing Guideline-Compliant Control Rooms

In the past decades, multiple methods, guidelines, best practices, and regulations have been developed that include human factors and ergonomics aspects in design (e.g., Johnsen, Bjørkli, Steiro, Fartum, Haukenes, Ramberg, & Skriver, 2011; O'Hara, Brown, Lewis, & Persensky, 2002). These human-centered requirements contribute to a safer operation and also towards the well-being of the operators. They refer to aspects such as the distance from the operator to controls and displays, the proximity between operators, the placement of furniture to facilitate communication, the field of view of individual operators, among many others. When designing a new control center or modifying an existing one, all of these aspects should be included in the control center design possibilities and tested multiple times to assure compliance with HF guidelines and regulations before the control room is physically built. To address these issues, specialized tools such as the IFE Halden Virtual Reality Centre CREATE (HVRC CREATE, 2017) have been presented by HFE and design experts. CREATE uses 3D visualization technologies to facilitate rapid prototyping and evaluation of alternative designs, incorporating information from HF guidelines and recommendations. It combines tools for layout design, planning, and formal review of the control room. It supports the visualization of different options, such as the location of work stations, size of computer displays, etc., providing manikins to evaluate ergonomics for the target population and tools for simulating field of view, reach envelopes, etc. Some of these checks can even be automated (Louka, 2015). Since it is used to produce a virtual mock-up, multiple versions of control rooms can be tested rapidly. Such a tool enables testing against different sets of guidelines that might be specific for certain industries, accelerating the design process since the designers do not need to wait for the completion of a physical mock-up before detecting aspects that will need to be modified (Drøivoldsmo & Louka, 2011). These visually oriented kinds of tools are especially useful as mediators in the communication between management, designers, reviewers, and the end-users (HVRC CREATE, 2017).

Safety Review of a Nuclear Power Plant Control Room

In this section we will describe a possible application of the verification and validation method in an industry case of safety review of a nuclear power plant control room. The goal of this HF safety review was to assess whether the HFE design fulfills requirements such as design principles, guidelines, and standards, allowing the control room crew to achieve the plant's safety and operational goals in an efficient way (O'Hara et al., 2012). The evaluated plant was operating since the 1980's and had modernized the control room by including digital interfaces. Multiple parameters in international and national standards as well as internal plant documentation were considered for the test (e.g., O'Hara et al., 2012; O'Hara, Stubler, Higgins, & Brown, 1995). In this case, performance evaluations from previous ISVs were also available as baseline data. An evaluation team was created with external members leading the evaluation and consulting with internal members such as training instructors and operators. A first step was the definition of a set of control room requirements for how the interfaces, procedures, and overall way of work should support the team. Prior to the test, the evaluation team identified non-acceptable performance criteria for the evaluation both by identifying the minimum acceptable ratings in the different measures and using the previous findings from ISV as baseline values. These criteria are termed human engineering discrepancies (HEDs; O'Hara et al., 2012), and refer to aspects of human performance that the crews/operators are expected to fulfill but are not verified in the test. A test was then conducted in the plant's training simulator, where four crews were presented with four complex, realistic accident scenarios. Measures of task performance, situation awareness, cognitive workload, teamwork, and usability were taken. The overall performance results showed that the control room supported safe operation and that the performance tended to improve in comparison with previous evaluations. Nonetheless, four different HEDs were identified, referring to low scores in performance measures' items in specific scenarios. Two of the HEDs related to low scores by one of the four crews in verification of safety functions and actions for loss of offsite power. The other two HEDs resulted from items in the performance measures where the crews performed below expectations (identification of the intermediate position of a valve) and where only one crew had satisfactory performance (understanding the status of a bypass valve and stuck flow measurement). This lead to a HED resolution plan, discussed with plant staff and training instructors. We concluded that the control room function was acceptable, since most identified HEDs could not be attributed to HSI characteristics and were not systematic (there was no item in which all crews performed below the acceptability threshold). The specific HEDs were assessed as non-acceptable as they were. These were all related to the detection of status in safety classified components and linked with an increase in detection time (when compared with previous assessments). These items required further investigation by the plant staff.

Assessing Manning Level for a New Field (IO MTO)

The IO MTO method was used for the purpose of assessing minimum manning requirements for normal operations of a new field. The new field was developed with a high level of instrumentation allowing for greater condition-based maintenance. The company already had extensive experience with integrated operations and adhered to a philosophy where administrative tasks are mainly done by the onshore organization. For this particular new field, the purpose of the IO MTO method was to determine whether manning for normal operations could be leaner than the standard operating model suggests, especially given the high level of instrumentation and by having offshore operators with combined technical competences. The analysis was performed based on interviews with the roles flagged by the company as particularly relevant. Through the interviews, a mapping of tasks for each role was performed, including at what time the task is performed, time flexibility, duration, location, participants, complementarity with other tasks, preconditions, and health, safety, and environment (HSE) criticality including competence requirements. The data was analyzed per role for both day and night shifts to estimate the amount of time that could potentially be freed given tasks that could be allocated to another role were removed from the current role. Next, the overall allocation of tasks was assessed, verifying that the proposed re-allocations did not result in too much work being assigned to any role, and whether there would be enough work to justify a full position or if the role could be combined with another. The IO MTO analysis showed that manning level could be significantly reduced given that the instrumentation replaces some manual tasks, some tasks are simplified, and combined technical competence for several offshore operators allows them to perform a greater set of diverse corrective maintenance jobs, while the main preventive maintenance work is performed by a campaign organization.

FUTURE TRENDS

HUMAN AS ASSET, NOT HAZARD

Initially, the main goals of HFE analysis and recommendations were to prevent human error, designing a "human-proof" system in process control. Underlying this approach was the acceptance of the human as a potential hazard. "Human error" was a concept that enabled detailed analysis and suggested which questions should be asked, for instance when improving human-system interfaces (e.g., Senders & Moray, 1991). Nonetheless, the focus on errors was clearly not enough to explain the many possible interactions and outcomes in complex systems, and a framework for enhancing adaptive capacity was needed (e.g., Woods, Dekker, Cook, Johannesen, & Sarter, 2010). The importance of the human element as a barrier for risk and not only as a potential risk source represents an important shift in the HFE approaches, leading professionals to identify critical decision-making contexts and activities, including success factors and learning from success stories. The analysis of operations often highlights the positive contribution of various human and organizational aspects that mitigate numerous instances of abnormality, quickly addressing existing issues, solving problems, and returning the process to normality (Hollganel, Woods, & Leveson, 2006). In this context the human is seen as an asset with unique capabilities that enable problem solving in expected and particularly unexpected situations (Woods et al., 2010).

Automation and Human-Robot Collaboration

Automation has been one of the central topics in HFE for many years (e.g., Parasuraman & Riley, 1997) and is becoming more relevant as automation evolves and generalizes. Automation topics are relevant both in large-scale projects such as the design of control centers, subsea exploration in the petroleum industry, or the operation of unmanned platforms both in petroleum and in renewable energy industries; but also to understand the implications of introducing smaller autonomous systems into currently existing interfaces. For example, being able to activate an automatic sequence for maintenance work at a power plant. Another rising area for the energy industry is the introduction of robots into the workflows, bringing human-robot collaboration aspects into consideration. Robots have been used in clean-up operations after accidents (Moore, 1985; Kawatsuma, Mimura, & Asama, 2017), but other examples include the use of robots for monitoring and maintenance in high-radiation areas, preventing radiation dosage to the workers (Dolan, 2013). Also, the use of autonomous underwater vehicles in the offshore petroleum industry is rising (e.g., Niu, Lee, Husain, & Bose, 2009) allowing monitoring of environmental conditions and status of underwater components. Likewise, the use of unmanned aerial vehicle systems has been explored, being considered promising in areas such as pipeline inspection (Sadovnychiy, 2004) or safety and security management (Cho, Lim, Biobaku, Kim, & Parsaei, 2015).

Augmented Cognition

In recent years HFE practitioners became able to collect relevant online data on the operators, by using physiological measurements such has heart rate or eye-tracking data and using them to better understand the status of the operators. Its potential on process control settings is also starting to be explored. For instance, Fallahi, Motamedzade, Heidarimoghadam, Soltanian, and Miyake (2016) used heart rate variability (HRV), electroencephalogram (EEG), and electrocardiogram (ECG) measures, together with a self-reported workload rating, to access the mental workload of operators in different settings, concluding that these measures were all sensitive to variations in mental workload demand. These types of application can eventually be used in connection with the process control system that could read the operator status and provide them with feedback, cuing ideal break times, or re-distributing tasks when one of the operators is perceived as overloaded. The use of eye-trackers to analyze how operators search information in screens (e.g., Ha, Byon, Baek, & Seong, 2016) is the most common example, and has been used in relation to control room work (e.g., Tran, Boring, Dudenhoeffer, Hallbert, Keller, & Anderson 2007; Fernandes, Renganayagalu, & Eitrheim, 2017). The use of augmented reality applications in the energy sector has also been growing, focusing in the presentation of procedures, and maintenance work (e.g., Johnsen & Mark, 2014; Ishii, Bian, Fujino, & Morishita, 2007), allowing the field operators to have online assistance from the control room while keeping the hands free for task completion.

From the Energy Workers to the Energy Consumers

There is an anticipated rise in "prosumers"—a term initially coined in media studies (Toffler, 1980) and borrowed in recent years in the energy industry (e.g., IEA-RETD, 2014) to refer to consumers that are also partially producing energy, for instance by adding solar panels to the household and selling the excess energy they are producing. This trend is anticipated to change the energy industries and thereby HFE for energy. For instance, the increase of digitalization and the potentials from integrating all the consumers/users in a dynamic network might imply the construction of dedicated controls centers, changing the traditional energy monitoring and control tasks. Other aspects being affected by this new trend relate to economical, behavioral, and technological issues, as well as the national frameworks that will set the conditions for prosumer growth. Prosumers are as such expected to have the capacity to alter the energy market, affecting electricity prices and contributing to an increase of renewable energy use, but the conditions for moving from a consumer to a prosumer role are very complex (e.g., Bremdal, 2013) and will surely require the contribution of HFE knowledge and experience.

CONCLUSION

HFE has proven to be a crucial discipline for the energy industry. Its contributions in understanding the contexts and predictors of accidents and incidents have contributed extensively to increased safety in operations, particularly in the nuclear and oil and gas industries. In recent years, associated with the quick development of new technology, new challenges have been presented to the HFE community related with both the digitalization of the energy industry and the environmental concerns surrounding the non-renewable energy sources. Being able to produce energy more efficiently, and in a sustainable manner, entails the involvement of not only the energy producers, but also consumers, who gain an active role in the energy production as well. The systemic approaches in HFE are particularly relevant in this context, allowing an integrated and complex analysis of how the use and production of energy can be optimized, involving the communities towards a more sustainable way of living.

KEY TERMS

Control room (or control center). A physical space where instrumentation and control tools and interfaces are gathered, enabling the monitoring and operation of a large set of components that are distributed in different locations (in the same building, other buildings in the same area, or even more remote distances).

Human reliability analysis. A framework that attends to the possibilities of error and success in human-machine systems, focusing on the potentials for error prevention and improvement.

Integrated operations. A concept used in the petroleum industry referring to the integration of technology and competences in new cross-disciplinary work processes and ways of working to optimize production and preserve safety.

Process control. Used in this chapter as the application of HFE principles to the actions and monitoring aspects required to operate complex production systems and to maintain them within productivity and safety boundaries.

Verification and validation. The set of independent methods and procedures required to assess a control room system to assure it will support the safety of operations.

Resilience engineering. The capacity to operate in a way that allows a system/organization to integrate changes into its processes and to react to undesirable events/failures maintaining its basic tasks and processes.

Risk assessment. The identification of hazards in a specific operation; a quantitative and/or qualitative description of risk linked to a concrete task/situation.

REFERENCES

Besnard, D. and Albrechtsen, E. (2013). *Oil and Gas, Tehnology and Humans: Assessing the Human Factors of Technological Change*. Dorchester: Ashgate.

Braseth, A. O., Veland, Ø., and Welch, R. (2004). Information rich display design. In *Proceedings of the Fourth American Nuclear Society International Topical Meeting on Nuclear Power Plant Instrumentation, Controls and Human-Machine Interface Technologies* (NPIC&HMIT 2004). Columbus, OH, September.

Bremdal, B. A. (2013). The impact of prosumers in a smart grid-based energy market. IMPROSUME Final Report. Norwegian Centres of Expertise: Smart Energy Markets, Norway.

Burns, C. M. and Hajdukiewicz, J. R. (2004). *Ecological Interface Design*. Washington, DC: CRC Press LCC.

Bye, A., Laumann, K., Taylor, C., Rasmussen, M., Øie, S., van de Merwe, K., Øien, K., Boring, R., Paltrinieri, N., Wærø, I., Massaiu, S., and Gould, K. (2017). The Petro-HRA Guideline (IFE/HR/E-2017/001). Institute for Energy Technology, Halden, Norway.

Cho, J., Lim, G., Biobaku, T., Kim, S., and Parsaei, H. (2015). Safety and Security Management with Unmanned Aerial Vehicle (UAV) in Oil and Gas Industry. *Procedia Manufacturing* 3, 1343–1349. http://doi.org/10.1016/j.promfg.2015.07.290.

Cooper, M. D. (2000). Towards a model of safety culture. *Safety Science* 36, 111–136.

Cullen, L. (1990). *The Public Inquiry into the Piper Alpha Disaster.* London: Stationery Office.

Curtis, B., Hefkey, W. E., and Miller, S. A. (2001). People Capability Maturity Model (P-CMM), version 2. Carnegie Mellon Software Engineering Institute, Pittsburg.

Davies, J. B., Ross, A., Wallace, B., and Wright, L. (2003). *Safety Management: A Qualitative Systems Approach*. London: Taylor and Francis.

Dekker, S. and Woods, D. (2002). MABA_MABA or ABRAKADABRA? Progress on Human Automation Co-ordination. *Cognition, Technology and Work* 4, 240–244.

Dolan, J. (2013). First of a kind IVVI Robot: Case Study. Hitachi leaflet.

Drøivoldsmo, A. and Louka, M. N. (2011). Virtual Reality Tools for Testing Control Room Concepts. In B. G. Liptak (ed.), *Instrument Engineers' Handbook: Process Software and Digital Networks* (Volume 3, 4th Edition). Boca Raton, FL: CRC Press. ISBN 9781439817766.

Drøivoldsmo, A. and Nystad, E. (2016). The Man, Technology and Organisation MTO Function Allocation Method. SPE Intelligent Energy International Conference and Exhibition, Aberdeen, United Kingdom. SPE-181102-MS.

Drøivoldsmo, A., Reegård, K., and Farbrot. (2014). The Capability Approach to Integrated Operations. IFE/HR/F-2014/1591. IO Center.

Edwards, A., Mydland, Ø., and Henriquez, A. (2010). The Art of Intelligent Energy – Insights and Lessons Learned from the application of iE. SPE Intelligent Energy International Conference and Exhibition, Utrecht, the Netherlands. SPE-128669.

Fallahi, M., Motamedzade, M., Heidarimoghadam, R., Soltanian, A. R., and Miyake, S. (2016). Assessment of operators' mental workload using physiological and subjective measures in cement, city traffic and power plant control centers. *Health Promotion Perspectives* 6(2), 96–103.

Fitts, P. (1951). *Human engineering for an effective air navigation and traffic control system.* Columbus, OH, Ohio State University.

Fernandes, A., Renganayagalu, S. K., and Eitrheim, M. H. R. (2017). Using eye tracking to explore design features in nuclear control room interfaces. Proceedings of the Human Factors and Ergonomics Society Europe chapter 2016 Annual Conference. Prague, Czech Republic.

Gertman, D., Blackman, H., Marble, J., Byers, J., and Smith, C. (2004). The SPAR-H Human Reliability Analysis Method (NUREG/CR6883). United States Nuclear Regulatory Commission, Washington, DC.

Ha, J. S., Byon, Y. J., Baek, J., and Seong, P. H. (2016). Method for Inference of Operators' Thoughts from Eye Movement Data in Nuclear Power Plants. *Nuclear Engineering and Technology* 48, 129–143.

Henderson, J., Hepsø, V., and Mydland, Ø. (2012). What is a Capability Platform Approach to Integrated Operations? An Introduction to Key Concepts. In Rosendahl, T. and Hepsø, V. (eds), *Integrated Operations in the Oil and Gas Industry: Sustainability and Capability Development.* IGC.

Hendrick, H. W. and Kleiner, B. M. (2001). *Macroergonomics: An introduction to work system design.* Proceedings of the Human Factors and Ergonomics Society Conference.

Hollnagel, E. (1998). *Cognitive Reliability and Error Analysis Method (CREAM).* Oxford: Elsevier Science Ltd.

Hollnagel, E. and Woods, D. D. (1983). Cognitive systems engineering: New wine in new bottles. *International Journal of Man-Machine Studies* 18(6), 583–600.

Hollnagel, E., Woods, D. D., and Leveson, N. (2006). *Resilience engineering: Concepts and Precepts.* Aldershot: Ashgate Publishing Limited, UK. ISBN 0-7546-4641-6.

HVRC-CREATE. (2017). User Guide for HVRC CREATE: Overview. Institute for Energy Technology, Halden, Norway.

IAEA: International Atomic Energy Agency. (2015). The Fukushima Daiichi Accident: Report by the Director General. IAEA, Vienna.

IEA-RETD (International Energy Agency – Renewable Energy Technology Deployment). (2014). Residential Prosumers Drivers and Policy Options, IEA, Utrecht.

IO Center: Integrated Operations Center. (2008). Annual Report 2008. Norwegian University of Science and Technology, Center for Research Based Innovation (SFI), Norway.

Ishii, H., Bian, Z., Fujino, H., Sekiyama, T., Nakai, T., Okamoto, A., and Shimoda, H. (2007). Augmented reality applications for nuclear power plant maintenance work. In Proceedings of the International Symposium on Symbiotic Nuclear Power Systems for the 21st Century (ISSNP).

ISO-DIS-9241-11. (1998). Ergonomic requirements for office work with visual display terminals (VTDs), Part 11. Guidance on usability.

Johnsen, S. O., Bjørkli, C., Steiro, T., Fartum, H., Haukenes, H., Ramberg, J., and Skriver, J. (2011). CRIOP: A scenario method for crisis intervention and operability analysis (SINTEF A4312). SINTEF Technology and Society, Trondheim, Norway.

Johnsen, T. and Mark, N.-K. (2014). Virtual and Augmented Reality in the Nuclear Plant Lifecycle Perspective. In H. Yoshikawa and Z. Zhang (eds), *Progress of Nuclear Safety for Symbiosis and Sustainability: Advanced Digital Instrumentation, Control and Information Systems for Nuclear Power Plants* (pp. 243–255). Tokyo: Springer Japan. http://doi.org/10.1007/978-4-431-54610-8_25.

Kantowitz, B. H. and Buck, J. R. (1982). Feedback and Control. In B. H. Kantowitz and R. D. Sorkin (eds), *Human Factors: Understanding People-System Relationships*. New York: John Wiley & Sons Inc.

Kawatsuma, S., Mimura, R., and Asama, H. (2017). Unitization for portability of emergency response surveillance robot system: Experiences and lessons learned from the deployment of the JAEA-3 emergency response robot at the Fukushima Daiichi Nuclear Power Plants. *ROBOMECH Journal* 4(1), 706. http://doi.org/10.1186/s40648-017-0073-7.

Kemeny, J. G., Babbitt, B., McPherson, H. C., Haggerty, P. E., Peterson, R. W., Lewis, C., Pigford, T. H., Marks, P. A., Taylor, T. B., Marrett, C. B., Trunk, A. D., and McBride, L. (1979). *Report of the President's Commission on the Accident at Three Mile Island*. Washington, DC.

Kirwan, B. and Ainsworth, L. K. (1992). *A Guide to Task Analysis* (1st ed). London: CRC Press.

Kleiner, B. M. (2006). Macroergonomics: Analysis and design of work systems. *Applied Ergonomics* 37(1), 81–89.

Kleiner, B. M., Hettinger, L. J., DeJoy, D. M., Huang, Y.-H., and Love, P. E. D. (2015). Sociotechnical attributes of safe and unsafe work systems. *Ergonomics* 58(4), 635–649.

Louka, M. N. (2015). Using Virtual Mock-ups and Automated Verification Assistance to Support Human Factors Engineering Evaluation Activities for Control Rooms Layouts. In *Proceedings of the Ninth American Nuclear Society International Topical Meeting on Nuclear Plant Instrumentation, Control and Human-Machine Interface Technologies, NPIC&HMIT 2015, Charlotte, North Carolina, February 2015*. LaGrange Park, IL: American Nuclear Society.

Moore, T. (1985). Robots for nuclear power plants. *IAEA Bulletin*.

Nachreiner, F., Nickel, P., and Meyer, I. (2006). Human factors in process control systems: The design of human–machine interfaces. *Safety Science* 44(1), 5–26. http://doi.org/10.1016/j.ssci.2005.09.003.

Niu, H., Adams, S., Lee, K., Husain, T., and Bose, N. (2009). Applications of Autonomous Underwater Vehicles in Offshore Petroleum Industry Environmental Effects Monitoring. *Journal of Canadian Petroleum Technology* 48(5), 12–16.

Nystad, E., Drøivoldsmo, A., Lunde-Hanssen, L., and Heimdal, J. (2014). *IO MTO Handbook*, Rev. 4. IO Center: Trondheim, Norway.

O'Hara, J. M., Brown, W. S., Lewis, P. M., and Persensky, J. J. (2002). *Human-System Interface Design Review Guidelines* (NUREG-0700, rev. 2). Washington, DC: United States Nuclear Regulatory Commission.

O'Hara, J. M., Higgins, J. C., Fleger, S. A., and Pieringer, P. A. (2012). *Human Factors Engineering Program Review Model* (NUREG-0711). Washington, DC: United States Nuclear Regulatory Commission.

O'Hara, J. M., Stubler, W., Higgins, J. C., and Brown, W. (1995). *Integrated System Validation: Methodology and Review Criteria* (NUREG/CR-6393). Washington, DC: United States Nuclear Regulatory Commission.

Oxstrand, J., Kelly, D. L., Shen, S.-H., Mosleh, A., and Groth, K. M. (2012). A Model-Based Approach to Human Reliability Analysis: Qualitative Analysis Methodology. Proceedings of the International Conference on Probabilistic Safety Assessment and Management (PSAM 11). Helsinki, Finland.

Parasuraman, R. and Riley, V. (1997). Humans and Automation: Use, Misuse, Disuse, Abuse. *Human Factors* 39, 230–253.

Reason, J. (1995). Understanding adverse events: Human factors. *Quality in Health Care* 4, 80–89.

Reegård, K., Drøivoldsmo, A., and Rindahl, G. (2015). Strengthening the HF/E value proposition: Introducing the capability approach. *Proceedings of the Human Factors and Ergonomics Society 59th Annual Meeting.*

Reegård, K., Drøivoldsmo, A., Rindahl, G., and Fernandes, A. (2014). *Handbook: The Capability Approach to Integrated Operations.* IO Center: Trondheim, Norway.

Rosendahl, T. and Hepsø, T. (2013). Preface. In: T. Rosendahl and V. Hepsø (eds), *Integrated Operations in the Oil and Gas Industry: Sustainability and Capability Development.* Hershey, PA: IGC

Sadovnychiy, S. (2004). Unmanned aerial vehicle system for pipeline inspection. In *Proceedings of the World Seas Conference.* Athens, Greece.

Senders, J. W. and Moray, N. P. (1991). *Human Error: Cause, Prediction, and Reduction.* Hillsdale, NJ: Lawrence Erlbaum Associates.

Swain, A. D. and Guttman, H. E. (1983). *Handbook of human reliability analysis with emphasis on nuclear power plant applications* (NUREG/CR-1278). Washington, DC: United States Nuclear Regulatory Commission.

Toffler, A. (1980). *The Third Wave: The Classic Study of Tomorrow.* New York: Bantam.

Tran, T. Q., Boring, R. L., Dudenhoeffer, D. D., Hallbert, B. P., Keller, M. D., and Anderson, T. M. (2007). Advantages and disadvantages of physiological assessment for next generation control room design (pp. 259–263). Presented at the 2007 IEEE 8th Human Factors and Power Plants and HPRCT 13th Annual Meeting, IEEE.

Wickens, C. D. and Hollands, J. G. (2000). *Engineering Psychology and Human Performance.* Prentice Hall, Upper Saddle River.

Woods D. D., Dekker S., Cook R., Johannesen L., and Sarter N. (2010). *Behind Human Error* (Second Edition). Farnham: Ashgate Publishing Ltd.

Case Study 9
Human-Machine Interface Innovations

Samantha J. Horseman, Steven Seay, and Colin M. Sloman

INTRODUCTION

The e-factory is an innovation lab, as seen in Figure CS9.1, that specializes in human-machine interfaces (HMI) to enhance the safety, well-being, and performance of the Saudi Aramco workforce. It is aligned with Saudi Aramco's corporate objectives of intensifying the focus on safety, and preparing the workforce for the future. The e-factory has created a cutting-edge intellectual property portfolio called "intellisense." Unlike any other traditional human factors engineering (HFE) program, the "intellisense" innovation collects and collates undetected human signaling patterns from the body and brain and provides profound insights into workforce health status and performance. Intellisense systems provide a game-changing platform to detect and manage real-time signals across the HMI. Intellisense provides both hardware and software solutions to make sense of human signals so that organizations can outperform their competitors with an energized, engaged, and empowered workforce. One such innovation is predictive personal protective equipment (PPPE) and this case study will demonstrate how HMI innovations are advancing HFE.

THE STATE OF HUMAN FACTORS ENGINEERING IN THE GCC

The integration of applying both HMI innovations using HFE principles is a new methodology for the region. Undoubtedly, there is a very impressive wealth of intellectual and innovation capital in the technical and industrial knowledge in the oil and gas sector. However, combining both the HFE sciences with HMI to enhance safety, well-being, and performance in the workplace is still an immerging scientific discipline. Culturally, with the billion-dollar wearable technology industry disrupting the norms in collecting and customizing data, we are seeing a shift in the workplace. Workers are more aware and engaged in wearable technologies and aligned with the "connected quantifiable self" concept. This emerging interest and excitement in the workplace paves the way for new industrial standards and policies. It also provides an important dataset that can contribute to advanced machine leaning

FIGURE CS9.1 All projects in the e-factory are Saudi Aramco patents that have been taken through accelerated and rapid prototyping and licensing/commercialization sectors. The PPPE is a SAO patent and prototype created at the e-factory.

analytics and future health, safety, and environment (HSE) trends in the region. This will be important for HSE polices and ongoing research for the region.

GOALS AND APPLICATION

The application of PPPE measures human sensory central and peripheral signals, via human-computer interfaces—namely brain signals measured by electroencephalography (EEG) and biometrics (heart rate, stress response, temperature, body position, and location). This data is measured by specialized sensors built into personal protective equipment (e.g., hard hats, safety glasses, gloves, and belts), as seen in Figure CS9.2.

The methodology includes specialized sensors built into existing personal protective equipment (PPE). The measurements from these sensors then alert, in real time, both the end-user (biofeedback system) and safety personnel onsite via Wi-Fi signals. The sensoring system consists of the below elements (Table CS9.1).

As examples, the real-time outputs from the sensoring systems produce specific motion and biomechanical patterns that are risk factors contributing to potential injuries (see Figures CS9.3 and CS9.4).

This system determines risk and alert levels associated to worksite tasks involving personnel, equipment, and the environment. This system redefines the industry's current thinking through five core value propositions: situational awareness, knowledge and skill retention, biofeedback loops, predictive analytics, and safety alert systems. The back-end engine driving this is the EMC^2 infrastructure consisting of a data stream pipeline collecting real-time data at specific physiological time stamps from the end-user sensoring system embedded in the PPE (see Figure CS9.5), as explained in Table CS9.1.

Furthermore, this architecture will allow for the development of machine learning algorithms across multiple sites. Based on the initial baseline data, this

FIGURE CS9.2. PPPE human sensoring systems for hard hats, safety glasses, belts, gloves, and shoes.

TABLE CS9.1
The Critical Components of the PPPE System

Body–Computer Interface Measurement	Sensor System
Heart rate (bpm)	HR sensors are embedded into gloves and glasses
Temperature (Celsius)	Surface temperature sensors are embedded into gloves
Galvanic skin response (coherence values)	Finger conductance and resistance are embedded into gloves
Biomechanical stress: wrist, spine, and hip	GPS accelerometers and gyro-flex sensors are embedded into gloves, belts, and shoes (and the measurement of wrist and hip angles is done through belts and gloves)
Digital processing: blink rate, fatigue, and sleep deprivation	Via safety glasses (digital processors)
Brain signals: alpha, beta, delta, theta, and gamma Monitors alertness, fatigue, distraction, boredom	Dry electrode electrocephalography (EEG) embedded into safety helmets

infrastructure selected the EMC^2 hardware (rack) that was installed to collate data pools and data lakes, in real time, for large amounts of data streaming. This intelligent human sensory system rises to the challenge in developing an innovative platform capable of providing early detection and prescriptive analytics, and further advances HFE capabilities.

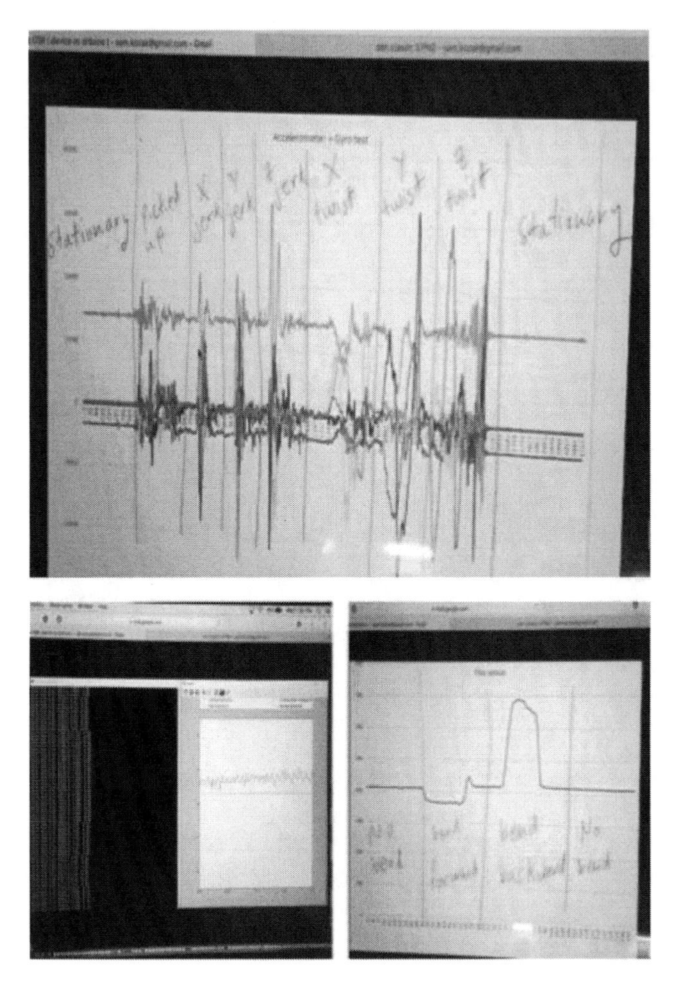

FIGURE CS9.3 Analyzing biomechanical patterns to determine predictive risk of injury.

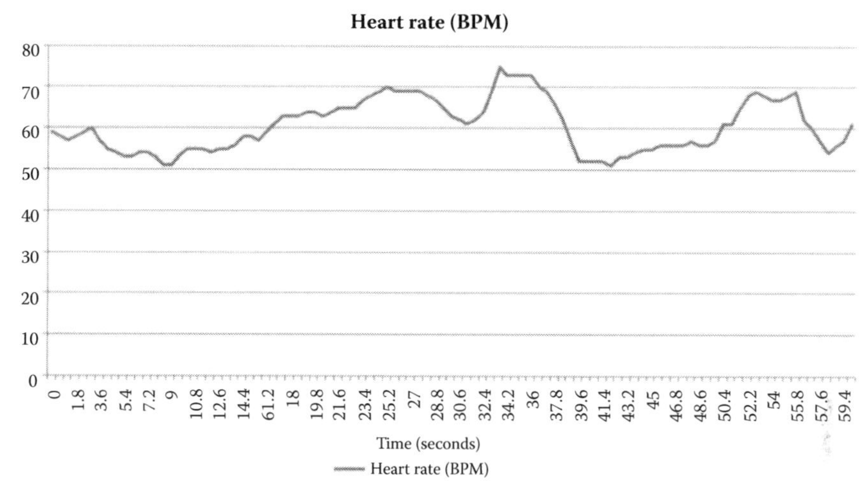

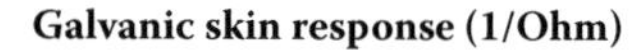

Placement of legends in the figure was modified to increase text font size. Please check if OK.

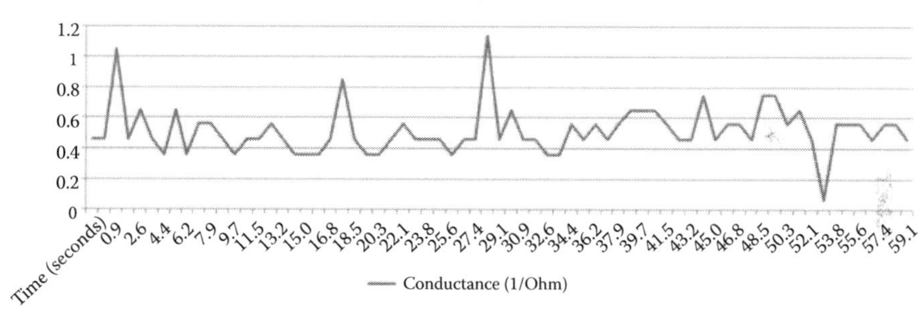

FIGURE CS9.4 Data collation: flex, GSR, HR, gyro, and accelerometer. Collected over a specific time-stamp gathered from the gloves and belt. *(Continued)*

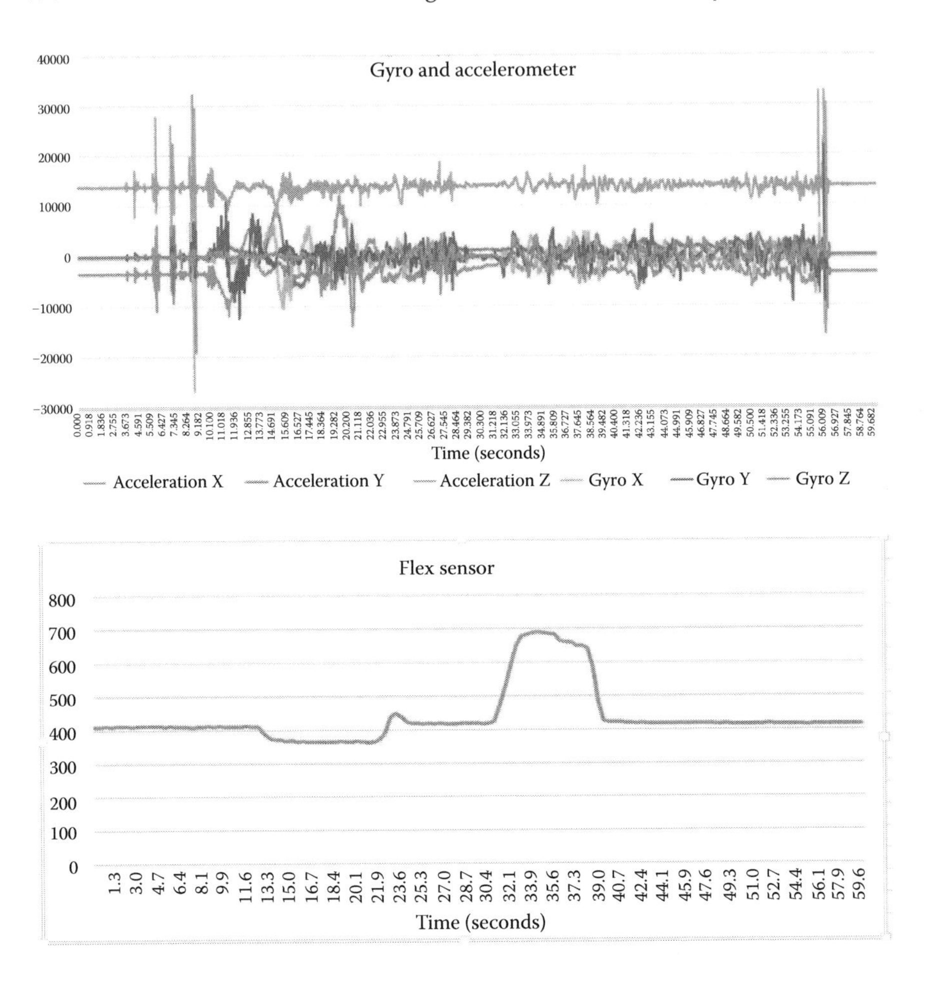

FIGURE CS9.4 (CONTINUED) Data collation: flex, GSR, HR, gyro, and accelerometer. Collected over a specific time-stamp gathered from the gloves and belt.

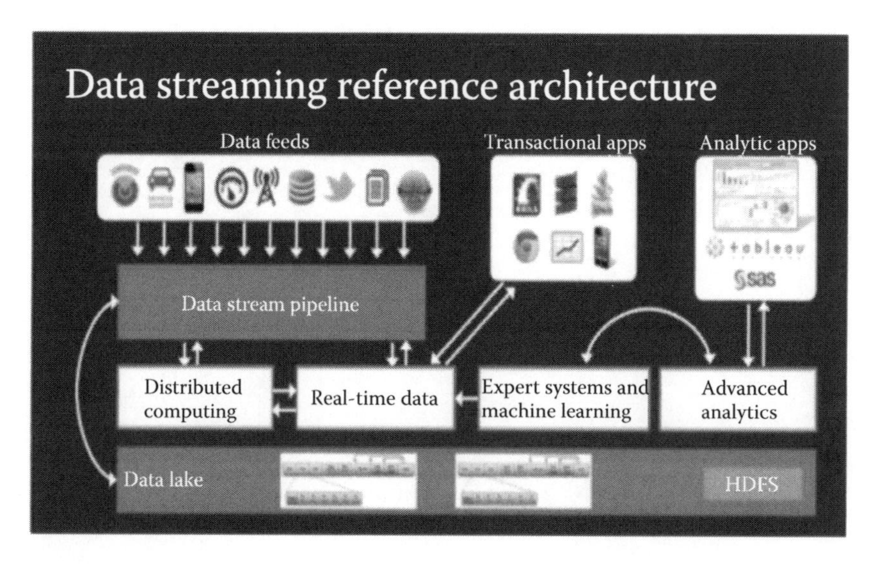

FIGURE CS9.5 Information technology data streaming and analytics: EMC[2].

10 Surface Transportation

Guofa Li and Ying Wang

CONTENTS

lease check ection levelling f correct.

INTRODUCTION

Surface transportation, as defined by the Human Factors and Ergonomics Society (HFES), refers to all forms of transit outside the aerospace sector. It contains numerous modes for conveying humans and resources on the earth, including both on- and off-road passenger/commercial/military vehicles, mass transit, maritime transportation, rail transit, pedestrian and bicycle traffic, and highway and infrastructure systems (HFES, 2017). A typical surface transportation system (e.g., a passenger vehicle) contains three integral components: the operator (e.g., driver), the vehicle (e.g., car), and the environment (e.g., roadway).

The majority of operators in current surface transportation systems are humans. Humans make errors and errors lead to accidents and failures. Some errors are made by the operators with intentional violations. Statistics from the World Health Organization indicate that the total number of annual road traffic deaths around the world is 1.25 million (WHO, 2016). Among the leading contributors, speeding violation is the primary cause. It leads to 9,262 fatalities, about 28% of all road accidental deaths in the USA in 2014 according to the National Highway Traffic Safety Administration (NHTSA, 2017). Driving under the influence (DUI, e.g., drunk and drugged driving) and distracted driving caused by cell phone use also contribute greatly to driving risks. It has been reported that driver distraction has become a prominent issue because of the overwhelming increase in the use of cell phones, communication devices, and the emerging technologies that are in portable vehicles (Tango & Botta, 2013). In 2014, 3,179 drivers died and 431,000 were injured in vehicle crashes involving distracted drivers in the USA; and 71% of young drivers who were killed in road accidents were reported to have experience of message texting while driving (NHTSA, 2017). Currently, 46 states in the USA have banned message texting while driving for all drivers, but many more potential uses of these devices during driving have not yet been legislated. Enforced laws, regulations, and management activities can effectively reduce drivers' unsafe behavior. For example, aggressive policies and other activities (e.g., education) were implemented to reduce alcohol consumption and improve road safety in Botswana. The overall crash rate declined by 22% in 2010–2011 compared to the number in 2004–2009 (Sebego et al., 2014). In addition, by monitoring, guiding, and restricting drivers' behavior based on personalized customization with sound human factor principles (e.g., gender and age considerations, road conditions, etc.), intelligent driving technology has been contributing to reduce the above-mentioned risky behavior greatly in recent years.

Some errors are made by the operators without intentional violation (Liang and Lee, 2014), such as gradually emerged fatigue, the insufficient experience of novice drivers, and the slow reaction of older drivers to road emergency. A national sample of serious US crashes showed that driver error was the critical reason for 97% of crashes involving older drivers, among which inadequate surveillance and gap misjudgment errors were more prevalent (Cicchino & McCartt, 2015). Understanding the operators' individual differences and physical/psychological limitations in policy making and technology development can help to avoid and reduce this type of error. For example, hazard perception training and vision screening can effectively improve older drivers' ability in preventing crashes (Horswill et al., 2015).

Other errors are not made directly by the operator, but due to design defects of equipment, infrastructures, or procedures (Dewar and Olson, 2007; SAE J2944, 2015).

For instance, the lack of lighting equipment was found very contributive to pedestrian–motor vehicle accidents in Guangzhou, China (Zhang et al., 2014). Increased prevalence of new in-vehicle information systems (IVIS) could occupy significant spare glance of drivers and create workload issues if the demand of IVIS was not carefully managed (Birrell & Fowkes, 2014). Exploring the limitation and comfortable zone of operators' capability while holding the vehicle can help to reduce this type of errors.

Therefore, it is important to bring HFE knowledge and experience to the domain of surface transportation to effectively prevent accidents caused by human errors and improve system performance. Besides, driving comfort, travel efficiency, and fuel economy can also be acquired by applying HFE (Hancock & Parasuraman, 1993). As Figure 10.1 shows, taking road transportation systems as an example, students, policy makers, practitioners, and other professionals can learn basic theories, research methods, and typical applications of HFE knowledge in surface transportation systems from this chapter (not all details in Figure 10.1 were discussed; further learning by readers is recommended).

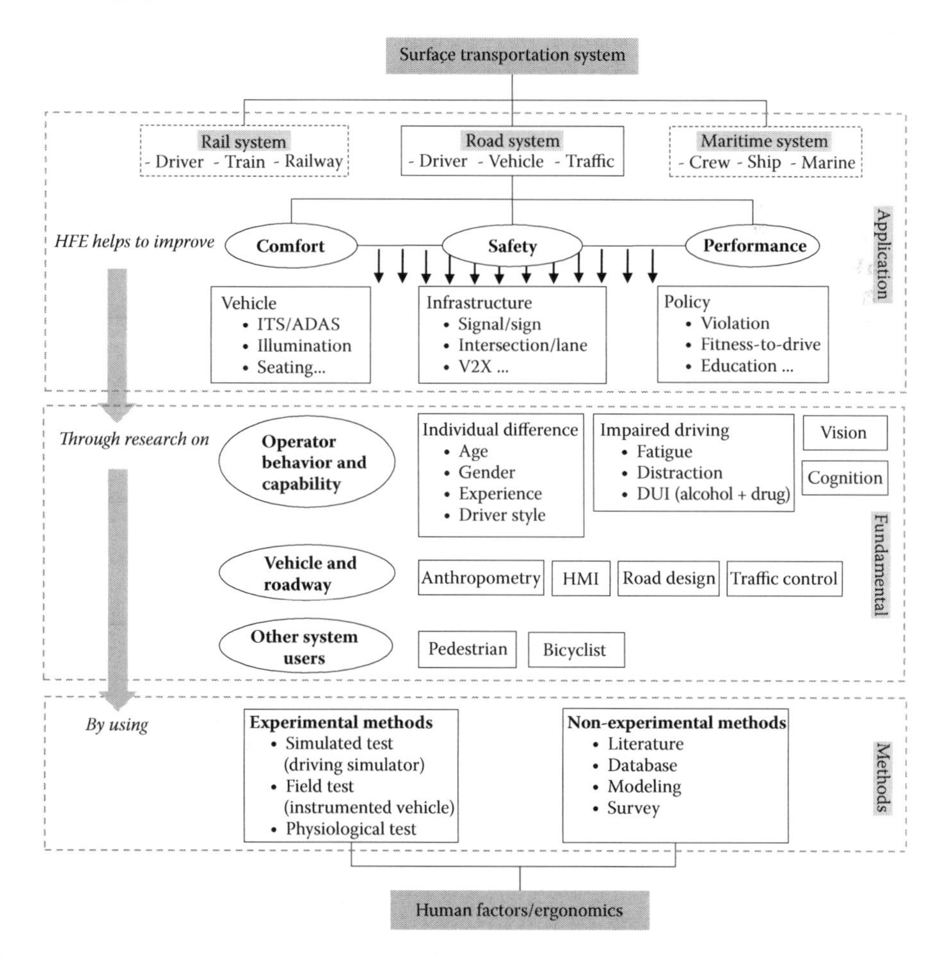

FIGURE 10.1 How HFE serves a surface transportation system.

FUNDAMENTALS

Can these be combined into one subheading?

Operator Behavior and Capability

Individual Difference

Age and Experience

Crash statistics reveal that both young (under 25) and older (65 plus) drivers are at greater risk than others on the road (Dewar and Olson, 2007). As compared to mature drivers, younger drivers perform higher violation rates, tend to underestimate the risks of various violations, have a lower level of motivation to follow traffic rules, and are overly involved in running red lights. In 2013, young male drivers involved in fatal collisions were twice as likely to be speeding as a male driver aged from 35 to 44 (NHTSA, 2015). Differently, the higher probability of older drivers to be involved in crashes at intersections or between multiple vehicles corresponds to their reduction in vision, difficulties in attention, slow decision-making, etc. (Anstey et al., 2005). Thus, the lack of experience for younger drivers and physiological defects for older drivers contribute to their high crash probability.

Experience is an even more critical factor for driving safety compared to age. Drivers with more experience can deal with various situations in a more timely and effective manner than those without. Take workload for example. Workload involves an interaction between the task and the capability of the driver, and thus affects driver behavior and driving safety. It can be either too high or too low in the same situation for different drivers. Driving on an urban street may involve a high workload for inexperienced drivers (or impaired drivers), while the same task may involve a low workload for experienced drivers. Although young drivers are usually inexperienced, not all inexperienced drivers are young. In developed countries, driver experience grows with a driver's age, since most people start to learn to drive during their teenage years. However, in developing countries, people start to drive at any possible age, including through their 70s. Experience can play confounding effects with age in influencing driving safety (Tao et al., 2017).

Gender

With respect to gender, male drivers tend to perform more aggressively than, or at least as aggressively as, female drivers (Farah et al., 2014). Male drivers are more likely to be involved in sensation seeking and in committing unsafe driving actions such as speeding and driving while intoxicated. This is because of their underestimation of hazards in different driving situations, higher confidence in their driving skills than females, and lower levels of motivation to follow traffic rules.

Impaired Driving; Fatigue

Fatigue has been recognized as a major cause of traffic accidents. More than 30% of drivers in the USA drive with fatigue at least once a month (Stutts et al., 2003). Driver fatigue mainly arises from the stress of high workload, long working hours, and lack of sleep during night driving. Driving with fatigue degrades the level of situation awareness and vehicle control; thus, driving performance will get worse, affecting steering movements (fewer and larger), variability of lane position and speed, and longer response time to road dangers. This degraded performance increases the risk of a crash. Current technologies for the detection of driver fatigue include: (1) detection from visual cues like eyelid movement, gaze movement, head movement, and facial expression (Jo et al., 2014); (2) using driving performance characteristics to recognize fatigue (Forsman et al., 2013); and (3) equipping physiological sensors (e.g., EEG) to detect fatigue (Fu et al., 2016).

Distraction

Similar to fatigue, driver distraction is identified as another major concern of traffic safety. About 25% of police-reported cases list driver distraction as a involved factor (Liao et al., 2016). It is defined as a momentary or transient redirection of attention from the task of driving to a thought, object, activity, event, or person, which is often caused by in-vehicle secondary tasks like cell phone use and in-vehicle technologies (e.g., navigation) (Sheridan, 2004). Distracted drivers tend to decrease their awareness of critical information for safe driving and degrade their driving performance, which makes them prone to causing severe car accidents. NHTSA has categorized driver distraction into four types, i.e., visual, auditory, biomechanical, and cognitive distraction (Ranney et al., 2000), among which visual distraction has been identified as the most significant. Dynamically, a driver's perception of a changing environment demands high cognitive workload (Wickens et al., 1998). Although distractions such as simple conversation can mitigate driving boredom and fatigue, driving safety will be threatened when cognitive workload is too high or the driving environment changes dramatically. To date, there have been many approaches to detecting driver visual distraction; usually these use camera-based systems to monitor drivers' eye movements and fixations (Tango & Botta, 2013; Liu et al., 2016). Different from visual distraction, the detection of cognitive distraction is more challenging since it happens inside the brain without apparent exterior features (Horrey & Wickens, 2006). Although difficult, approaches have been tried in the detection of cognitive distraction based on physiological measures (e.g., EEG) (Sonnleitner et al., 2014) and/or driving performance indicators (Liang and Lee, 2014; Liao et al., 2016). The final two distraction types, auditory and biomechanical, have the least impact on driving, but inexperienced drivers may still be affected more than experienced ones.

Driving Under the Influence

The use of alcohol or drugs is severely harmful to driving safety (Strand et al., 2016). In the USA, nearly 11,000 deaths were caused by alcohol-impaired driving in 2009 (Compton & Berning, 2009), accounting for 31% of all traffic-related fatalities (NHTSA, 2013). Alcohol and drugs reduce drivers' awareness level of dangers, reasoning, and muscle coordination, all of which are essential to safely operating a vehicle. The risk of a crash increases significantly when a driver takes alcohol and drugs together (Movig et al., 2004). Other than the factors mentioned above, driving style also significantly affects safety (Li et al., 2017).

Vision

Vision is essential to guiding a vehicle along its path, following cars, detecting roadway hazards, and making judgments about turning maneuvers, as well as seeing and identifying traffic control devices. Eye movement provides possibly the most beneficial information of where and to what extent drivers attend to various objects in and outside of their cars (Shinar, 2007). The allocation of a driver's fixation is the first clue that we have of verifying what drivers attend to, and how much time they devote to different objects. An important concept in understanding driver attention is useful field of view (UFOV), which includes the total visual field from which characteristics of the target can be acquired when head and eye movements are excluded (Rayner, 1998). The extent of UFOV depends on how well a driver can divide his/her attention, ignore distractions, and select relevant information from the environment. In a study of visual/cognitive factors in accidents, it was found that drivers with UFOV problems had 4.2 times more collisions and 15.6 times more intersection accidents than those without UFOV problems (Owsley et al., 1991).

An alert driver spends most of his/her time looking ahead of the vehicle. However, a good deal of information comes from peripheral vision. Thus, it is necessary to divide attention between what is ahead and what is off to the side. This is especially important at intersections and in heavy traffic. A healthy human vision system extends to about 180 degrees in a horizontal direction, reduced to about 140 degrees by age 70 (Owsley et al., 1991). High vehicle speed, heavy traffic, high workload, and rain and snow also reduce the UFOV of drivers.

Cognition

Driving is essentially a perceptual-motor task, so the importance of driver perception and information processing (driver cognition) cannot be overemphasized. A well-known decision-making model in dynamic systems proposed by Endsley (1995) shows that drivers' situation awareness to the state of the environment, information processing mechanism, current stress, and workload are all critical factors to the accuracy and speed of decision-making and following performance. Situation awareness involves being aware of the traffic environment, understanding what is happening, and knowing how to deal with perceived information to keep safe in the near future (Endsley, 1995). Situation awareness can be degraded by many environmental, vehicle, and personal factors, such as insufficient illumination, monotonous roads, unclear in-vehicle information, driver fatigue, and inattention. A typical scenario of degraded situation awareness is driving with fatigue in darkness, where the lack of light limits a driver's perception capabilities. Such a degraded level of situation awareness leads to frequent lane departures and even crashes. Interestingly, automation is not always good for situation awareness. Drivers in highly automated vehicles were found to lose more situation awareness compared to normal vehicles with less automation technology (Lu et al., 2016).

Driver workload also needs to be considered. Prolonged heavy workload driving leads to degraded situation awareness, but this doesn't mean that workload should always be reduced to relieve drivers from driving tasks. Studies show that decreased workload may cause a driver to direct his/her attention away from the main driving task and thereby affect his/her ability to retain control of the vehicle in emergency situations (Stanton et al., 1997). Thus, driver workload should be controlled within an appropriate range to keep an optimal level of arousal for safety (Reimer & Mehler, 2011).

It is clear that optimizing the design of roadway system (road, signs, signals, etc.) and vehicle (human-machine interface, automation, etc.), as well as controlling driver stress and workload to an appropriate level, are important to enhance driver situation awareness and information processing capability. From another perspective, measuring driver workload and situation awareness could be an effective way of evaluating the design of roadways and in-vehicle devices. Detection response task (DRT) has been a popular tool in recent years. It requires drivers to respond to a stimulus as soon as it appears in their visual field by hitting a response button. The stimulus could be visual, audio, or tactile. Long detection-response times of DRT suggest high cognition-demanding tasks and low situation awareness levels, based on which the design of roadways and in-vehicle devices could be further improved.

VEHICLE DESIGN

ANTHROPOMETRY

Vehicle design impacts driving performance and traffic safety. Human factor issues that need to be considered in vehicle design include anthropometry (physical dimensions of the human body), visibility of the traffic environment from inside the vehicle, the design and placement of controls and displays, noise, vibration, occupant protection, and the use of electronic devices. For details, see Dewar and Olson (2007). Among these factors, anthropometry design aims to ensure that drivers: (1) can see the road, traffic signals, and other vehicles outside of the car; (2) can see controls and displays inside the car; (3) are able to reach controls; and (4) find controls comfortable and convenient for operating while driving. The hip-point (H-point, the origin on the human body for vehicle car design) and the eyellipes (range of the driver's eye position) are two primary concepts in anthropometry design, which guide the design of driver seat, position of the steering wheel, layout of the vehicle cabin, etc. If a vehicle is not designed based on sound human factors principles, both physical and mental workload will increase, which often leads to driver fatigue, distraction, and the likelihood of driver error (Cheng and Trivedi, 2010). Examples of poor design include small and too-similar symbols in entertainment and climate control systems, poor visibility of the road in front, and insufficient space for drivers to operate.

HUMAN-MACHINE INTERACTION

The demand of in-vehicle human-machine interaction (HMI) has increased dramatically in recently years, according to the boom of advanced driver assistance systems

(ADAS). ADAS are electronic systems that are designed to support drivers in their driving tasks (see Figure 10.2 for the development history of ADAS). This support ranges from simple information presentation to advanced assisting and even taking over the driver's tasks in critical situations. Typical ADAS include ACC (adaptive cruise control system), FCW (forward collision warning system), and LCA (lane-change assistance system) (Bengler et al., 2014). Although the purpose of ADAS techniques is to generate a positive effect on traffic safety, several human factors issues need to be considered seriously. One ubiquitous issue is the design of warnings that promote appropriate responses from the driver. A badly designed and overly sensitive system can increase driver workload and thereby decrease situation awareness, comfort, and safety (Vahidi & Eskandarian, 2003). For example, if the alarm timing of an ADAS is designed to be correct, inappropriate responses such as ignoring collision warning signals, failing to brake when essential, or sharp braking maneuvers that jeopardize traffic safety for both the driver and other vehicles on the road are reduced (Lee et al., 2004). This may also lead to poor driver trust and acceptance of these systems. Excessive trust in a system creates dependence that can reduce vigilance while lack of trust in a system can affect a driver's notice of warnings or even cause a driver to shut down the system if possible.

Driver workload is another consideration in HMI. Studies have shown reduced mental workload of drivers when using ACC due to the driver being relieved from some elements of the driving task. Other research (Stanton et al., 1997) showed that this decreased workload may cause the driver to direct his/her attention away from the driving task when using ACC and thereby affect the drivers' ability to retain control of the vehicle in an emergency situation. The authors also pointed out that the

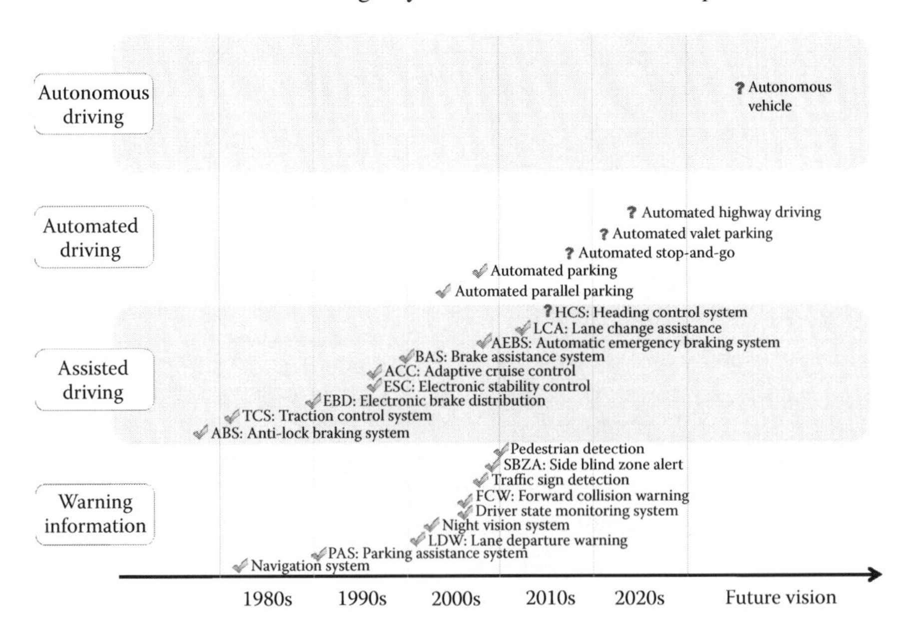

FIGURE 10.2 Development history of ADAS (Beiker, 2012).

improvements in secondary task performance while using ACC may lead to unexpected increases in accidents caused by driver distraction when performing more in-vehicle secondary tasks.

ROAD AND INFRASTRUCTURE DESIGN

Roadway Design

Drivers and pedestrians are often at a disadvantage due to inadequate roadway designs. It is essential for highway designers to be aware of road user limitations since the interaction between the road's physical characteristics and road users' expectations and limitations are important to traffic safety. A number of design countermeasures are available to increase driver and pedestrian safety. Below are some safety countermeasures for a specific highway design (Dewar & Olson, 2007; SAE J2944, 2015):

- wider medians: less headlight glare; reduced risk of head-on collisions
- fewer intersections and longer spacing between them: reduced cognitive load and fewer decisions to be made by drivers; less frequent lane changes; less speed variance
- reduction of roadside vegetation (e.g., trees, bushes, tall grass): increased visibility at curves and intersections

Traffic Control Devices

The purpose of traffic control devices (e.g., signals, signs, pavement markings) is to provide information to regulate, warn, and guide road users. The devices must fulfill some information need, command attention, convey a clear and simple meaning, command respect, and give adequate time for a proper response. Criteria for the design of effective traffic control devices include (Summala & Näätänen, 1974; Dewar & Olson, 2007):

- higher conspicuity: more likely to attract drivers' attention
- longer legibility distance: earlier detection, allowing longer reaction times for drivers
- clearer glance legibility and increased ease of understanding: easier to read and discern

METHOD

Simulated Testing

Simulation studies involve "driving" a mockup of a real vehicle inside a laboratory. This is achieved by projecting a computer-generated driving scene on a screen in front of the car and by having the driver control the apparent movement of the scene via the vehicle's pedals and steering wheel. Various equipment (e.g., eye tracker, cameras, etc.) can be instrumented in the simulator to collect drivers' physiological and behavioral data and traffic environment details.

There are different reasons why a study can best be conducted in a simulator. Some situations are dangerous to study in a controlled fashion on the road and are difficult to replicate in a valid manner in a rudimentary laboratory test. Some situations occur only rarely on the road and it is therefore expensive to collect sufficient data. Simulation studies are also uniquely applicable for those situations that do not yet exist in the real world, such as those involving innovative safety devices. Experiments using a driving simulator are time-efficient, repeatable, controllable, and do not expose subjects to the risk of real injury in a crash.

Field Testing

To collect driving data in real traffic environments, or to verify the effectiveness of developed systems, or to develop further techniques for autonomous driving, an instrumented vehicle is needed. Equipment that can be instrumented on the vehicle mainly includes: video recorders to collect front-road images, side images, driver images, etc.; radar or ultrasonic sensors for distance detection; and global positioning systems (GPS) for location and time information. In the USA, the NHTSA conducted a 100-instrumented-car naturalistic driving study (Guo & Fang, 2013). Driver behavior, traffic environment, and vehicle performance data in accidents and near-accident incidents were collected for analysis. To date, naturalistic driving studies and driving simulator studies have become powerful tools in analyzing driving behavior, assessing effectiveness, and identifying problems in intelligent transportation systems.

Physiological Testing

Apparatus commonly used in simulators or vehicle studies includes eye trackers and physiological sensing systems (e.g., electroencephalogram (EEG), electrocardiogram (ECG), galvanic skin response (GSR)). Eye trackers measure a driver's gaze point (where he/she is looking), blinks, eye closures, pupil variation, and even head motions. They can also be used to detect driver fatigue and distraction. EEG and ECG measure the electrical activities of the brain and the heart, respectively, while GSR measures the electrical resistance between two electrodes attached to the skin. These physiological signals strongly indicate changes in driver status while driving. For example, the heart rate of a driver will increase when irritation is induced by other drivers' rage behavior or when encountering traffic jams (Denson et al., 2011).

APPLICATION

Applications in Vehicle

The center high-mounted stop lamp (CHMSL) is a truly successful application in the history of human factors development in surface transportation. Research in the 1970 and 1980s found that the inclusion of an additional central brake lamp in a location closer to the line of sight of drivers directly behind a vehicle reduces the frequency and severity of rear-end accidents, alerting the following driver with a more visible warning of braking by adding another light source. The effectiveness of the CHMSL design found that cars were 17% less likely to be hit from the rear, preventing chain collisions of three or more vehicles, and preventing an estimated 126,000 accidents, avoiding 80,000 non-fatal injuries, and reducing total property damage by approximately $1 million (Sanders & McCormick, 1993). From the success of the CHMSL, other improvements were proposed to provide additional cues (e.g., a flashing brake light to warn of emergency deceleration of the lead vehicle) to warn drivers of impending dangers with shorter perception-response time (Li et al., 2014).

annually?

Research on automotive human factors reached made a turning point in 1990 with the introduction of intelligent transportation systems (ITS). ITS apply advanced technologies to resolving the problems in surface transportation. The main objective of ITS is to improve travel efficiency and mobility, enhance safety, and conserve energy. Human factors considerations play an important role in ITS development and application, from the design of in-vehicle navigation displays and controls, to the safe and efficient transition from manual to automatic control in highly automated systems (Hoogendoorn et al., 2014; Lu et al., 2016). For example, the advanced traveler information system (ATIS) provides real-time in-vehicle information to drivers regarding route guidance and navigation, entertainment services, and hazard warnings. A human factors challenge in the design of these in-vehicle displays and controls may be of concern. For instance, various in-vehicle functions and the extensive diversity of the driving population make this particularly challenging. In complex driving situations, multiple warning signals may occur simultaneously. In such cases, the driver may become confused and be unable to respond to the warnings or may not react appropriately. Therefore, how drivers can be kept informed of the driving situation with ATIS equipped should be addressed (Barfield & Dingus, 2014).

Applications in Road Infrastructures

Applications of human factors principles in road infrastructures include stop sign assist, signalized left-turn assist, red-light violation warning, curve speed warning, etc. (Richard et al., 2015). Signalized left-turn assist (SLTA) is a an example that is extremely useful at intersections with a high density of traffic participants. Thus, the decision to provide SLTA information should consider intersection characteristics and the safety record of the specific intersection. The SLTA information can be presented using a DII (driver-infrastructure interface), which is available to all drivers making a left turn. The display location should be integrated with locations where drivers would typically scan. If SLTA information is presented on DII and in-vehicle display simultaneously, the activation timing of both displays should be the same. The elements and symbols used on the displays should also be consistent with the allowable movements. Therefore, SLTA could be an effective countermeasure to assist in the coordination of vehicle-to-infrastructure (V2I) and vehicle-to-vehicle (V2V) displays and messages to safely support driver information needs and decision-making.

Applications in Policy Making

HFE has helped policy makers to define/change violation criteria, implement effective driver education, and regulate drivers' fitness-to-drive for years. Citing technology's continual progress, lawmakers in California just expanded the state's existing law of prohibiting drivers from texting behind the wheel to all forms of smart device operations unless the device can be mounted on the dashboard/windshield and can be activated by only one finger tap or swipe (Nichols, 2016). Human factors studies can support these legislations with solid evidence from investigating the best locations of cell phone placement in cars (Harwood et al., 2014) to addressing the impact of new forms of cell phone use while driving (Hassani et al., 2017). In another instance, qualified professional drivers can be screened by assessing their cognitive capabilities related to potential risky driving behaviors (Wang et al., 2016), which is very helpful in improving road safety since professional drivers spend much more time on the road.

FUTURE TRENDS

The Challenge of Autonomous Vehicles

Both NHTSA and the Society of Automotive Engineers (SAE) recommended a classification system based on six automation levels, ranging from none (fully operated by drivers) to fully automated driving (SAE, 2014). The current automation technology is still under development in level 2~3, and the fully autonomous vehicle that is 100 percent safe any time on any street in any weather condition is still a decade or more down the road (Truett, 2016). As the level of automation increases, it is important to keep in mind that humans will still remain essential for many tasks, especially at levels lower than full automation. Human factors will therefore play an essential role in the future of technological advances, where people and technology are being integrated more closely and more intensively than ever before. Human and automation will both participate in the control through some sort of partnership. What's the correct partnership? Who should own the final authority of control if the driver and computer disagree? How should the authority of control transfer between the driver and computer safely and smoothly? All these questions require addressing in the future development of autonomous vehicles.

Should this be 2–3?

The Role of Operators

In current surface transportation systems, drivers are in charge of all activities including strategic planning, tactical maneuvering, and vehicle operation, as well as maintaining situation awareness and engaging in secondary tasks (Li et al., 2017). However, in future well-developed transportation systems, drivers will be relieved from most of these activities. Tactical maneuvering and vehicle operation can be realized by employing intelligent control algorithms and mechanical systems. Environment perception will be replaced by multiple sensors functioning as a human's vision system, and situation awareness will be achieved through the

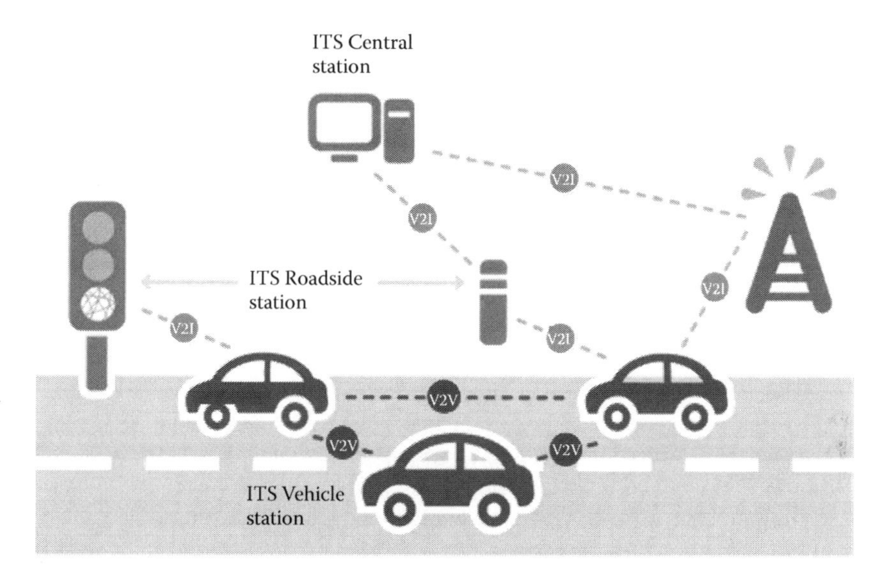

FIGURE 10.3 A simple framework for future ITS.

analysis of intelligent algorithms. Drivers will still be charge of the strategic planning and secondary tasks, for example, an email about work. They will manage the information from the driving vehicle, other traffic participants, and traffic management systems or road infrastructures to decide on an optimized plan, and then guide the vehicle to the destination. The role of drivers will gradually change from a perceptual-motor machine to a manage-plan employer, but the job, responsibility, and challenges of this employer need to be redefined in the near future.

The Fusion of All

A simple framework of a future ITS is shown in Figure 10.3. The vehicles are connected via vehicle-to-vehicle (V2V) technology, and the vehicles are connected with roadside infrastructures or a management center via vehicle-to-infrastructure (V2I) technology. Considering the characteristics of the driver (e.g., gender, age) and the surrounding traffic situation (e.g., highway or city street), the best way for a vehicle to communicate with another one or with infrastructures needs to be figured out to improve driving safety. Also, the previously mentioned human factors issues need to be considered for effective communications between the driver and the vehicle.

CONCLUSION

This chapter clearly demonstrates how HFE helps to improve safety, comfort, and system performance of surface transportation systems. From exploring physical and psychological individual differences, the causes and results of violated behaviors, and drivers' visual and cognitive limits, HFE contributes greatly to reducing accidents and failures caused by human and system errors, as well as enhancing human-friendly designs of in-vehicle devices and roadway elements. Along with rapid progress in vehicle intelligence, HFE will still play an important role and face many unknown challenges, since automation may relieve human operators from tremendous tail-end physical and low-level cognitive jobs, but increase head-end supervising and high-level cognitive jobs simultaneously.

KEY TERMS

Human factors and ergonomics (HFE). Consideration of HFE in design helps to reduce human error and improve performance in surface transportation.

Impaired driving. Driver fatigue, distraction, and driving under the influence (DUI) are the most critical challenges transportation safety faces to today.

Individual differences. Driver age, gender, experience, and driving style play important roles in transportation safety management.

Situation awareness. Involves being aware of the traffic environment, understanding what is happening, and knowing how to deal with the perceived information to keep safe in the near future, which can be degraded by insufficient illumination, monotonous roads, unclear in-vehicle information, driver fatigue, inattention, etc.

Vision and cognition. Driver visual and cognitive capacities should always be considered in the design of transportation systems.

Workload. Either too high or too low a workload can degrade drivers' situation awareness; thus, workload should be kept at an appropriate level for safety concerns.

Vehicle and road/infrastructure design. Sound HFE principles contribute to road safety improvement and driver acceptance.

Future surface transportation systems. As man-machine co-driving is the future, HFE must develop alongside trends in the development of autonomous vehicles and intelligent transportation systems (ITS).

REFERENCES

Anstey, K. J., Wood, J., Lord, S., and Walker, J. G. (2005). Cognitive, sensory and physical factors enabling driving safety in older adults. *Clinical Psychology Review* 25(1), 45–65.

Barfield, W. and Dingus, T. A. (2014). *Human Factors in Intelligent Transportation Systems*. Psychology Press.

Beiker, S. A. (2012). Legal aspects of autonomous driving. *Santa Clara Law Review* 52(4), 1145–1156.

Bengler, K., Dietmayer, K., Farber, B., Maurer, M., Stiller, C., and Winner, H. (2014). Three decades of driver assistance systems: Review and future perspectives. *IEEE Intelligent Transportation Systems Magazine* 6(4), 6–22.

Birrell, S. A. and Fowkes, M. (2014). Glance behaviours when using an in-vehicle smart driving aid: A real-world, on-road driving study. *Transportation Research Part F: Traffic Psychology and Behaviour* 22, 113–125.

Cheng, S. Y. and Trivedi, M. M. (2010). Vision-based infotainment user determination by hand recognition for driver assistance. *IEEE Transactions on Intelligent Transportation Systems* 11(3), 759–764.

Cicchino, J. B. and McCartt, A. T. (2015). Critical older driver errors in a national sample of serious US crashes. *Accident Analysis & Prevention* 80, 211–219.

Compton, R. and Berning, A. (2009). *2007 National Roadside Survey of Alcohol and Drug Use by Drivers: Alcohol Results*. NHTSA Report DOT HS 811 175. US Department of Transportation, Washington, DC.

Denson, T. F., Grisham, J. R., and Moulds, M. L. (2011). Cognitive reappraisal increases heart rate variability in response to an anger provocation. *Motivation and Emotion* 35(1), 14–22.

Dewar, R. E. and Olson, P. L. (2007). *Human Factors in Traffic Safety*. Tucson, AZ: Lawyers & Judges Publishing Company.

Endsley, M. R. (1995). Toward a theory of situation awareness in dynamic systems. *Human Factors* 37(1), 32–64.

Farah, H., Musicant, O., Shimshoni, Y., Toledo, T., Grimberg, E., Omer, H., and Lotan, T. (2014). Can providing feedback on driving behavior and training on parental vigilant care affect male teen drivers and their parents? *Accident Analysis & Prevention* 69, 62–70.

Forsman, P. M., Vila, B. J., Short, R. A., Mott, C. G., and Van Dongen, H. P. (2013). Efficient driver drowsiness detection at moderate levels of drowsiness. *Accident Analysis & Prevention* 50, 341–350.

Fu, R., Wang, H., and Zhao, W. (2016). Dynamic driver fatigue detection using hidden Markov model in real driving condition. *Expert Systems with Applications* 63, 397–411.

Guo, F. and Fang, Y. (2013). Individual driver risk assessment using naturalistic driving data. *Accident Analysis & Prevention* 61, 3–9.

Hancock, P. A. and Parasuraman, R. (1993). Human factors and safety in the design of intelligent vehicle-highway systems (IVHS). *Journal of Safety Research* 23(4), 181–198.

Harwood, L., Klauer, S., and Doerzaph, Z. (2014). Cell Phone Resting Locations: Use of the 100-Car Naturalistic Driving Study to Determine the Most Frequent In-Vehicle Cell Phone Placement and Containers. *Transportation Research Record* 2434, 63–71.

Hassani, S., Kelly, E. H., Smith, J., Thorpe, S., Sozzer, F. H., Atchley, P., and Vogel, L. C. (2017). Preventing distracted driving among college students: Addressing smartphone use. *Accident Analysis & Prevention* 99, 297–305.

Hoogendoorn, R., van Arem, B., and Hoogendoorn, S. (2014). Automated driving, traffic flow efficiency, and human factors: Literature review. *Transportation Research Record* 2422, 113–120.

Horrey, W. J. and Wickens, C. D. (2006). Examining the impact of cell phone conversations on driving using meta-analytic techniques. *Human Factors* 48(1), 196–205.

Horswill, M. S., Falconer, E. K., Pachana, N. A., Wetton, M., and Hill, A. (2015). The longer-term effects of a brief hazard perception training intervention in older drivers. *Psychology and Aging* 30(1), 62.

Human Factors and Ergonomics Society (2017). Surface Transportation Technical Group, www.hfes.org/web/TechnicalGroups/STTG.pdf, accessed January 28, 2017.

Jo, J., Lee, S. J., Park, K. R., Kim, I. J., and Kim, J. (2014). Detecting driver drowsiness using feature-level fusion and user-specific classification. *Expert Systems with Applications* 41(4), 1139–1152.

Lee, J. D. and See, K. A. (2004). Trust in automation: Designing for appropriate reliance. *Human Factors* 46(1), 50–80.

Li, G., Li, S. E., Cheng, B., and Green, P. (2017). Estimation of driving style in naturalistic highway traffic using maneuver transition probabilities. *Transportation Research Part C: Emerging Technologies* 74, 113–125.

Li, G., Wang, W., Li, S. E., Cheng, B., and Green, P. (2014). Effectiveness of Flashing Brake and Hazard Systems in Avoiding Rear-End Crashes. *Advances in Mechanical Engineering* 1–12.

Liang, Y. and Lee, J. D. (2014). A hybrid Bayesian Network approach to detect driver cognitive distraction. *Transportation Research Part C: Emerging Technologies* 38, 146–155.

Liao, Y., Li, S. E., Wang, W., Wang, Y., Li, G., and Cheng, B. (2016). Detection of driver cognitive distraction: A comparison study of stop-controlled intersection and speed-limited highway. *IEEE Transactions on Intelligent Transportation Systems* 17(6), 1628–1637.

Liu, T., Yang, Y., Huang, G. B., Yeo, Y. K., and Lin, Z. (2016). Driver distraction detection using semi-supervised machine learning. *IEEE Transactions on Intelligent Transportation Systems* 17(4), 1108–1120.

Lu, Z., Happee, R., Cabrall, C. D., Kyriakidis, M., and de Winter, J. C. (2016). Human factors of transitions in automated driving: A general framework and literature survey. *Transportation Research Part F: Traffic Psychology and Behaviour* 43, 183–198.

Movig, K. L. L., Mathijssen, M. P. M., Nagel, P. H. A., van Egmond, T., de Gier, J. J., Leufkens, H. G. M. and Egberts, A. C. G. (2004). Psychoactive substance use and the risk of motor vehicle accidents. *Accident Analysis and Prevention* 36, 631–636.

NHTSA (2012). *Traffic Safety Facts 2010: A compilation of motor vehicle crash data from the Fatality Analysis Reporting System and the General Estimates System*. Tech. Rep. DOT HS 811 659. US Department of Transportation, Washington, DC.

NHTSA (2013). *Alcohol-Impaired Driving: Traffic Safety Facts.* Tech. Rep. DOT HS 811 870. US Department of Transportation, Washington, DC.

NHTSA (2015). *Speeding Traffic Safety Facts Sheet: 2013 Data.* Tech. Rep. DOT HS 812 162, US Department of Transportation, Washington, DC.

NHTSA (2017). Risky driving, www.nhtsa.gov/risky-driving, accessed February 3, 2017.

Nichols, C. (2016). New California Law: Keep Your Hands Off Your Smartphone While Driving, www.capradio.org/articles/2016/12/28/new-california-law-keep-your-hands-off-your-smartphone-while-driving/, accessed February 8, 2017.

Owsley, C., Ball, K., Sloane, M. E., Roenker, D. L., and Bruni, J. R. (1991). Visual/cognitive correlates of vehicle accidents in older drivers. *Psychology and Aging* 6(3), 403–415.

Ranney, T. A., Mazzae, E., Garrott, R., and Goodman M. J. (2000). NHTSA driver distraction research: Past, present, and future. In Proceedings of Driver Distraction Internet Forum, 1–11.

Rayner, K. (1998). Eye movements in reading and information processing: 20 years of research. *Psychological Bulletin* 124(3), 372.

Reimer, B. and Mehler, B. (2011). The impact of cognitive workload on physiological arousal in young adult drivers: A field study and simulation validation. *Ergonomics* 54(10), 932–942.

Richard, C. M., Morgan, J. F., Bacon, L. P., Graving, J. S., Divekar, G., and Lichty, M. G. (2015). Multiple sources of safety information from v2v and v2i: Redundancy, decision making, and trust—Safety message design report. Tech. Rep. FHWA-HRT-15-007. Federal Highway Administration, McLean, VA.

Sebego, M., Naumann, R. B., Rudd, R. A., Voetsch, K., Dellinger, A. M., and Ndlovu, C. (2014). The impact of alcohol and road traffic policies on crash rates in Botswana, 2004–2011: A time-series analysis. *Accident Analysis & Prevention* 70, 33–39.

Sheridan, T. B. (2004). Driver distraction from a control theory perspective. *Human Factors* 46(4), 587–599.

Shinar, D. (2007). *Traffic Safety and Human Behavior.* Amsterdam: Elsevier.

Society of Automotive Engineers (2014). Automated Driving: Levels of Driving Automation Are Defined in New SAE International Standard J3016, www.sae.org/misc/pdfs/automated_driving.pdf, accessed February 8, 2017.

Society of Automotive Engineers (2015). Operational Definitions of Driving Performance Measures and Statistics (Recommended Practice J2944). Warrendale: Society of Automotive Engineers.

Sonnleitner, A., Treder, M. S., Simon, M., Willmann, S., Ewald, A., Buchner, A., and Schrauf, M. (2014). EEG alpha spindles and prolonged brake reaction times during auditory distraction in an on-road driving study. *Accident Analysis & Prevention* 62, 110–118.

Stanton, N. A., Young, M., and McCaulder, B. (1997). Drive-by-wire: The case of driver workload and reclaiming control with adaptive cruise control. *Safety Science* 27(2/3), 149–159.

Strand, M. C., Gjerde, H., and Mørland, J. (2016). Driving under the influence of non-alcohol drugs–an update. Part II: Experimental studies. Forensic Science Review 28(2), 79–101.

Stutts, J. C., Wilkins, J. W., Osberg, J. S., and Vaughn, B. V. (2003). Driver risk factors for sleep-related crashes. *Accident Analysis & Prevention* 35(3), 321–331.

Summala, H. and Näätänen, R. (1974). Perception of highway traffic signs and motivation. Journal of Safety Research 11, 150–154.

Tango, F. and Botta, M. (2013). Real-time detection system of driver distraction using machine learning. *IEEE Transactions on Intelligent Transportation Systems* 14(2), 894–905.

Tao, D., Zhang, R., and Qu, X. (2017). The role of personality traits and driving experience in self-reported risky driving behaviors and accident risk among Chinese drivers. *Accident Analysis & Prevention* 99, 228–235.

Truett R. (2016). Fully Autonomous Vehicles Won't Arrive for a Long Time, www.autonews.com/article/20161010/OEM06/310109972/fully-autonomous-vehicles-wont-arrive-for-a-long-time, accessed February 8, 2017.

Vahidi, A. and Eskandarian, A. (2003). Research advances in intelligent collision avoidance and adaptive cruise control. *IEEE Transactions on Intelligent Transportation Systems* 4(3), 143–153.

Wang, H., Mo, X., Wang, Y., Liu, R., Qiu, P., and Dai, J. (2016). Assessing Chinese coach drivers' fitness to drive: The development of a toolkit based on cognition measurements. *Accident Analysis & Prevention* 95, 395–404.

Wickens, C., Lee, J., and Liu, Y. (1998). *Introduction to Human Factors Engineering*. Upper Saddle River, NJ: Prentice-Hall.

World Health Organization. Global status report on road safety 2015, www.who.int/violence_injury_prevention/road_safety_status/2015/en, accessed January 28, 2017.

Zhang, G., Yau, K. K., and Zhang, X. (2014). Analyzing fault and severity in pedestrian–motor vehicle accidents in China. *Accident Analysis & Prevention* 73, 141–150.

Case Study 10
Human Factors in Police Training

Mostafa Aldah

THE DUBAI POLICE

Road users across the world take risks when using transport systems. While these risks vary by region and transportation mode (Aldah, 2010), all share similar outcomes when accidents occur. Efforts to control these risks that examine every part of the transportation system include the four "E"s: education and training of road users; engineering safe environments for travel; enforcement of laws and regulations ensuring a common system is followed by everyone; and providing emergency services for when things go wrong (Mace et al., 2001).

The police's role varies in many jurisdictions, but among its many duties it typically involves the enforcement of the law and the detection of illegal behavior. The Dubai Police manages crime scenes and accidents, as well as running a quick intervention search-and-rescue services that intervenes in cases of entrapment and reported missing persons. It is also responsible for air ambulance services when land transport is not viable.

The Police is a service provider and is on call 24 hours a day, 365 days a year. It is imperative that the response provided in all cases, and especially in emergencies, is optimized. Dubai is the most cosmopolitan city in the world, with 83% of the population foreign-born and 90% being expatriates from more than 200 nationalities. This unique environment of varied and changing population drives Dubai Police officers to constantly be prepared with the knowledge, skills, and attitudes required to fulfill their duties. It is common for other government authorities in the United Arab Emirates to call on the knowledge and experience of Dubai Police experts in special fields to share their best practices.

DUBAI POLICE TRAINING

One method the Dubai Police uses to sustain high standards in its operations is by heavily investing in its employees via personnel training, and providing them with the latest technologies to assist them in performing their duties. As human capital is a key asset to the force, training is part of the annual appraisal of employees and continuing education is strongly encouraged. The Dubai Police has a high number of officers with advanced degrees (M.S., Ph.D.) in various fields of current and future interest. The training of personnel working in the areas of traffic patrol and safe driving has been ongoing for decades, with the establishment of a dedicated Traffic Institute in 2009 to unify the provision of training across the force in this area. The Traffic Institute aims to offer training programs to improve human-machine interaction with police equipment, and training for excellent customer experience in human-human interaction.

Human-machine interaction is critical when operating advanced equipment and interacting with emergency services. The equipment used in traffic patrols varies and is constantly updated due to improved technology systems. Police officers are constantly required to acquire a range of knowledge and skills to operate this old and new equipment, regardless of the age or make of the patrol car they drive. For instance, there are electronic in-car and mobile systems for reading number plates (ANPR; automated number plate recognition) and issuing traffic fines. Given the wide range of age and education levels, all Police officers must go through the same training courses that fit their job description. Officers are given on-the-job training along with printed reference guides for operating data systems. Officers must be proficient in operating the Police band handheld radio and the emergency warning systems (flashing light bar, beacons, sirens, and horns) fitted to cars and motorcycles.

Safe driving training is provided to Police officers to preserve property, lengthen the service life of vehicles, maximize the longevity of components to ensure the maximum availability of vehicles, and reduce costs to the government. Driver training takes place in class and on the road. Case study scenarios and video recordings of incidents are also used to demonstrate how things have gone wrong in the past and how to minimize risk in dealing with such situations. Assessment is provided with experienced instructors along with a driver record monitoring system to offer a long-term evaluation of training effectiveness. The Dubai Police car fleet has historically used famous car models like Lamborghinis and a host of twenty-first-century "supercars" to reflect its modern image. This helps to break barriers between the Police and the public. These vehicles require special training to ensure they are used safely and efficiently and this is provided on a closed track dedicated for this purpose.

In addition, human-human interaction training is given to all Police officers. Following the visionary leadership of the UAE in making happiness a core mission to all government departments, all customer-facing police employees are empowered to provide the best customer service when interacting with the general public. Training is performed to ensure a uniform high level of customer satisfaction, which is measured independently and centrally by the local government. Current legislation is actively used in training officers to set the foundation for operations. The specific language terms are standardized and embedded in experiential learning. Successful methods and examples from service industries are illustrated with regard to the local customs and conservative society of the region to ensure best practices while preserving the national identity.

TRAINING AND HUMAN FACTORS

Training programs are taught using blended learning techniques of in-class instruction, virtual training and assessment, and live exercises. Most of the virtual training has been developed in-house, starting with a serious game-based learning environment (Binsubaih et al., 2006, 2009) that takes place in a purpose-built laboratory with dedicated workstations. The video game simulation programs mimic real-life situations and many are based on real cases that were handled by the police. The virtual training includes the whole journey of an accident investigator from the point of dispatch to the scene of the collision, including managing the scene and following the required steps in the right order.

The evaluation of training is integral to assessing the effectiveness of any program. When the trainee goes through the virtual training program, each action is scored for evaluation. Dubai Police has a dedicated department that oversees the content and effectiveness of training programs throughout all divisions. Industry-leading systems such as the Kirkpatrick (2016) model are used to evaluate training provision to ensure fitness for purpose and maximum effectiveness.

REFERENCES

Aldah, M. K. (2010). Causes and consequences of road traffic crashes in Dubai, UAE and strategies for injury reduction. Ph.D. Thesis. Loughborough University.

Binsubaih, A., Maddock, S., and Romano, D. (2006). A serious game for traffic accident investigators. *Interactive Technology and Smart Education* 3(4), pp. 329–346, https://doi.org/10.1108/17415650680000071.

BinSubaih, A., Maddock, S., and Romano, D. (2009). Developing a Serious Game for Police Training. In R. Ferdig (ed.), *Handbook of Research on Effective Electronic Gaming in Education* (pp. 451–477). Hershey, PA: IGI Global. doi:10.4018/978-1-59904-808-6.ch026.

Kirkpatrick, J. D. and Kirkpatrick, W. K. (2016). *Kirkpatrick's Four Levels of Training Evaluation.* Virginia: Association for Talent Development. ISBN: 978-1607280088.

Mace, S. E., Gerardi, M. J., Dietrich, A. M., Knazik, S. R., Mulligan-Smith, D., Sweeney, R. L., and Warden, C. R. (2001). Injury prevention and control in children.

Annals of Emergency Medicine 38(4), 405–414. http://dx.doi.org/10.1067/mem.2001.115882.

11 Aerospace Human Factors

Barrett S. Caldwell

CONTENTS

INTRODUCTION

The history of human performance in the aerospace setting is well into its third century of research and innovation. Since the 1780s, researchers have been concerned about issues of aviation safety, including the ability to survive and perform complex tasks in extreme environment while controlling aircraft (Grant, 2007). Many advances in aerospace human factors are tied to military advances, which also include concerns about training, awareness of changing events and threats, and human capabilities in even more dangerous environmental conditions. Much of the history of human factors and ergonomics (HFE) dates to the challenges of standard information presentation and stimulus-response compatibility within and across aircraft controls in World War II ("Of Men and Machines," 1962; Casey, 1993). In an early demonstration flight of the Boeing model 299 in October 1935, the aircraft crashed, with no obvious technical breakdown or engineering failure (Meilinger, 2004). This accident helped highlight the important role of human factors researchers in improving human safety, performance, and capability in the aerospace environment.

As a complex sociotechnical system, there are several significant HFE components to aerospace systems engineering design encompassing technical, human, and organizational aspects (NASA, 2007). The applications of human factors research in aerospace include emphasis on multiple aspects of the *vehicle*, the *environment*, the *task*, and the *human*. As a result, human factors researchers frequently represent combinations of engineering, psychology, and medical domains. "Aviating, navigating, and communicating" are primary task requirements of aerospace human performance, requiring effective presentation of environmental information and vehicle control interfaces. As a result, aerospace HFE has been a long-standing domain to address applied research and technology development issues in human performance, human error, teamwork, and human-technology interactions in safety-critical systems.

Successful human performance in extreme environments starts with human survival. As altitude increases, threats to human physiology increase. Limited availability of ambient oxygen, pressure, and temperature all represent potentially fatal issues that need to be considered when designing a vehicle. The below sections discuss examples of human factors research related to *vehicle* design, the physical *environment*, critical aviation *tasks*, and performance of the *humans as individuals and teams involved.*

FUNDAMENTALS

Vehicle

Aircraft and spacecraft must operate in a challenging physical environment, and still safely protect the health and functional capacity of human crew and passengers. Vehicle design in aerospace engineering has changed substantially in the past 25 years, based on advances in increasingly more powerful computers to enable analysis and modeling of mechanical component design, fluid dynamics, and vehicle structures (Page and Olsen, 2015). Research in the design of materials and engines for advanced supersonic aircraft continues to address the challenge of vehicle response to high stresses, temperatures, and power requirements (Hudi and Edi, 2013). However, more research in the design of civil aircraft (including passenger airliners) focuses on safety, efficiency, and environmental quality, such as reducing engine noise (Lockhard and Lilley, 2004; Papamoschou, 2004) and greenhouse emissions (Melis, Silva, and Sylvestre, 2017), and even air quality within the aircraft (Spicer et al., 2004; Strøm-Tejsen, Wyon, Lagercrantz, and Fang, 2007). As a result, human factors research has provided design engineers with important data regarding human performance affected by environmental quality (air and temperature), safety, and even passenger loads for commercial aircraft (Melis, Silva, and Clothier, 2015).

From an engineering perspective, aircraft and spacecraft are known to have very limited safety margins (Grant, 2007). While buildings, bridges, or other engineering structures may have a margin of safety of 2.0–5.0 (the structure is built to withstand loads two to five times larger than expected), aircraft and spacecraft may have margins of only 1.4–1.5 (the structure can only withstand loads 40–50% larger than expected) (Modlin and Zipay, 2014). Vehicle design for very light jet aircraft does require tradeoffs between material selection and anthropometry of passenger and crew size affecting the number of persons that can be carried; differences in population size between countries may influence vehicle design options (Dahari, et al., 2010). Human factors research may actually be more influential in the early stages of vehicle design, where engineers are using computer-aided design tools to test design concepts that can reduce weight without compromising structural strength or passenger/crew safety (Page and Olsen, 2015). However, larger passengers mean heavier aircraft in flight and greater fuel costs and resulting impacts on greenhouse gas emissions, a constraint that cannot be addressed just by changing aircraft materials or reducing safety margins (Melis, Silva, and Clothier, 2015; Melis, Silva, and Sylvestre, 2017).

From a vehicle design perspective, a pilot unable to perform a critical task due to insufficient time, strength, or information represents not a human error, but a breakdown in effective engineering systems design. Prior to World War II, engineering design of aircraft instruments (including presentation of altitude, direction, speed, and engine conditions) was not consistent in where, or how, instruments would show critical information essential for safe and effective flight to the pilot (Landry, 2009). Human factors research created a standardized instrument design across aircraft as it became evident that pilots became confused moving from one aircraft type to another (Casey, 1993; Grant, 2007).

Environment

Documented cases of human survival at temperatures and pressures equivalent to commercial aviation altitudes (7000–10,000 meters) or above is very limited. Although French balloonists during a flight in 1875 carried supplemental oxygen, two of three died due to lack of oxygen (Harding, 2002). Human factors researchers studying physiology have determined that cognitive performance may be unpredictable and erratic in hypoxic conditions (Fowler, Paul, Porlier, Elcombe, and Taylor, 1985; Fowler, Elcombe, Kelso, and Porlier, 1987). Aircraft can vary greatly in cabin pressure during flight, which can result in dangerous reductions in oxygen levels for passengers with health conditions (Gendreau and DeJohn, 2002; Humphreys, Deyermond, Bali, Stevenson, and Fee, 2005).

Human factors researchers also study the effects of blood flow to the brain in high-performance aircraft and spacecraft. With accelerations in the vertical direction of 6–12 times the force of gravity ("G force"), consciousness is reduced, limited, and shut down when blood is forced *away from* the head ("black-out") or *towards* the head ("red-out") (Seedhouse, 2012). Another challenge in the aerospace environment is the way in which information about direction (whether the vehicle is ascending or descending, or flying straight or turning) can be confusing or unavailable to the pilot due to clouds or darkness. Important human factors work has addressed the causes and possible reduction of pilot confusion or motion sickness that occur in aerospace settings (see also David, 2005).

The vestibular and visual systems are challenged both due to missing perceptual cues (e.g., atmospheric cues to distance), and to differences in movement in microgravity and lunar gravity (where walking becomes a type of "kangaroo hop") (Cernan and Davis, 1999; Mindell, 2008). The longer-term exposure to these environments may also have impacts of bone loss and muscle degradation as time in space increases, where the level of loss of calcium and muscle tone begins to simulate the impacts of old-age symptoms of osteoporosis and extended bed stay (Kluger, 2016).

Therefore, both pilots and astronauts may have difficulties with true motion in three dimensions. Sensory confusion between visual, vestibular (sense of movement), and kinesthetic senses (body location and orientation), particularly in cloudy or nighttime weather, can give the pilot divergent cues about the true orientation of "straight and level" flight (Grant, 2007). Even with accelerations less than those that create acute physiological effects, confusing cues from ground features (e.g., slowly rising terrains) or a lack of any cues (e.g., heavy overcast, fog, nighttime landscapes over water) can lead to perceptual disturbances that can lead to poor interpretation of aircraft orientation or direction (Landry, 2009). The spaceflight environment also highlights these problems of interpretation, both because of confusing distance cues and the lack of gravity providing orientation information to the vestibular system in the inner ear (Larimer, 2017). Although definitive population estimates are difficult to confirm, approximately 40–50% of spacefarers experience some type of "space adaptation sickness" (SAS) due to disorientation; those who do not receive the conflicting information from their vestibular system, due to conditions such as meningitis, do not experience motion sickness or SAS (Larimer, 2017).

Task

While operational checklists and procedures are now common across a range of technical domains, the pilot's checklist has become one of the most iconic technical job performance aids (Fosshage, 2014). The pilot's checklist is the result of human factors research following the demonstration accident of the Boeing 299 (Meilinger, 2004). Increases in aircraft vehicle capability and complexity had resulted in engineering designs that were too complicated for a pilot to easily manage with the demands for rapid situation awareness and decision-making during combat or reconnaissance operations (Fosshage, 2014; Gawande, 2010). Therefore, standardized operating procedures for pilots were developed to help make sense of task requirements for aviation (and later space) environment, and to reduce the need for memory or intuition in decision-making and problem solving (Fosshage, 2014; Smilie, 1985).

In parallel with the development of technical procedures for aircraft pilots, the need has grown for clear, effective, and reliable procedures to support aircraft mechanics. The development of simplified English (SE) was created in response to challenges in international aviation contexts, where it could not be validated that mechanics could reliably read and understand maintenance documents written in English (STEMG, 2015). Aircraft manufacturers such as Boeing developed SE as a standard application of English with a drastically reduced vocabulary (approximately 1000 words) and where each word has one, unambiguous meaning: these factors increase ease of learning and clarity of meaning for non-English speakers

(Boeing IPM, 2016). The use of SE also increases understanding among fluent as well as non-fluent English speakers and drastically decreases the costs of translation (automated or human-expert) for a variety of contexts, including computer software and hardware interfaces (Boeing IPM, 2016; Mills and Caldwell, 1997).

Human

Human factors research in the aerospace context emphasizes both the physical and cognitive capabilities and limits of pilots/astronauts, passengers, and crew of aircraft and spacecraft. A focus on physical human factors research will highlight the interaction of the human body with the design of the aircraft. As discussed above, medical concerns for patient health include an awareness of the effective pressure within the aircraft cabin and the risk of reduced oxygen levels, particularly in passengers with health conditions (Gendreau and DeJohn, 2002; Humphreys, Deyermond, Bali, Stevenson, and Fee, 2005). The layout of aircraft seats, controls, and other features are also affected by attention to the physical sizes and movement patterns of passengers and crew members. Minimum size limits for aircraft seating, and maximum reach distances to pilots' controls, are defined through anthropometry research data collected for different populations. Recently, there has been greater recognition of the need to incorporate population-specific anthropometry data in the design of aircraft, or the procurement of aircraft already built, so that they are suitable for the user population (Al Wardi, Jeevarathinam, and Al Sabei, 2016; Trudgill and Harrigan, 2016).

While anthropometry measures are critical for the physical design of systems for human users, to ensure they can use the equipment safely and effectively, cognitive human factors research emphasizes the learning, training, and skill associated with performing aerospace-relevant tasks. The general concept of situation awareness (SA), formulated by Endsley (1995, 1999) is based on a much older concept of providing information to support pilot decision-making, highlights the need for considering the human and machine together as components of a complex system (Grant, 2007; Landry, 2009). The human operates in both a physical and information environment, and must effectively process, analyze, and act upon changing conditions and information to complete tasks in a timely way. SA incorporates three major phases: perception of elements in the physical environment; comprehension of those elements for appropriate understanding of the current state of the environment; and projection of those elements into the future (Endsley, 1995). Physiological factors can affect all three of these phases of SA, as can instrument-based information influenced by human-machine interface design (Landry, 2009). Design of technology interfaces influences how well the first two phases of SA (perception and comprehension) can be transferred to the pilot/astronaut. Expertise and awareness of task goals and strategies tie the third phase of SA (projection) to pilot/astronaut use of controls and displays to achieve skilled performance.

The challenges of SA and the interaction of physical and cognitive factors exist for pilots that are in a traditional cockpit or when piloting a UAV/drone (Ayaz et al., 2012; Drury, Riek, and Rackliffe, 2006). Thus, the human-machine interfaces and sources of physical and perceptual information are different whether the pilot is in the aircraft or relying on remote, sensor-based information displays for flight information (Ayaz et al., 2012; Drury, Riek, and Rackliffe, 2006).Unfortunately, delays in information presentation via weather display technologies have been shown to substantially degrade pilot decision-making and increase the risk of fatal outcomes in cases of adverse weather (Caldwell et al., 2015).

Team Coordination

The impacts of human performance have been applied to both individual and team performance contexts. A challenge occurred in civil aviation in the 1960s, when advances in civil aviation demand and aircraft design began to uncover mismatches between military command structures and the need for shared situation awareness in the cockpit environment (Helmreich, Merritt, and Wilhelm, 1999). Several high-profile aviation accidents attributed to breakdowns in effective communication, function allocation, or situation awareness, especially during critical phases of flight.

Team coordination and effective performance is not simply an aggregation of individual performances, but distinct capabilities in shared and coordinated expertise (Caldwell, 1997; Garrett et al., 2006). Although confusion remains regarding "shared situation awareness" (all members of the team have a common understanding of the task environment and required actions) and "team situation awareness" (members of the team have complementary understandings of each other's skills, task responsibilities, and mutual dependencies), it is recognized that situation awareness in the team environment does involve understanding of other team members' performance strengths, domains of relative expertise, and communication patterns.

METHODS

Improvements in aerospace vehicles and systems are due to the effective use of applied human factors research methods. Human factors researchers apply both physical and cognitive data collection techniques. Examples of physical human factors methods include detailed measurements of physical features (anthropometry) that represent one traditional (but still essential) approach to ensuring that cockpit controls and aircraft maintenance components can be effectively used by humans of a range of heights and strength capabilities. Physiological research on the effects of oxygen deprivation or motion sickness helps to develop appropriate physical equipment and medications to help maintain effective cognitive functioning and task performance by pilots and astronauts. These data collection methods often involve studies of professional pilots or maintenance workers performing tasks in "mock up" or other simplified prototype task settings (including computer simulations of aerospace vehicles, environments and tasks). A variety of methods (including surveys, observations, and expert analyses) can be combined to distinguish relative contributions of engineering, training, and other factors that might increase the risk of accidents or system breakdowns. Kroemer and Grandjean (1997) provide a classic overview of occupational ergonomics methods within the domain of physical human factors. General descriptions of three important areas of complex cognitive methods are described below.

Task Analysis

Systems engineering design processes often start with the question, "What is the goal of this system, and what is the definition of is function?" A version of this question for human factors researcher is, "What are the tasks that the human needs to successfully complete in this task environment?" Breaking down overall system goals and functions into required human performance tasks is an important and very widespread initial analysis step. Task analysis is conducted both when developing new aerospace tasks and functions and when analyzing existing tasks for improvements in accuracy, performance, or safety.

Task analysis focuses on the situation and strategic task elements of what the person needs to do in a task environment. The task analysis method includes a structured breakdown of the task activities and goals, and how the pilot or other operator is expected to complete those activities (see, for example, Raby and Wickens, 1994). A hierarchical task analysis (HTA) is the most general way of creating a step-by-step description of how to achieve the goals of the task (Stanton, 2005; Stanton, Salmon, and Rafferty, 2013). Based on these descriptions, a detailed examination of each step should be completed, including a set of decomposition categories (breakdowns of how the task is structured, and how the human operator is expected to use relevant equipment). Example lists of categories are found in Stanton (2005), including required inputs and outputs, controls and displays used, and success criteria.

Another approach to task analysis is known as critical task analysis (CTA) and focuses on the risks of system failure or other negative outcomes if the task is not performed as designed or intended (Seamster, Redding, Cannon, Ryder, and Purcell, 1993). Both HTA and CTA include descriptions of important interactions between the person and equipment. CTA can also include highlights of both intended and unintended consequences of human-machine interactions that are compatible or inconsistent with patterns of human behavior.

Human Error Analysis and Performance-shaping Factors

Research investigating human error in aviation is directly related to the complex interaction between vehicle, environment, and task factors. As aircraft mature in design capability and more information is available about aerospace environmental conditions, human error has come to represent the greatest risk of adverse aviation events such as crashes or other sources of damage, whether leading to injury or death (Grant, 2007).

Early studies of human error (see Miller and Swain, 1987; Swain, 1964) focused on two types of performance breakdowns: errors of omission (the pilot neglected to do something that s/he should have done) and errors of commission (the pilot did something that s/he should not have done). Additional research helped to identify factors such as order (doing the right tasks in the wrong order) and timing (doing things too quickly or slowly). Earlier studies of sources of human error do not necessarily distinguish physical HFE limitations (a display is too difficult to see or a switch is too far away to reach), cognitive HFE concerns (the display is too confusing for the pilot to understand the proper response). In these examples, aircraft design flaws or organizational training breakdowns may "set the user up for failure," or create situations where correct performance is difficult (or even impossible) to achieve. Thus, human factors research addressing human error (see Reason, 1990) recognizes that several interacting issues may influence the likelihood of adverse outcomes. Modern, systems-level error research often includes a discussion of performance-shaping factors (PSF). A PSF is any set of concerns from design and engineering, environmental, organizational and social, and task performance perspectives that affect a human's ability to correctly complete required tasks (Kirwan, 1998).

Task analysis helps to define human performance constraints as one type of PSF. Task procedures (and the engineering systems to which these procedures are to be applied) must be developed with a recognition of the limitations of pilot/astronaut performing complex physical, technical, and task environment (Fosshage, 2014; Smilie, 1985). From a technical design perspective, a pilot/astronaut (or even mechanic or flight attendant) unable to perform a critical task due to insufficient time, strength, or information does not represent a human error but a breakdown in effective engineering systems design.

Risk Assessments

Risk assessment and root cause analysis tools such as failure modes and effects analysis (FMEA) use a combination of descriptive and expert-driven techniques to define what could go wrong in a situation. FMEA is a technique for determining potential risks and is often used when designing a new engineering system or process. As the name of the technique suggests, the FMEA method includes developing descriptions of what could go wrong and how (failure modes), and if that failure mode occurs, what would be the outcomes in the system (effects analysis).

FMEA is sometimes described as an "expert-driven" method, since a comprehensive examination of complex failure modes is most likely based on domain expertise in a specific application area rather than simply from general human factors knowledge. Thus, FMEA for the aerospace domain will be based on knowledge of the interactions between the human, environment, and task or vehicle. An additional aspect of the FMEA method is the development of a scaling of relative risk for different types of failure modes. Relative risk is not just a function of the overall likelihood of a problem, but also the danger inherent in a failure mode not being appropriately considered or mitigated in the engineering design process. FMEA risk scores are often determined by a multiplier of three distinct features: probability (likelihood that this failure mode could occur); severity (net impact on overall system performance if this failure were to occur); and detectability (the likelihood that this failure could exist or worsen without being noticed). Experts develop qualitative risk assessments often described as "high," "medium," or "low," with scores (such as 1-3-5 for least to most severe) on each of these features, with a total risk score based on multiplying the three scores. The scores themselves remain a relatively coarse qualitative evaluation, but since the FMEA process may generate dozens or even hundreds of risk assessments, this technique is a relatively effective way of providing a rough prioritization of risks and interventions.

Applications

It is important to recognize that mechanisms of flight catastrophes are no longer "single-point" failures of equipment or human decision-making, but complex interactions of design, maintenance, human performance, communications, and environmental conditions (Grant, 2007; Caldwell, 2008). Complex interactions of thousands of aircraft in a nation's airspace, multinational space stations with multiple mission operations centers, and numerous aircraft combined with unpiloted aerial vehicles (UAVs/drones) all represent "systems of systems" problems of human-technology interactions at multiple levels of analysis (DeLaurentis and Fry, 2008; Sindiy, DeLaurentis, and Caldwell, 2010).

Crew Resource Management

Improving communications and decision-making between pilots or between pilots and air traffic control is known as crew resource management (CRM). A challenge began to be noted in civil aviation in the 1960s, when advances in civil aviation demand and aircraft design began to uncover mismatches between military command structures and the need for shared situation awareness in the cockpit environment (Helmreich, Merritt, and Wilhelm, 1999). Analysis of many fatal aviation accidents indicated causes attributed to breakdowns in effective team coordination rather than technical failure (Helmreich and Merritt, 2000; Helmreich et al., 1999). In response to these accidents, CRM training (originally known as "cockpit resource management"; see Lauber, 1984) was developed to ensure better information flow and knowledge-sharing between crew members in the cockpit (Kanki, Helmreich, and Anca, 2010). Over time, improvements in communications and safe operations were seen to result from providing training programs to flight attendants as well as pilots, extending the CRM concept from "Cockpit" to "Crew," with a great expansion of applications (O'Connor et al., 2008; Kanki, Helmreich, and Anca, 2010).

Effective crew performance relies on a number of critical areas of knowledge, skills, and abilities (KSAs) that are part of CRM training. CRM training includes emphasis on KSAs including situation awareness, team communication and task coordination, decision-making, shared mental models, and workload management (Kanki, Helmreich, and Anca, 2010; Waller, Gupta, and Giambatista, 2004). Team-level coordination and leadership using CRM skills have been shown to improve performance specifically in non-routine situations (Waller, Gupta, and Giambatista, 2004).

Spaceflight Mission Operations

The relative complexity of spacecraft behavior and small error margins across all ranges of spaceflight operations has resulted in the development of ground-based "mission control" facilities designed to monitor and support spacecraft and spaceflight crews. The most famous of these is the NASA Mission Operations Control Center at Johnson Space Center in Houston, Texas. Unlike air traffic controllers, whose responsibilities are limited to directing the activities of aircraft pilots who control their own aircraft, spaceflight mission operations personnel monitor the energy, life support, and other aspects of the spacecraft itself. Teams of flight controllers, separated into functional specialties, coordinate system monitoring and troubleshooting in a distributed human-machine "supervisory coordination" model, led by a flight director and crew liaison known as "CAPCOM" (from "Capsule Communicator," based on the first US manned flights in space "capsules") (Caldwell, 2000, 2005).

Current and future spaceflight missions combine increasing levels of engineering complexity and computational power with human oversight and strategic planning to ensure mission success (Perl, DeLaurentis, Caldwell, and Crossley, 2009; Sindiy, et al., 2010). This human oversight function is based on a well-established approach to human-machine interaction known as human supervisory control (Sheridan, 1992). Sheridan describes five general functions of *managing* a complex engineering process, in addition to the functions required to *execute* the process: plan; teach; monitor; intervene; and learn (Landry, 2009; Sheridan, 1992). See Figure 11.1 for a description of these supervisory control and task execution functions in an aerospace environment.

The task execution phase itself involves determining which task functions are completed by humans (individually or together), and one or more automation systems. This determination is known as function allocation; Sheridan describes "levels of automation" ranging from pure human control through complete automation (Sheridan, 1992). As computer and automation systems gain in power and sophistication, they are more able to take on the supervisory control management functions of planning, monitoring, and intervening to perform system functions. With increasing operational experience, both humans and automated systems are also able to use past data to teach and learn better procedures for improved operations (Garrett and Caldwell, 2002). As automation systems increase in capability, a broader question involves whether humans or automation (or some combination) are in control of the system at any given point to optimize system performance (Caldwell and Onken, 2011).

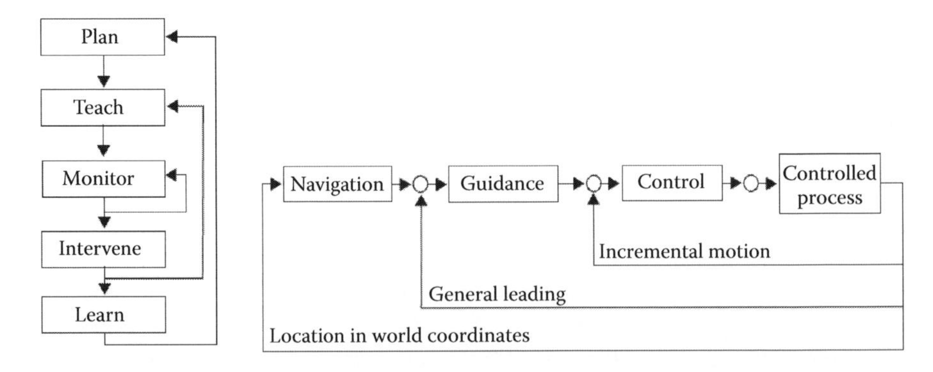

FIGURE 11.1 Supervisory control and aerospace navigation and task execution functions. (From Landry, S. J., Human–Computer Interaction in Aerospace. In *Human-Computer Interaction: Designing for Diverse Users and Domains* (pp. 197–216), Boca Raton, FL, CRC Press, 2009; adapted from Sheridan, T. B., *Telerobotics, Automation, and Supervisory Control*, Cambridge, MA, MIT Press, 1992. With permission.)

An additional challenge occurs when considering human space exploration to Mars or other interplanetary destinations, where light time communications delays are measured in minutes or hours, rather than seconds. Substantial advances in artificial intelligence and human supervisory coordination will be required to enable spaceflight crews to be able to plan, monitor, intervene, and optimize mission operations in real time with on-board capabilities, rather than with the substantial time delays associated with a ground-based control facility (Caldwell, 2014).

FUTURE TRENDS

AVIATION "NEXTGEN"

National and international commercial air travel continues to increase worldwide, with year-to-year increases in the 4–5% range in both the USA and Europe over most of the twenty-first century. Although commercial aircraft safety has been very good, there are concerns about maintaining both safety and efficiency of commercial passenger air travel. In the USA, the effort to modernize technology and upgrade the capacity of commercial air transport is known as "NextGen," a shortening of "next-generation air transportation system" (Kochenderfer, Holland, and Chryssanthacopoulos, 2012; Liddle and Millett, 2015). Commercial aircraft are especially subject to operating in potentially crowded airspaces, with needs to clearly communicate spacing between aircraft during takeoff, landing, and "en-route" (cross-country flights at steady altitudes), as well as appropriate operations for airport runways and gates. While the pilot and crew remain in control of a single aircraft, management of the overall aerospace environment (the "airspace") becomes the responsibility of air traffic control personnel who manage communication and coordination between aircraft; human factors research has been underway to ensure that new technology designs can help human air traffic controllers effectively manage NextGen airspace demands (Prevot, Lee, Callantine, Mercer, and Homola, 2010).

One design goal for NextGen is to enable commercial airlines to carry twice as many passengers in 2025 as were carried in 2008. Not only increased traffic volume, but safety, security, and environmental sustainability will be major goals of such a NextGen system. It is important to recognize that the NextGen system is not simply more commercial aircraft flying traditional routes, but a variety of aircraft types (including UAVs and drones) operating safely and effectively (DeLaurentis and Fry, 2008). This NextGen concept represents a growing need to consider coordination of "distributed expertise" (Caldwell, 2005; Cuevas, Fiore, Caldwell, and Strater, 2007) among pilots, air traffic controllers, aerospace engineers, and software designers.

New technologies in seemingly unrelated areas can significantly influence airspace management. For example, advances in aviation maintenance and automated aircraft vehicle health information are important elements of NextGen operations. For airlines to carry twice the passenger volume as 2008, maintenance delays and out-of-service aircraft must be reduced. One method of reducing cost and risk of maintenance-related outage is through condition-based maintenance (Asadzadeh and Azadeh, 2014). Thus, there must be strong information flow and task coordination between aircraft still in the air, and aviation maintenance crews at destination airports and central airline maintenance facilities, to reduce errors and improve responsiveness to aircraft maintenance needs (Fogarty, 2004). New technologies for automated information flow, such as radio frequency identification (RFID) and "internet of things" (IOT) capabilities, are being addressed to help improve information available to maintenance workers (Edwards, Bayoumi, and Eisner, 2016; Li and Liu, 2014). Human factors research in areas such as augmented reality can help to create new ways to assist aviation mechanics to detect and resolve anomalies on aircraft before they result in breakdowns or time-consuming repairs.

Spaceflight Human Factors

As discussed previously, spaceflight causes significant short-term as well as longer-term physical, cognitive, and physiological effects on human performance. Over the short term, many astronauts have expressed feelings of discomfort (like a severe head cold) as the lack of gravity reduces the drainage of blood and other fluids from the head (Seedhouse, 2012). It is still unknown how to predict who will suffer from SAS, although drug therapies, such as "ScopeDex," can be useful in some cases (David, 2005). Longer-term impacts of bone loss and muscle degradation do not have major consequences over the course of a few days; however, as time in space increases, the level of loss of calcium and muscle tone begins to simulate the impacts of old-age symptoms of osteoporosis and extended bed stay (Kluger, 2016).

Space Tourism

The first "space tourist" (paid passenger on a space vehicle) was Dennis Tito, who spent eight days in orbit in 2001 (Grant, 2007). While he (and other paid passengers) have paid many millions of US dollars for their opportunity for flight, there is a desire for commercial space flight operations that may take paid passengers to space, depending on the costs and durations of the flight (Freeland, 2005). One challenge is that spaceflight is still a very risky activity. However, the push for lower costs and higher participation in the space tourism industry may work to reduce the perceived risk of debilitating injury or death in space. Space tourists are more likely to be seeking a luxury experience than to have the skills and training as experimental test pilots or research specialists (Billings, 2006; Reddy, Nica, and Wilkes, 2012). It is possible that one model of trained astronaut crew members for space tourist flights is that of a "space attendant," like flight attendants in commercial aircraft. Originally, flight attendants on commercial airlines were selected to both instill confidence in the flying public, and respond to in-flight health emergencies; they still perform these roles on occasion (Caldwell, 2013; Grant, 2007; Peterson et al., 2013).

Dangerous health conditions on board commercial aircraft are low, and may be estimated in terms of emergencies per tens of thousands of passengers (Cocks and Liew, 2007), per billions of revenue passenger miles or kilometers (Sand, Bechara, Sand, and Mann, 2009), or per hundreds of flights (Peterson et al., 2013). However, the costs of an emergency to a scheduled airline flight requiring diversion may include schedule disruptions, dumping of fuel, and other costs (Sand et al., 2009). As the duration and distance of the spaceflight expands from suborbital to lunar and other missions, space attendants will need additional capabilities to provide medical interventions, and not simply stabilize the passenger until regular or emergency landing is available. Therefore, if space tourism remains an "extreme experience" rather than a "routine event" as is now the case with commercial aircraft flights, such incidence rates may be considered acceptable as "waiver" conditions absolving the provider of liability (Reddy, Nica, and Wilkes, 2012).

For suborbital or orbital flights, relatively few constraints exist for scheduling space traffic. As the distance and specificity of the destination increases, the "windows" for acceptable launch (and time between such windows) increase. For instance, Space Shuttle launch windows for travel to the ISS permitted only 5–15 minutes' leeway from planned schedule; if a window was missed, the next available launch opportunity was not until the next day (Goodman, 2006). Launch windows for Mars journeys may only allow for a few minutes' error over a period of two to three years. Therefore, pre-flight preparation, engineering check-out, and all necessary launch activities must operate on a highly regimented time schedule. This is not impossible as some airlines, and even some national train systems, do operate with second- to minute-scale precision over the course of a year.

CONCLUSIONS

Human factors and ergonomics research has made crucial contributions to the design, improvement, and increasing safety of aviation and spaceflight operations through the first two decades of the twenty-first century. Human factors research in aerospace has enabled pilots, passengers, and crew to survive, travel safely, and perform tasks in what would be an otherwise fatal environment. Physical and cognitive human factors methods have improved vehicle designs and controls, and the integration of perception, information, decision-making, and task performance described by the concept of situation awareness has been a major human factors research innovation over the past 20 years. Effective human factors design and system integration for both increased aviation travel volumes, and exploration of space destinations uncover a range of physiological, decision-making, and distributed expertise linking humans, artificial intelligence, and autonomous systems, and other engineering technologies.

KEY TERMS

Aerospace environment. A description of the physical conditions (such as air pressure, temperature, wind speed and direction, humidity, oxygen content, and other chemical, physical, and thermal properties) affecting aircraft and spacecraft during flight.

Airspace management. The processes where air traffic controllers regulate the spacing and pathways that aircraft use, in order to improve safety and efficiency of the use of airports and routes that aircraft take from takeoff to landing.

Crew resource management (CRM). A research-based team performance technique and training approach (originally known as "cockpit resource management") used to improve communication, task prioritization, and safety behaviors in aircraft operations during both routine flight and potentially dangerous conditions.

Failure analysis. The process of studying a new or existing engineering system or organization to determine potential problems (for new systems or existing systems that have not failed) or actual causes of failure (for existing systems that already have failed).

Human-machine interaction. The interdisciplinary study of how to design and build technologies that are easier, safer, and more effective for people to use.

Performance-shaping factors. Aspects of the physical environment, equipment design, work demands, organizational culture, and other issues that can affect the possibility of an operator error or system failure.

Supervisory control. A technology design where a human operator has responsibilities for controlling, managing, monitoring, and/or recovering from automated machine performance in a complex human-machine interaction system.

Task analysis. A human factors and ergonomics technique to study and document how a person completes a work activity.

Team coordination. Processes for how people work together and share information, knowledge, and tasks to complete shared goals.

Vehicle design. The engineering process of creating a new aircraft, spacecraft, or other built object to carry one or more objects or people from place to place with desired capabilities.

REFERENCES

Al Wardi, Y. M., Jeevarathinam, S., and Al Sabei, S. H. (2016). Eastern bodies in western cockpits: An anthropometric study in the Oman military aviation. *Cogent Engineering* 3(1), www.tandfonline.com/doi/full/10.1080/23311916.2016.1269384.

Asadzadeh, S. M. and Azadeh, A. (2014). An integrated systemic model for optimization of condition-based maintenance with human error. *Reliability Engineering & System Safety* 124, 117–131.

Ayaz, H., Çakir, M. P., İzzetoğlu, K., Curtin, A., Shewokis, P. A., Bunce, S. C., and Onaral, B. (2012, March). Monitoring expertise development during simulated UAV piloting tasks using optical brain imaging. *Aerospace Conference, 2012 IEEE* (pp. 1–11).

Billings, L. (2006). Exploration for the masses? Or joyrides for the ultra-rich? Prospects for space tourism. *Space Policy* 22(3), 162–164.

Boeing IPM (2016). Boeing Simplified English Checker. Website, www.boeing.com/company/key-orgs/licensing/simplified-english-checker.page, accessed October 17, 2016.

Caldwell, B. S. (1997). Components of Information Flow to Support Coordinated Task Performance. *International Journal of Cognitive Ergonomics* 1(1), 25–41.

Caldwell, B. S. (2005). Multi-team dynamics and distributed expertise in mission operations. *Aviation, space, and environmental medicine* 76(6), B145–B153.

Caldwell, B. S., Johnson, M. E., Whitehurst, G., Rishukin, V., Udo-Imeh, N., Duran, L., Nyre, M. M, and Sperlak, L. (2015). Impact of Weather Information Latency on General Aviation Pilot Situation Awareness. *International Symposium on Aviation Psychology 2015*, 1810.

Casey, S. (1993). Rental Car. In *Set Phasers on Stun, and Other True Tales of Design, Technology, and Human Error.* Santa Barbara, CA: Aegean Publishing.

Cuevas, H. M., Fiore, S. M., Caldwell, B. S., and Strater, L. (2007). Augmenting team cognition in human-automation teams performing in complex operational environments. *Aviation, space, and environmental medicine* 78(5), B63–B70.

Dahari, M., Yeo, S. E., Cho, W. K., Sabri, M. A. M., Mingo, M. Z., and Ong, W. Y. (2010). Design and Development of Aircraft Cabin for Very Light Jet (VLJ).

DeLaurentis, D. and Fry, D. (2008). Understanding the Implications for Airports of Distributed Air Transportation Using a System-of-Systems Approach. *Transportation Planning and Technology* 31(1), 69–92.

Drury, J. L., Riek, L., and Rackliffe, N. (2006, March). A decomposition of UAV-related situation awareness. In *Proceedings of the 1st ACM SIGCHI/SIGART conference on Human-robot interaction* (pp. 88–94). ACM.

Edwards, T., Bayoumia, A., and Eisner, L. (2016, October). Internet of Things–A Complete Solution for Aviation's Predictive Maintenance. In *Advanced Technologies for Sustainable Systems: Selected Contributions from the International Conference on Sustainable Vital Technologies in Engineering and Informatics, BUE ACE1 2016, 7–9 November 2016, Cairo, Egypt* (Vol. 4, p. 167). Springer.

Endsley, M. R. (1995). Toward a theory of situation awareness in dynamic systems. *Human Factors* 37(1), 32–64.

Endsley, M. R. (1999). Situation Awareness in Aviation Systems. In Garland, D. J., Wise, J. A., and Hopkin, V. D. (eds), *Handbook of Aviation Human Factors.* Mahwah, NJ: Lawrence Erlbaum Associates.

Fogarty, G. J. (2004). The role of organizational and individual variables in aircraft maintenance performance. *International Journal of Applied Aviation Studies* 4(1), 73–90.

Fosshage, E. (2014). The Effects of Job Performance Aids on Quality Assurance. *Proceedings of the Human Factors and Ergonomics Society 58th Annual Meeting*, 1959–1963.

Fowler, B., Paul, M., Porlier, G., Elcombe, D. D., and Taylor, M. (1985). A re-evaluation of the minimum altitude at which hypoxic performance decrements can be detected. *Ergonomics* 28(5), 781–791.

Fowler, B., Elcombe, D. D., Kelso, B., and Porlier, G. (1987). The threshold for hypoxia effects on perceptual-motor performance. *Human factors* 29(1), 61–66.

Freeland, S. (2005). Up, up and... back: The emergence of space tourism and its impact on the international law of outer space. *Chicago Journal of International Law* 6(1).

Gendreau, M. A. and DeJohn, C. (2002). Responding to Medical Events during Commercial Airline Flights. *New England Journal of Medicine* 346(14), 1067–1073.

Goodman, J. L. (2006). History of space shuttle rendezvous and proximity operations. *Journal of Spacecraft and Rockets* 43(5), 944–959.

Grant, R. G. (2007). *Flight: The Complete History.* New York: DK Publishing.

Guldenmund, F. W. (2007). The use of questionnaires in safety culture research–an evaluation. *Safety Science* 45(6), 723–743.

Kanki, B. G., Helmreich, R. L., and Anca, J. (eds) (2010). *Crew Resource Management.* Amsterdam: Academic Press/Elsevier.

Kirwan, B. (1998). Human error identification techniques for risk assessment of high risk systems—Part 1: Review and evaluation of techniques. *Applied Ergonomics* 29(3), 157–177.

Kochenderfer, M. J., Holland, J. E., and Chryssanthacopoulos, J. P. (2012). Next generation airborne collision avoidance system. *Lincoln Laboratory Journal* 19(1), 17–33.

Kroemer, K. H. E. and Grandjean, E., (1997). *Fitting the task to the human: a textbook of occupational ergonomics (5th Ed).* London: Taylor & Francis.

Landry, S. J. (2009). Human–Computer Interaction in Aerospace. In *Human-Computer Interaction: Designing for Diverse Users and Domains* (pp. 197–216). Boca Raton, FL: CRC Press.

Larimer, S. (2017). "I wanted to serve": These deaf men helped NASA understand motion sickness in space. *The Washington Post*, online edition, May 5, 2017. Available at: www.washingtonpost.com/news/retropolis/wp/2017/05/05/i-wanted-to-serve-these-deaf-men-helped-nasa-understand-motion-sickness-in-space/?utm_term=.a62c7ae9ae13.

Lauber, J. K. (1984). Resource management in the cockpit. *Air Line Pilot* 53, 20–23.

Li, Q., and Liu, H. L. (2014). RFID Network and its Application in Management of Aviation Maintenance. *Applied Mechanics and Materials*, Vols. 475–476, pp. 920–924.

Lockhard, D. P. and Lilley, G. M. (2004). The Airframe Noise Reduction Challenge. NASA/TM–2004–213013.

Please provide in text citation for Mearns and Flin.

Mearns, K. J. and Flin, R. (1999). Assessing the state of organizational safety—Culture or climate? *Current Psychology* 18(1), 5–17.

Melis, D. J., Silva, J. M., and Clothier, R. (2015). The changing size of the commercial aviation passenger and its potential impact on the aviation industry [online]. In: Asia-Pacific International Symposium on Aerospace Technology (APISAT 2015). [Canberra]: Engineers Australia: 319–328. Available at http://search.informit.com.au/documentSummary;dn=270660984722210;res=IELENG. ISBN: 9781922107480.

Melis, D. J., Silva, J. M., and Silvestre, M. A. (2017). Characterisation of the anthropometric features of airline passengers and their impact on fuel usage in the Australian domestic aviation sector [online]. In: 17th Australian International Aerospace Congress: AIAC 2017. Melbourne, Vic.: Engineers Australia, Royal Aeronautical Society: 529–537. Available at http://search.informit.com.au/documentSummary;dn=739578338940410;res=IELENG. ISBN: 9781922107855.

Meilinger, P. S. (2004). When the Fortress Went Down. *Air Force Magazine* 87, 78–82.

Miller, D. P. and Swain, A. D. (1987). Human error and human reliability. *Handbook of human factors*, 219–250.

Mills, J. A. and Caldwell, Barrett S. (1997). Simplified English for Computer Displays. In *Proceedings of the Seventh International Conference on Human-Computer Interaction*, pp. 133–136.

Modlin, C. T. and Zipay, J. J. (2014). The 1.5 & 1.4 Ultimate Factors of Safety for Aircraft & Spacecraft-History, Definition and Applications.

NASA (2007). *NASA Systems Engineering Handbook*. Washington, DC: NASA SP 2007-6105.

O'Connor, P., Campbell, J., Newon, J., Melton, J., Salas, E., and Wilson, K. A. (2008). Crew resource management training effectiveness: A meta-analysis and some critical needs. *The international journal of aviation psychology* 18(4), 353–368.

"Of Men and Machines" (1962). Film, Washington, DC: National Educational Television/American Psychological Association. Available online at https://archive.org/details/ofmenandmachines.

Page, J. and Olsen, J. (2015). Aerospace design - a new world thanks to digital technology [online]. In: Asia-Pacific International Symposium on Aerospace Technology (APISAT 2015). [Canberra]: Engineers Australia, 2015: 98–103. Available at http://search.informit.com.au/documentSummary;dn=261419030978143;res=IELENG. ISBN: 9781922107480.

Papamoschou, D. (2004). New method for jet noise reduction in turbofan engines. *AIAA Journal* 42(11), 2245–2253.

Peterson, D. C., Martin-Gill, C., Guyette, F. X., Tobias, A. Z., McCarthy, C. E., Harrington, S. T., and Yealy, D. M. (2013). Outcomes of medical emergencies on commercial airline flights. *New England Journal of Medicine* 368(22), 2075–2083.

Prevot, T., Lee, P., Callantine, T., Mercer, J., Homola, J., Smith, N., and Palmer, E. (2010). Human-in-the-loop evaluation of NextGen concepts in the Airspace Operations Laboratory. In *AIAA Modeling and Simulation Technologies Conference*, p. 7609.

Raby, M. and Wickens, C. D. (1994). Strategic workload management and decision biases in aviation. *The International Journal of Aviation Psychology* 4(3), 211–240.

Reason, J. (1990). *Human Error*. Cambridge University Press.

Reddy, M. V., Nica, M., and Wilkes, K. (2012). Space tourism: Research recommendations for the future of the industry and perspectives of potential participants. *Tourism Management* 33(5), 1093–1102.

Sand, M., Bechara, F. G., Sand, D., and Mann, B. (2009). Surgical and medical emergencies on board European aircraft: a retrospective study of 10189 cases. *Critical Care* 13(1), R3.

Seamster, T. L., Redding, R. E., Cannon, J. R., Ryder, J. M., and Purcell, J. A. (1993). Cognitive task analysis of expertise in air traffic control. *The International Journal of Aviation Psychology* 3(4), 257–283.

Seedhouse, E. (2012). *Pulling G: Human Responses to High and Low Gravity.* Chichester, UK: Springer Praxis Publishing.

Sheridan, T. B. (1992). *Telerobotics, Automation, and Supervisory Control.* Cambridge, MA: MIT Press.

Smillie, R. J. (1985). Design Strategies for Job Performance Aids. In Duffy, T. M. and Waller, R. W. (eds), *Designing Usable Texts* (pp. 213–241). Orlando, FL: Academic Press.

Spicer, C. W., Murphy, M. J., Holdren, M. W., Myers, J. D., MacGregor, I. C., Holloman, C., and Zaborski, R. (2004). Relate air quality and other factors to comfort and health symptoms reported by passengers and crew on commercial transport aircraft (Part I), ASHRAE Project 1262-TRP, American Society for Heating, Refrigerating and Air Conditioning Engineers. Atlanta.

Stanton, N. (2005). *Human Factors Methods: A Practical Guide for Engineering and Design.* London: Ashgate Publishing.

Stanton, N., Salmon, P. M., and Rafferty, L. A. (2013). *Human Factors Methods: A Practical Guide for Engineering and Design (revised ed.).* London: Ashgate Publishing.

STEMG (2015). "ASD-STE100." Website accessed October 17, 2016 from www.asd-ste100.org/.

Strøm-Tejsen, P., Wyon, D. P., Lagercrantz, L., and Fang, L. (2007). Passenger evaluation of the optimum balance between fresh air supply and humidity from 7-h exposures in a simulated aircraft cabin. *Indoor Air* 17(2), 92–108.

Swain, A. D. (1964). Some problems in the measurement of human performance in man–machine systems. *Human factors* 6(6), 687–700.

Trudgill, M. J. A. and Harrigan, M. J. (2016). "Anthropometry and Aircrew Integration." In Gradwell, D., and Rainford, D. J. (eds), *Ernsting's Aviation and Space Medicine (5th ed.).* New York: John Wiley/CRC Press.

Waller, M. J., Gupta, N., and Giambatista, R. C. (2004). Effects of adaptive behaviors and shared mental models on control crew performance. *Management Science* 50(11), 1534–1544.

Wiegmann, D. A. (2002). *A synthesis of safety culture and safety climate research.* University of Illinois at Urbana-Champaign, Aviation Research Lab.

Case Study 11
Flying High in the Gulf Region: A Case Study of Aviation Human Factors at Emirates Airline

Nicklas Dahlstrom, Human Factors Manager

Please confirm if affiliation should be deleted (author name only) for consistency to other chapter author by line.

INTRODUCTION

At least one claim to the birth and foundation of Human Factors is made by aviation. This goes back to the pioneers of aviation more than a century ago, who were followed by the revelation of the importance of basic physiological factors (i.e., pilots flying at higher altitude suffering from hypoxia), the realization that not everyone is suitable to become a pilot (the emergence of pilot selection procedures), and the recognition of aircraft design (with the most celebrated example being Alphonse Chapani's design invention of putting a round item on the gear lever and a flat item on the flap lever). Human Factors in aviation expanded to include cognitive skills, including workload management, situation awareness, and decision-making as well as cooperative skills such as communication, teamwork, leadership, and cultural awareness. Today, applied aviation Human Factors is present in regulations, operational procedures, and training, specifically in the form of crew resource management (CRM) and its different derivatives of training provided to pilots, air traffic controllers, engineers, dispatchers, cabin crew, and other professional groups.

EMERGENCY AND URGENCY: THE FIRST TRACES OF HUMAN FACTORS IN THE GULF

The origins of aviation Human Factors in the GCC region can be linked to the arrival of CRM training in the late 1980s. Pilots who were in the region remember their first training on this new topic and that CRM started to be included in manuals at the time. In the mid-1990s, Human Factors gained traction, with the first Human Factors Manager positions emerging and CRM courses being more widely implemented. Still, the "C" in CRM was at that time more focused on "cockpit" than "crew"; pilots and cabin crew did not have any combined CRM training, engineers received no Human Factors training nor did any other aviation professionals. A turning point for Human Factors and CRM in the region came with the accident of Gulf

Air 072 in August 2000, when a fully operational aircraft hit the waters by Bahrain International Airport. That accident made it impossible for anyone to not recognize the critical role of skills other than the traditional piloting ones. It was also undeniable that the operational risks highlighted by that accident were not unique to the airline it happened to. Consequently, selection, operations and training—specifically CRM training—in all airlines in the region were developed towards improved safety, with an increasingly important role for Human Factors.

THE RISE OF A REGION AND OF HUMAN FACTORS

Recently, airlines in the GCC region have risen to world prominence and become major players in the aviation industry. At the same time, Human Factors and CRM have increased in importance in the aviation industry as well as in the airlines of the region. The "Big Three" airlines of the region (Emirates, Etihad, and Qatar) have Human Factors Managers positions, responsible for the development and delivery of CRM training and for many other Human Factors needs in their organizations. At Emirates airline, the author is working as Flight Training Human Factors Manager (who runs the Human Factors Office together with three Human Factors Specialists). Together with the Engineering Human Factors Manager the author represents Human Factors throughout the organization. Human Factors is also represented in the form of safety managers and line managers, for which Human Factors is an integrated aspect of their work. This work is coordinated through a Human Factors Safety Action Group as well as with the safety department of the Emirates Group and integrated into the larger framework of the organization's safety management system.

PRESENT POTENTIAL AND POSSIBLE PROGRESSION

The main mission for Human Factors in Emirates is to develop and deliver CRM training for pilots, cabin crew, and engineers as mandated by regulations. This is the foundation of how knowledge and practice of Human Factors is shaped, spread, and applied in an airline. The scope and aim of CRM training at Emirates goes beyond meeting regulatory requirements, focusing on applying research and industry best practice to meet organizational needs beyond mandatory initial, recurrent, upgrade CRM training and other Human Factors courses. Examples include:

- integrated recurrent CRM training for pilots, with CRM and technical topics combined to increase relevance of the training and facilitate transfer of training to line operations
- combined CRM training for pilots and cabin crew based on scenarios and performed in a full flight cabin simulator, including briefing and debriefing
- command coaching, a course planned a year ahead for pilots coming up for their training course to become captains, which prompts thinking and preparation for the new role
- cooperation with other professional groups such as dispatchers, air traffic controllers, network control staff, and engineers to make CRM "go both ways," including training for Crisis Directors

Beyond the core of CRM training, the Human Factors Office delivers coaching sessions for some pilots who may have failed a training/checking session, supports and drives development of trainers, develops Human Factors training for other parts of the organization (i.e., cargo, airport services, etc.), and is consulted in investigations and other organizational challenges. Additionally, the Human Factors Office is in contact with researchers and academia to stay at the forefront of new and relevant knowledge for the aviation industry, including supervising masters and Ph.D. students. The Human Factors Office is also currently working together with partners in the aviation industry on eye-tracking in relation to pilot monitoring as well as with development of mid-fidelity simulation. Furthermore, we have numerous visits by other airlines and aircraft manufacturers to observe our CRM training. Outside of the aviation industry, we cooperate and share our CRM training with representatives of Human Factors in other fields, such as in the medical, rail, and nuclear industry.

In conclusion, the development of CRM training in Emirates airline and the expansion of Human Factors within the organization, together with the contacts and interest from other industries, point to the great potential for Human Factors to continue to grow across the GCC region and contribute to the safety and efficiency of all industries.

12 HFE in the GCC
From Missed Opportunities to Promising Possibilities

Shatha Samman

CONTENTS

INTRODUCTION

The etymology of the word ergonomics comes from the Greek word "ergo," defined as work, and "nomos," meaning law. Ergonomics essentially denotes the science of work, where work can involve anything that we do to achieve a purpose, not necessarily specific to a job. Consequently, human factors and ergonomics (HFE) impacts us daily at work, home, and play, whether we recognize it or not, to enhance our performance and optimize our well-being. The theory and practice of HFE has been used in nearly all economic sectors. This was revealed in the previous chapters demonstrating how work processes, products, organizational activities, and physical environments are designed around human capabilities in a wide range of domains.

HFE is regarded as one of the first truly multi-, inter-, and cross-disciplinary fields, with interactions including human-system, human-environment, and human-human (Wilson, 2000). These interactions were illustrated throughout the book in diverse domains emphasizing how humans interact with systems such as products/technologies (product design, ICT, and cybersecurity), transportation (surface transportation, aerospace), complex processes and systems (healthcare, energy), in organizational structure and human interaction (macroergonomics and safety), and within the physical environment (environmental design).

HFE's Three Characteristics

HFE's scientific principles are based on three main characteristics (Dul et al., 2012). First, HFE takes a *system approach*. The system can be the environment, a product/technology, task/process, organization, or another human interaction. The level of interaction can take a holistic perspective ranging from a micro-level interaction (human completing a task on a computer), meso-level (individual interacting with a team on a technical process), or a macro-level (person is part of a greater organization performing a complex mission) (Rasmussen, 2000). It is essential to define the specific elements and scope of interactions (e.g., reducing operator workload of a collaborative complex task with multinational teams that are geographically distributed).

Second, HFE is *design-driven*. HFE professionals analyze and assess human-system interaction problems in various domains. They are involved throughout the design process stages including-planning, designing, developing, implementing, evaluating, and redesigning. Using an iterative process and the various methodologies discussed throughout this book (i.e., task analysis, behavioral-based assessment, usability testing) are some of the tools HFE specialists use to continuously improve the human-system design compatibility.

Third, HFE focuses on two main interconnected outcomes, *performance* and *well-being* of the individual. These outcomes are dependent on a project's goal (e.g., efficiency, productivity, safety, and enjoyment). The strongly connected outcomes are optimized when there is a great "fit" in human-system interaction (Dul et al., 2012).

HFE's Socio-Technical System Approach

HFE is a socio-technical system approach that aligns people, technology, work systems, and the environment using a design-driven system approach that aims to optimize performance and well-being (see Chapter 2). Cultural context is a significant factor of the environment, and thus is essential for optimal compatibility in any human-system interaction. The need to align the technology, work processes, and overall environment with the user's cultural background is an important step towards an optimal outcome (Thatcher, Waterson, Todd, & Moray, 2017).

The world is increasingly interdependent economically and geographically. With almost half of the GCC population being expatriates (see Table 12.1), it is customary to work in this region with a culturally diverse workforce that deal with systems designed internationally. As such, the HFE's social-technical system assists to understand how to design for the local context to meet both human-centric and culture-centric attributes (Moray 1995).

The chapter starts by considering the current state of HFE in the GCC and missed opportunities. This is followed by a discussion of some thriving platforms to activate HFE potential in several domains. The chapter concludes with a proposed future roadmap of promising possibilities for HFE to play a meaningful role and prosper in the GCC.

CURRENT STATE: HFE MISSED OPPORTUNITIES IN GCC DEVELOPMENT

The discipline and profession of HFE in the GCC is a case of missed opportunities that were skipped during the fast-paced development of the region. GCC countries may have overlooked HFE mainly due to its unfamiliarity to policy-makers, academic institutions, researchers, professionals, and the private sector. The GCC frequently implements technological advancements with little HFE research support. The developed countries from where most of those technologies come employ a strong evidence-based HFE practice that disseminates its findings into practical applications to various HFE domains. However, these technologies might not be optimal to use in the GCC due to different user needs and cultural backgrounds. By advocating the tremendous power of HFE, stakeholders will recognize its benefit and strong alignment to the GCC's development goals.

HFE typically goes unnoticed in the GCC due to its multidisciplinary nature. Unclear communication of how it uniquely fits in any respective project or domain marginalizes its distinct contributions to performance and well-being. The real value proposition of HFE is realized when the benefits are adequately explained to stakeholders using relatable language from their industry that motivate their actions and address their needs. By identifying and defining the stakeholder's priority and how they connect to HFE, stakeholders will appreciate the economic, social, and legal outcomes.

Limited GCC HFE Professional

Expert HFE professionals are scarce in the GCC countries. The case studies in this book highlight the pockets of experts dispersed throughout the region. HFE is a highly specialized field throughout the globe, with few professionals spread across many industries; this scarcity is more profound in the GCC. This underutilization is

TABLE 12.1
Total Population and Percentage of Nationals and Foreign Nationals in GCC Countries (2010–2016)

						% in Total Population	
	Date/Period	Total Population	Date/Period	Nationals	Foreign Nationals	Nationals	Foreign Nationals
Bahrain	Mid-2014	1,314,562	Mid-2014	630,744	683,818	48.0	52.0
Kuwait	31 March 2016	4,294,171	31 March 2016	1,316,147	2,978,024	30.6	69.4
Oman	20 April 2016	4,419,193	20 April 2016	2,412,624	2,006,569	54.6	45.4
Qatar	April 2015	2,404,776	April 2010	243,019	2,161,757	10.1	89.9
Saudi Arabia	Mid-2014	30,770,375	Mid-2014	20,702,536	10,067,839	67.3	32.7
United Arab Emirates	Mid-2010	8,264,070	Mid-2010	947,997	7,316,073	11.5	88.5
Total[a]		**51,467,147**		**26,253,067**	**25,214,080**	**51.0**	**49.0**

Source: National Institutes of Statistics, latest year or period available as of April 20, 2016 (http://gulfmigration.eu/).

[a] Total provides the sum of population numbers at different dates. It is not exactly the total population at any of these dates.

evident by the limited job opportunities and unavailable HFE degreed education and research in the region. Given the rapid development in the GCC countries, a greater presence would be advisable as a proactive measure to economic improvement.

Adaptability of GCC End-User

Part of GCC culture is its *adaptable* nature. The values of trust, compromise, flexibility, and adaptability reflect some cross-cultural differences between the East and West, where Westerners tend to be more structured and directive in their interactions. GCC locals are accustomed to adapting to "fit" into the system using selection, training, and/or trial-and-error instead of demanding that systems are designed to be compatible to their needs and requirements. With immense imports of products and technologies from around the world, GCC consumers are used to adapting to learn how to use these systems.

GCC Policy-Makers

Internationally, commercial human-system interaction is monitored and has legal responsibilities to the concerned government or private entity, and is regulated by specific standards and codes. Examples of US and UK government agencies that adopted HFE practices to enact regulations, policy requirements, and standards in many fields include: the US Food and Drug Administration (FDA) and medical device design, the UK National Health Service (NHS) and patient safety, the Federal Aviation Administration (FAA) and aviation human performance, and the National Aeronautics and Space Administration (NASA) for spaceflight human-system integration (Pew & Mavor, 2007).

The importance of international standards has not been ignored by GCC regulators and stakeholders. Indeed, standardization and regulating entities exist throughout the GCC. What might be lacking is a wealth of regional research to support their mission. Typically, in developed countries regulators can call on scientists and researchers from universities, research institutes, industry, and government labs to advise them on the best policies and practices to draft and update regulations. A similar process is recommended to encourage home-grown research and rigorous HFE evaluations suitable for local, cultural, and environmental conditions.

Furthermore, the GCC region has been late to initiate consumer protection and awareness efforts (Eid, 2017). Strong legal accountability and a comprehensive regulatory framework that protects end-users from international sellers, especially related to HFE policies, are missing. Immediate actions are needed to safeguard GCC consumers from harm and possible threats from bad design. Sufficient legal regulations are essential to ensure that design and work processes are appropriately designed for the GCC end-user.

HFE Integration in the GCC

As evident from the preceding discussion, HFE may not be sufficiently utilized in the GCC countries and many initiatives are urgently needed. Emphasis should be on avoiding problems before they escalate, and maximizing the benefits by early adoption of HFE and introducing the discipline and profession to policy-makers, academics, professionals, and end-users. Published books, professional associations, research and development, promoting academic inclusion of HFE, and integrating evidence-based practices of HFE in government, university, and industry are steps to mitigating these shortcomings. The mantra of "good ergonomics is good economics" (Hendrick, 1996) needs to be adopted by all stakeholders emphasizing HFE's importance to the GCC's economic development. With that conviction, a paradigm shift will occur for end-user advocacy. The next section explores how HFE can be applied in a variety of domains for the betterment of the GCC society.

HFE THRIVING PLATFORMS IN THE GCC

The GCC transformational vision will have a major impact on the region and the world at large. The economic diversification towards developing high-productivity industries and services aims to create more jobs and sustainable economies. The member countries' strategic programs strive to reduce dependence on oil revenues and expand the economy and guide human, social, economic, and environmental developments.

HFE will play a pivotal role in addressing the GCC's current challenges and aid in preventing future obstacles. By asking the relevant questions and focusing on the problems that impact human-system interaction (i.e., human, technology, organization, and environment) and the methodologies/tools that may assist in addressing these issues will likely provide practical solutions. In each chapter of this book, the reader can gather principles, concepts, methodologies, and applications that may resolve these challenges using HFE. Below is a discussion of the challenges and opportunities in various GCC sectors that will benefit from HFE using design-driven system approach propositions.

GCC Children, Youth, and the Aging Population

In 2017, the Human Factors and Ergonomics Society (HFES) established a new technical group that focuses on children and youth in relation to products and the environment (www.hfes.org). Any product that a child interacts with to promote safety and enjoyment is a topic of interest (e.g., toys, playground equipment, nursery products). These issues are of great interest to the GCC as, in July 2017, a new policy in the UAE obliges drivers to provide child car seats for children under the age of four (Shahbandari, 2017). This is an example of steps in the right direction in the GCC to establish regulatory standards and policies to protect children in case of car accident. Car injuries are a leading cause of death among children in the USA and may be prevented using the appropriate car seat (Zackowitz, 2016). Furthermore, car seat misuse in terms of installation, restraining a child in the seat, and following installation instruction may reduce protection in event of a crash. Data found the rate of misuse of car seats to be approximately 59%, with parents being overly confident in their knowledge of how to use car seats (Greenwall, 2015). HFE experts play a pivotal role to ensure that car seat design and installation are done properly for the safety of children. This example illustrates how HFE can positively improve children's environments at home and in public.

Youth (under age of 25) comprise more than 50% of total GCC population. HFE provides a unique opportunity as technology continues to advance and permeate youth's daily lives. Research found that the sharing economy (exchange of goods/services directly between individuals using online platforms) is promising, with GCC consumers spending US$10.7 billion online in 2016 and generating an estimated US$1.7 billion in revenues ("The GCC can benefit," 2017). The rising use of the Internet and the development of regional mobile applications to provide online business opportunities (i.e., Careem, a regional private taxi application) is encouraging. It allows youth and underutilized segments of the population (e.g., women, people living in rural areas) to develop start-up innovative businesses to accommodate their lifestyle (i.e., flexible hours, freedom to work from anywhere at any time). This is aligned with all of the GCC national transformation plans to address the rise in youth unemployment and to drive economic growth by promoting innovative entrepreneurial ecosystems. A HFE design-driven system approach is a foundation to promote the concept of *designing for the youth by the youth.* Applying HFE evidence-based principles and methodologies encourages youth to think from a holistic multidisciplinary perspective, which is vital to meet job market requirements and build necessary competencies. It is important to note that exposing youth to HFE at an early stage will also contribute to promoting science, technology, engineering, the arts, and mathematics (STEAM) curriculum by demonstrating its usefulness in our everyday life.

The GCC countries also have growing elderly populations. The average life expectancy of adults in the GCC was found to approach 77 years, which is in accordance with other affluent countries and above the global average (Ingham, 2017). HFE can greatly improve the quality of life and increase the independence of the older GCC population. Age-related changes are accompanied by a decline in visual and auditory acuity, a decay in motor skills and agility, and a deterioration in cognitive processing such as memory (Fisk, Rogers, Charness, Czaja, & Sharit, 2009). To improve the quality of life and health of older adults in the GCC requires that the knowledge of the aging community be applied to the design of products and environments specific to the cultural and environmental conditions of this region. Issues include: how to design environments to maximize the safety and functional abilities of older adults while ensuring social connection with family and friends; designing ICT and products that accommodate their physical and cognitive restrictions (e.g., large buttons for better dexterity); and developing effective training programs to teach older adults necessary skills (e.g., mobile applications). Importantly, the issues of privacy and trust in technology are critical, especially as seniors tend to be vulnerable to identity theft, hacking, and cybersecurity extortions.

GCC Multicultural Communities

The majority of the GCC population consists of expatriates (approximately 50%) that are interacting with different cultural backgrounds and with a host of complex technologies. National culture influences human attitude, behavior, and cognition

(Samman, 2010, 2016) and has contributed to major incidents and catastrophic accidents (Noort, Reader, Shorrock, & Kirwan, 2016; Strauch, 2010; see Case Study 3). Also, products designed for one culture do not necessarily transfer well to other cultures. For instance, anthropometric data standards used to design cockpits for Western pilots may not be suited for Eastern pilots (Al-Wardi et al., 2016). This is especially critical with GCC workforce nationalization indicatives. These findings will encourage GCC policy-makers to customize anthropometric international standards to reflect the GCC population and use it when designing and purchasing products/systems. This is a small example of the need to customize and update standards to improve compatibility with GCC nationals to ensure ideal human-interaction. Moreover, multicultural aspects in occupational safety and health (OSH) and managing cultural diversity are essential to ensure a robust safety culture. Safety culture is a product of both national and organizational culture (Helmreich & Merritt, 1998), and the implications on safety management systems within the GCC region will greatly benefit from HFE to develop safety assessment programs that indicate the impact of culture and provide outcomes that optimize safety, performance, and overall well-being.

GCC Service and Customer Experience Design

The GCC region is highly reliant on the service economy, with businesses working with customers to provide particular services including banking, government, retail, healthcare, transportation, food services, etc. According to the CIA World FactBook, the contribution of the service sectors to the GCC economy is estimated to be an average of 50% in 2016 (Bahrain: 62.3%; KSA: 54%; Kuwait: 40%; Oman: 53.9%; Qatar: 52%; and UAE: 40.1%). Findings from customer experience research predicted that by 2018 more than 50% of organizations globally will redirect their investments to customer experience innovations (Meulen & Rivera, 2015). Results found that currently 86% of buyers will pay more for a better customer experience, and by 2020 customer experience will be the key brand differentiator compared to price and production (Kulbytė, 2017). Omni-channel customer experience ensures that human interaction is consistently unified across multiple channels to provide a unified brand experience for customers to easily switch between channels and have the same customer experience (see Case Study 4). One iconic example of omni-channel customer experience and service interaction design is Disney. Disney is renowned for its customer-centric services, which anticipate the needs of customers to design an interaction with exceptional attention. Research found that organizations that provide consistent customer experience service across multiple channels will retain 89% of their customers (Lemzy, 2017). As Steve Jobs said: "you've got to start with the customer experience and work back toward the technology, not the other way around" (Solomon, 2014).

Human-service interactions are developed and designed from a customer's perspective using HFE design-driven systematic principles resulting in customer satisfaction. Service design (Lynn Shostack, 1982) is improving the experience of both the customer and employee by designing and aligning operations to support customer interaction. Customer experience design (CXD; Verhoef et al., 2009; Hartson

& Pyla, 2012; Longanecker, 2016) is designing for the whole experience a customer has whenever they interact with not only the product but the organization as a whole. Both are centered around the needs of the user experience and culturally relevant interaction design. Hence, both are design processes based on foundational theories, principles, methodologies, and processes of HFE using the framework of user-centered design (see Chapter 4). The premise is having an ecosystem that practices service and customer experience design to ensure that this is the responsibility of the whole organization, from an internal process system involving stakeholders and senior managers to external front-end processes where staff members interact directly with the customer.

GCC Health Care Digital Transformation

The health care sector in the GCC countries is confronted with a rise in health risks and chronic diseases such as obesity and diabetes, and unhealthy habits such as smoking. GCC governments have identified health care as a key factor in their transformation vision. Total GCC health care spending is expected to rise to US$69 billion in 2020, with GCC governments and private companies both investing to improve health care services (Issac, 2017a). Part of Saudi Arabia's National Transformation Program is to increase private sector contribution to healthcare by 35% by 2020 (Kerr, 2016). One of the strategic objectives set for the Saudi Ministry of Health is to "improve the efficiency and effectiveness of the health care sector through the use of IT and digital transformation" (Giguashvili, 2017). A budget of SAR5.9 billion from government sources for "electronic health" initiatives over the next five years was allocated (Lalchandani, 2017). The UAE also aims to create a world-class health care system by 2021 in their national agenda (see Case Study 8). The goal is to develop a health care system that is patient-centered, employing the best global standards and innovative technology to enable strategic health care objectives.

HFE has been recognized by the health care sector to play a pivotal role in patient safety and improvement in health care services (see Chapter 8). Implementing HFE in GCC health care services will assist in designing environmental evidence-based rooms that are based on safety design principles, mitigating and reducing medication error, improving the design and implementation of health IT, and enhancing design and implementations of technologies, work processes, and task-flows with effective team performance (Carayon, 2017).

GCC Surface Transportation

The GCC will invest US$121.3 billion in the expansion of its transportation infrastructure ("GCC To Invest," 2014). A population increase and the occurrence of two major events in the region (Dubai Expo 2020, Qatar FIFA World Cup 2022) demands an improvement in the transportation infrastructure in the GCC countries. Furthermore, the investment of US$200 billion in railway network projects that will link all GCC countries in 2021 is promising (Issac, 2017b). Roads, bridges, metros, and railways are being designed and developed to meet the number of commuters and vehicles on the road. As traffic increases, road congestion problems arise that will be followed by an increase in accident rates. Traffic accidents bring heavy losses to the GCC countries, losing approximately 1–2.5% of the national income generated ("GCC loses," 2017). For instance, on average, a car accident occurs every minute in Saudi Arabia, totaling more than 460,000 crashes per year (Lemon, 2017).

Road accidents are a serious problem globally. Unfortunately, the scale of injuries and fatalities are amplified significantly in the GCC countries. HFE plays an essential role in road safety by addressing the integral components of the transportation system including the human (driver), vehicle, and environment (roadway) (see Chapter 10). The goal is designing roadways and environments (traffic management centers, signage, and traffic signals) that ensure the operator's safety, efficiency, and optimum driver performance, while safeguarding pedestrian protection. Furthermore, HFE principles can be used to improve passenger services in public transportation to promote environmental sustainability (e.g., buses, metro, and rail). Lastly, HFE provides evidence-based design approaches to improve training and safety management systems (safety culture).

Promoting Sustainability in the GCC

Sustainability can be defined as "the development that meets the needs of the present without compromising the ability of future generations to meet their own needs," and includes attention to natural and physical resources (the planet), but also attention to human and social resources (people), in combination with economic sustainability (profit) (Dul et al., 2012, p. 9). The UN's Sustainable Development Goals (SDGs) propose that the three dimensions of sustainable development are economic, social, and environmental. All dimensions are critical to GCC countries and are part of their transformation vision plans. For instance, one of the core themes of the EXPO 2020 plans in Dubai is sustainability, and Dubai has the objective of becoming one of the most sustainable cities in the world.

To address the global challenges of sustainability, in 2008 the International Ergonomics Association (IEA) created the Human Factors and Sustainable Development technical group (www.iea.cc). As HFE is design-driven, it can contribute to sustainable development by employing design strategies in workplace sustainable environments and designing products using user-centered sustainable design approaches. Unfortunately, studies have shown that ergonomic considerations are rarely, if ever, taken into account when designing sustainable environments (Hedge & Dorsey, 2013). A research study found usability flaws with the use of control designs for flexible lighting scenes in a LEED-certified workplace relating to consistency, feedback, visibility of status and function, and discoverability of functions (Lee, Carswell, Seidelman, & Sublette, 2013). If these functionalities were designed properly, they could have increased energy-conscious behaviors.

Sustainability cannot be accomplished with design alone, but through knowledge and motivation to promote conserving habits such as motivation to reduce energy cost, mitigate global climate change, and maximize user comfort and well-being (Nickerson, 2012; Sethumadhavan, 2014a,b). Several behavioral interventions have been found successful in raising awareness and promoting energy conservation habits. Well-designed feedback may be used to make the invisible energy visible through designing systems that facilitate changing behavior and learning to conserve through goal-relevant information (Flemming, Hilliard, & Jamieson 2008; Trinh & Jamieson, 2014). Another behavioral intervention is that of comparison with either oneself (past consumption behavior) or with others (neighbor's behavior) to motivate conservation habits (Fisher, 2008). Finally, rewards and incentives (or disincentives) provide extrinsic motivation to learn new behavior or maintain conserving habits (McKenzie-Mohr, 2011).

Other than HFE design principles, the extensive collection of methodologies used in HFE (some described in the previous chapters) have also been applied in sustainable development. Stevens and Salmon (2015) propose the use of cognitive work analysis as a methodology to understand the socio-technical system approach of urban planning and potentially design sustainable cities. Other methodologies can also be used to design sustainable urban planning environments that advocate walkable urban areas with community facilities that are close to public transportation.

While all GCC governments advocate for sustainability, they cannot act alone and are in dire need to align with effective partners in the private and academic sectors to drive the sustainability agenda in the region. A survey report focusing on Saudi Arabia, Qatar, and the UAE found that 86% of decision-makers share a sense of urgency to address the region's social, economic, and environmental sustainability challenges (Rung et al., 2016). This is promising to move the sustainability vision forward, but great efforts are needed to ensure that evidence-based design sustainability principles are implemented rather than experimental design developed by foreign companies unaware of the GCC's cultural and environmental conditions (see Case Study 7).

GCC Cybersecurity

The increasing number of cyber-attacks on the GCC countries are of great concern, especially as Saudi Arabia and the UAE were found to be the two most targeted Middle East and North Africa (MENA) countries for ransomware attacks (Diwakar, 2017). The Shamoon virus was deployed against Saudi Aramco and Qatar's RasGas in 2012, and more recently on the Saudi Labor Ministry in 2017, infecting networks in at least 22 organizations ("What is the Shamoon," 2017). These incidents highlight the urgency for all GCC countries to address this problem. Cybersecurity cooperation efforts between and within the GCC countries will protect attacks that will potentially target critical infrastructures such as government entities, airports, banks, or any telecommunication networks. To provide solutions to cybersecurity within the GCC, HFE's systematic approach can contribute to resolve both the vulnerability of the end-user being exposed to cyber attacks as well as understanding the threats of cyber hackers that are targeting this region (see Chapter 5).

The GCC cyber security market is estimated to grow to over US$10.41 billion by the end of 2022 and GCC network security spending to reach US$1 billion by 2018, positioning the GCC region among the global leaders in cyber security preparedness (Bou Samra, 2017). Research initiatives employing HFE methodologies and tools can be used to improve security measures within the GCC cultural environment. The current estimation of Internet users in the GCC region is approximately

40 million. Many of these users are youth and elderly who are susceptible to threats and are complacent in protecting themselves digitally. GCC nationals tend to be over-trusting, sharing vital information with their colleagues, subordinates, and acquaintances, which may be exploited by social phishing tactics. Furthermore, GCC system analysts may not have the required skills in situation awareness, expertise, and teamwork training, nor the tools to assist with deciphering cyber enemies (see Chapter 5). Hence, adding more security technologies for security analysts will not protect any organization from evolving threats without improving their competencies for analytic analysis. Therefore, under a national security roadmap and a GCC cyber defense strategy, it is proposed to invest in cyber security HFE research and development to investigate the factors that jeopardize security from an end-user perspective and provide evidence-based solutions while providing security analysts with the tools necessary to combat these attacks and mitigate the risks of operating safely in a digitally connected smart-region.

Artificial Intelligence, Robotics, and Automation in the GCC

The digital evolution index ranked the UAE as one of three countries in the world that stands out for being a leader in digital advancement while exhibiting high momentum for driving innovation and intelligent transformations (Chakravorti, BhallaRavi, & Shankar, 2017). Robots are beginning to replace humans in customer care services, with the Dubai Electricity and Water Authority (Dewa) recruiting five robots ("Robots begin," 2017) to staff their customer happiness centers (centers developed to boost community happiness via customer service). The Dubai Autonomous Transportation Strategy aims to have 25% of all its transport to be driverless by 2030 (Achkhanian, 2016). Furthermore, the Dubai Roads and Transport Authority (RTA) introduced the first concept of an Autonomous Air Taxi (AAT), a vehicle that will be used as a self-flying taxi service in the near future (Baldwin, 2017).

While advancements in technology are seen across sectors assisting our daily work and leisure activities, it may likely be beyond user capabilities. Twenty years ago, Parasuraman and Riley (1997) discussed the use, misuse, disuse, and abuse of automation and these issues are still applicable to current conditions. Taking driverless vehicles as an example, we will likely *use* the driverless vehicle if it is consistently reliable and predictable. We will *misuse* its automation when we put excessive trust into the vehicle, which may lead us to rely on the automated system without adequately monitoring the driving process, thus, removing ourselves from the driver interaction and taking a passive rather than an active role in driving. We will *disuse* the vehicle when our trust is reduced because of continuous false alarms that alert constantly. Lastly, *abuse* of the driverless vehicle occurs when the automated functions are designed to disregard our authority as a driver over to the automated vehicle to eliminate (falsely) any human error (Sethumadhavan, 2017).

Findings from driverless car research in the USA by the American Automobile Association (AAA) found that 78% of respondents said they were afraid to ride in autonomous vehicles. Another study by AIG insurance found that 41% did not want to share the road with driverless cars (Hutson, 2017). With autonomous vehicle accidents reported in the news recently, this is not surprising; Uber's driverless car ran a red light in San Francisco, Google's driverless car bumped into a bus in Mountain View, and Tesla's semiautonomous car was on autopilot and hit a truck and killed its driver in Florida. With these outcomes, consumer distrust is on the rise and researchers in industry and academia in the USA are examining how people interact with these driverless cars (i.e., how pedestrians and other drivers react to driverless cars, how drivers/passengers interact with the vehicle). Therefore, it is crucial that further research and development is performed in the GCC with local residents in their environment to assess the usability of these autonomous systems. The goal is a changing focus from imported technology-push innovations to user-centered design-driven innovations. This will ensure that the region is employing the technologies to fulfill GCC population needs, cultural context, and local environment.

GCC Manufacturing and the Fourth Industrial Revolution

The GCC countries are looking at the manufacturing sector to enhance their economic growth. The region's manufacturing industries include steel, chemicals, aerospace, shipbuilding, plastics, electronics, clothing, and food processing. The petrochemical industry in the GCC is the second-largest manufacturing sector in the region, producing over US$108 billion worth of products in 2016 ("EQUATE sponsored," 2017). Also, manufacturing is Dubai's third-largest economic contributor. The UAE's Department of Economic Development (DED) expects around US$19 million to be spent on research and development by manufacturing companies, as the sector expands from a current value of US$11.2 to US$16.1 billion by 2030 (Talhouk, 2017).

The manufacturing industry is changing rapidly, with the fourth industrial revolution driving greater digitization and intelligent automation. Industrial robotics is one of the main technologies being developed. Consequently, HFE will play an essential role in manufacturing design processes as production lines become more complex with increasing levels of robotics, sensors, informatics, and high levels of automation (Charalambous, Fletcher, & Webb, 2016). To prepare factory workers for the future of manufacturing, it is important to understand workers' capabilities and limitations and the best integration between human and robot. The GCC's factory workers are multicultural, with varying abilities. Thus, it is critical to fully understand how to best design and operationalize both worker and industrial robot functionalities for optimal cooperation. This will assist to increase efficiency, lower cost production, and optimize operation processes. Other than designing for optimal worker-robot interaction, HFE can assist in developing training programs to introduce the new work process and aid workers to learn the new manual and cognitive skills required in the redesigned production system. Iain Wright, UK's chair of Parliament's Business, Energy and Industrial Strategy Select Committee, states that "While technology has the ability to increase productivity and reduce costs, it's human factors that will enable us to fully integrate our supply chains and enable differentiation" ("The role of," 2016).

GCC Training with Virtual/Mixed/Augmented Reality

Virtual reality (VR), mixed reality (MR), and augmented reality (AR) are in great demand in the GCC region for their potential use across various sectors (Buller,

2017). According to a digital consumer survey performed in 26 countries, demand for AR and VR is high (68% and 55%, respectively) in the UAE, more than their global counterpart averages ("UAE consumers," 2017). In the USA and Europe, VR, MR (merging real and virtual environments where physical and digital objects mix and interact in real time), and AR (superimposing a digital image on the user's view of the real world) have been instrumental tools to effectively create realistic synthetic training environments (i.e., military, transportation, health care) (Vincenzi, Wise, Mouloua, & Hancock, 2009). When these technologies first started in the 1980s, the emphasis was on the technology. The technology push approach lead to failures as the user was neglected. The GCC region can learn from these mistakes and move from a technology-focused perspective to a human-centered approach. HFE maximizes human performance efficiency in these virtual worlds for effective design and optimal performance, safety, and well-being.

The importance of understanding HFE when developing these interactive technologies is vital to address the requirements of the stakeholders and end-users. Also, human-centered design using cultural context is essential when developing and evaluating these technologies to create a more realistic multimodal training experience that optimizes training transfer into the real environment. The incorporation of HFE in design and implementation of these synthetic environments has been recognized by the International Organization for Standardization (ISO 9241) for ergonomics of human-system interaction and design for interactive systems.

GCC Space Ambitions

Within the GCC, the UAE has the most ambitious space program, aiming for a mission to Mars in 2021, building a settlement by 2117, and developing a space program that will have a manned spaceflight program with Emirati astronauts (see Case Study 12). Back in 1985, Saudi Arabia pioneered the way by sending Prince Sultan bin Salman Al-Saud to space on the NASA space shuttle Discovery, which made him the first Arab Muslim and a Saudi royal family member to travel in space. In addition to Saudi Arabia's active satellite program, it recently announced interest in investing in the space tourism of Virgin Galactic, which plans a space mission in 2018. Saudi Arabia also signed an agreement with Russia to build another International Space Station by 2023 (Khatib, 2017). Furthermore, Kuwait reported similar endeavors to establish a space program as it has been working with NASA on jet propulsion projects with its Kuwait Institute for Scientific Research. Qatar has also been present in space activities, with its commercial communication satellite company. Oman and Bahrain are also exploring options on how to advance their space interests.

There are pronounced scientific advancements in the region. The GCC countries have used research and development programs and harnessed the required technologies of space systems related to satellite manufacturing capabilities. However, HFE contributions to space flight in the GCC region are in their infancy. NASA found that in order to undergo a future journey to Mars it needs to understand the effects of spaceflight on human performance and the role of HFE in the design of complex systems and human-system integration (Martin, 2015). HFE plays a significant role in spaceflight in several areas, including space vehicle design, the tasks astronauts will perform and the extreme environment that surrounds them, and the influences on their physical, cognitive, physiological, behavioral, and social well-being in terms of biomechanics, anthropometrics, workload, situation awareness, and overall performance (see Chapter 11). A journey to Mars is likely to take several months; hence, HFE factors related to the needs of the astronaut will be of great concern. For instance, physical, psychological, and cognitive effects are likely to influence the astronaut's performance (Fitts, 2016). Consequently, HFE principles, methodologies, and applications in spaceflight are used to improve design, augment astronaut decision-making, and provide training processes to enhance safety and reduce risks of accidents, failures, and any other adverse events.

THE FUTURE OF HFE IN THE GCC: PROMISING POSSIBILITIES

> The solutions to the problems of the 21st Century absolutely require the redesign of society to change human behavior. (Moray, IEA 1994, p. 4)

The most significant need for the GCC in future is addressing the human component. Many of the challenges and developments within the GCC cannot be solved by technological innovation alone but by employing human-centered design principles to change, when necessary, human behavior. The previous section demonstrated how all the challenges and opportunities in the GCC have a human component and how HFE can play a unique role in providing solutions such as reducing driver accidents, mitigating medication error, and consuming less electricity for sustainability. If more technologies are designed with HFE principles, then the aggregate contributions could amount to major improvements in the GCC region. However, it is important to note that HFE cannot solve these problems alone; however, its unique contribution is critical because it deals with the human dimension of these complex challenges.

The transition of change in the GCC occurs in the context of complex sociotechnical systems (see Chapter 2), where HFE functions to optimize performance and enhance well-being across human, social, economic, and environmental development. This progress will certainly influence human-system design and how GCC locals interact with such systems (human-human, human-system, and human-environment interactions). To be effective, tremendous opportunities for HFE in the GCC have been proposed, exhibiting numerous application areas that will benefit society. HFE's design-driven system approach forms a catalyst to cross-utilization of research and development (R&D).

For HFE to take its rightful place within GCC regional development, it requires a strategic plan for a sustainable HFE ecosystem composed of a strong foundational base in advocacy using a quadruple helix model with university, industry, government, and general GCC society (see Figure 12.1). HFE will facilitate a collaborative relationship towards human-centric innovations among all four tracks. GCC universities develop HFE programs and conduct HFE R&D to produce, transmit,

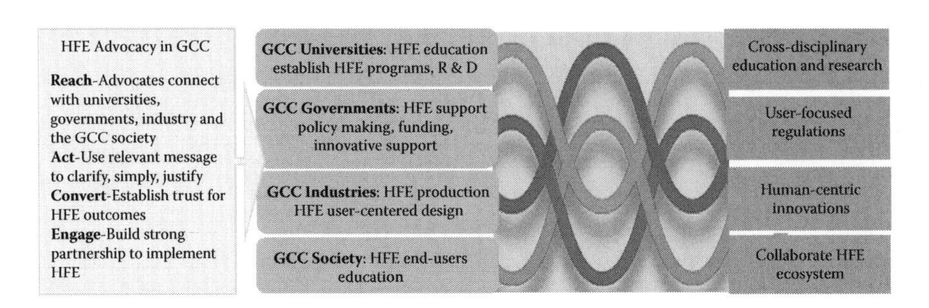

FIGURE 12.1 HFE Roadmap in the GCC.

and transfer knowledge. Industry works with HFE professionals to create and innovate technologies that are home-grown and designed to align with the needs and requirements of the GCC locals. Government attends to policy formation, funding, and innovative support. Finally, the general public are the end-users that value the importance of user-centered design, safety, and well-being. These informed consumers will appreciate how employing HFE will result in good design that is compatible with human needs, capabilities, and limitations. Along these intersections, the hope is that GCC countries will create cross-disciplinary university curriculums and research programs, develop user-focused regulations and standards, build human-centric innovations, and produce a collaborative HFE ecosystem.

HFE Advocacy

HFE advocacy aims to influence decision-makers across all sectors for the betterment of all GCC countries in terms of human, economic, social, and environmental development (optimal human performance, improved well-being). HFE advocacy is a shared responsibility of international and regional associations (i.e., IEA, HFES, CIEHF, HFES GCC Chapter) along with HFE professionals globally and in the GCC countries. Also, identifying allies and partners that are not necessarily HFE experts may support this process with their participation. What is needed is to focus on decision-makers' interests and how HFE is significant to their work. It is crucial for HFE advocates to create strong trusting relationships with governments, industries, universities, and the general public, and to build strong partnerships that strengthen HFE understanding and utilization. HFE advocacy can include many activities, including public-awareness and media campaigns, speaking at professional forums, and publishing evidence-based research projects.

Regardless of medium, the first step is introducing HFE to all stakeholders (government, industry, university, general public). Developing a multidimensional communication format (e.g., top-down, bottom-up, diagonal, and horizontal) throughout any organization will assist in greater outreach. Advocating HFE using effective communication requires creating a defined message, developing a communication strategy, and customizing the message to the respective stakeholder. A clear, memorable, and meaningful message framed around HFE evidence-based practices and outcomes will assist stakeholders to focus their attention on how HFE relates to their economic, social, and environmental motivation and priorities. Building an evidence-based advocacy message that has clear HFE value proposition includes: a tailored relevant message (explaining how its principles, methodologies, and applications solve practical problems); delivering specific benefits in quantified value (performance quality and reliability, low operating cost, better decision-making and innovation, overall greater profitability in the organization); and demonstrating the unique differentiation of HFE (design-driven system approach using socio-technical system analysis). To illustrate the potential impact, Honeywell, a US company, implemented a company-wide HFE program that led to both direct and indirect cost savings of US$2 million. Other benefits included zero repetitive-strain and musculoskeletal injuries, along with a 24% improvement in worker productivity (Budnick & Osborne, 2012).

Universities: HFE Education

Throughout national and international universities across the GCC, only a limited number of HFE courses are offered in undergraduate- and graduate-level programs. Regrettably, no HFE degree exists in the GCC region teaching the required competencies for HFE professionals. This is concerning, as HFE is crucial to any human-system interaction. For example, while engineering is the application of scientific knowledge that meets technical human and societal needs, rarely do you find HFE courses in engineering colleges. It is necessary to move towards developing academic institutions in the GCC that break out of college silos and cut across disciplines, allowing students to employ holistic problem-solving approaches. This will be accomplished via innovative university studies and programs designed on the principle of multi-, inter-, cross-, and trans-disciplinary courses of study. One such program successfully implemented at Arizona State University is a hybrid HFE program that embeds HFE within engineering programs with rigorous research studies. The Fulton Schools of Engineering combined engineering and psychology to design technologies that work for humans, offering both undergraduate and graduate degrees (https://poly.engineering.asu.edu/hse/).

Additionally, universities provide tremendous support for HFE R&D. Unfortunately, while the GCC's GDP per capita has developed at global levels, its investment in R&D remains at developing-country levels. The experience of the developed countries can be used to fast-track the capacity of R&D by promoting research-based programs within and between the GCC countries. This will provide incentives to GCC nationals to take part in their transformational vision to ensure sustainability and relevance to the local GCC culture and environment. While R&D is viewed by some as a costly proposition with little return on investment (ROI), if viewed from its net effect it is clear to see its positive returns. When research findings are shared and implemented through different levels and applications, their incremental economic contribution can outweigh their initial cost. This will lead to establishing research hubs that are viewed as a source of financial contribution and economic activity. Creating strong engagements from academia, industry, and government sectors in funding HFE R&D

will help develop research infrastructures for basic and applied research. The GCC countries cannot rely on imported research and knowledge production that have been generated without GCC local input; thus, investing in building human capacity to develop knowledge and skills of local researchers is essential.

The thriving platforms identified earlier demonstrate how GCC challenges and opportunities are complex topics that require cross-disciplinary partnership. Innovative solutions stem from a synergy of collaborative researchers from different disciplines. As HFE is considered to be one of the first cross-disciplinary fields, it is essential that research integrates projects employing multidisciplinary (examining topics from multiple perspectives without efforts to systematically integrate disciplinary views); interdisciplinary (examining issues from multiple perspectives leading to systematic efforts to integrate alternative views into a unified coherent framework); cross-disciplinary (examining a matter that is typically suitable to one discipline through the lens of another discipline, such as how engineers explore cognition); and trans-disciplinary (integrating different unrelated academic disciplines and non-academic participants to research a common goal and create a new theory of knowledge).

Governments: HFE Support

Government is the driving force in creating policy decisions, funding opportunities, and innovative support. Experiences from governments in developed countries that employed HFE in policy-making can be duplicated to build a science ecosystem in the GCC. For instance, the US National Highway Traffic Safety Administration (NHTSA) was established in 1970 in response to public outcry to increase vehicle safety issues, such as promoting seat belts, and the development of US safety regulations (Pew & Mavor, 2007). At NHTSA, the HFE labs aim to improve safety and enhance operations throughout the highway transportation system with considerations of driver behavior. HFE labs are created throughout many government sectors to align with their strategic vision to create policies and regulations based on HFE standards. They also often collaborate with academia and industry to conduct HFE research projects.

Government-supported programs can work with universities and industry to enable that societal challenges are resolved around HFE evidence-based practices. Governments can help plan the development of HFE research labs in sectors that address challenges in society and take priority according to their policy-making agenda from creation, implementation, and revision requirements. The government's role can also be central to the involvement of educational institutions to aid HFE knowledge-sharing with industry through consulting, contracting research, developing user-centered products, and creating industry incubators for research-based projects. These activities will be sustainable when creating local national capacity with home-grown local HFE researchers that foster product and service-based innovation. Government HFE labs can create innovative technologies to improve GCC society with the support of SME (small and medium-sized enterprises) to stimulate knowledge-based economies. Programs such as the Small Business Innovative Research (SBIR) and Small Business Technology Transfer (STTR) are government-funding mechanisms supporting academia and industry to resolve US government challenges. Likewise, if applied in the GCC, products and services from government-funded programs can reach successful commercialization in partnership with academia and industry.

Industries: HFE Production

It is essential that GCC industries develop products employing a HFE design-driven system approach. This will ensure that HFE evidence-based practices incorporating research, design, and evaluation are implemented for easy to use features. An aspiring model of employing HFE design is how the Food and Drug Administration (FDA) in the USA enforced regulations on medical device manufacturers to design their products more effectively with HFE principles (FDA, 2016). The FDA developed guidance documentations to assist industry in following appropriate HFE processes to maximize the likelihood that new medical devices are safe, effective, and easy to use. Premarket information for HFE device design, along with post-market evaluation for device surveillance and reporting, guides industry to ensure that devices meet government regulations and standards.

Industry can also build strategic knowledge partners with universities to develop industry incubators that support HFE technology transfer, patent sharing, and licensing of innovative technologies. More importantly, this collaboration will help develop local innovators to create home-grown technologies and services that meet GCC needs and requirements. Furthermore, providing support to universities in terms of funding, mentoring, and production commercialization will aid developing human capital ready for the labor market.

GCC Society: HFE End-User

Most of the general public are unfamiliar with HFE. Those that do recognize ergonomics will commonly link it to chairs, pens, or to office design and workstation layout. This narrow view of HFE needs to expand to realize the breadth and depth of the various fields covered in this book and beyond. It is important to take advantage of all opportunities to offer the general GCC community education about the importance of user-centered design, safety, and well-being. Providing educational material such as magazines, newspapers, and other media outlets (e.g., TV, social media) will likely reach a broader audience and have a greater impact. These educational sessions could explore common HFE evidence-based advocacy practices related to common themes such as patient safety in the health care sector (e.g., awareness of medication errors).

Undoubtedly, promoting HFE is important at any age. Hence, starting to teach HFE to school students will help them apply their STEAM knowledge to solve engineering problems and provide solutions that will have GCC societal impact. Teaching students the principles of engineering and human capabilities, along with research methodology skills, helps students conduct experiments and develop solutions that meet their culture-specific needs. This will not only cultivate a culture of problem-solving and experimenting but also a society that develops entrepreneurship and innovation and builds strong research skills at an early age. These young, savvy consumers will reject bad design and create solutions for better user-centered design.

CONCLUSION

The previous chapters demonstrated HFE principles, methods, and applications throughout numerous domains written by academic experts from around the world. The book also provided successful examples of case studies written by professionals to effectively illustrate the benefits of HFE in industry and governments around the GCC countries. This final chapter provides a preview of the current state of HFE in the GCC and missed opportunities, discusses thriving platforms to activate HFE potential in several domains, and proposes a future roadmap of promising possibilities for HFE to play a meaningful role and prosper in the GCC.

Taken together, the ultimate goal is to achieve social progress and add economic value through optimizing performance quality and reliability, lowering operation cost, improving decision-making and innovation, and gaining greater profitability in any organization. This can only happen when the HFE discipline and profession is promoted by advocating for its evidence-based value. By creating trust and building partnerships between universities, industry, government, and the general public, the HFE discipline will grow to address the challenges discussed in this chapter across the various sectors. As a result, the GCC region will reap great benefits from harnessing the scientific, creative, and economic potential of HFE that will stimulate the necessary knowledge and skills to participate in human, social, environmental, and economic growth.

REFERENCES

Achkhanian, M. (2016). 25% of all transportation in Dubai will be smart and driverless by 2030: Mohammad Bin Rashid. *Gulf News*. Retrieved November 20, 2017 from http://gulfnews.com/news/uae/transport/25-of-all-transportation-in-dubai-will-be-smart-and-driverless-by-2030-mohammad-bin-rashid-1.1810896.

Al Arabiya English (2017). *What is the Shamoon virus that has returned to hack Saudi networks?* Retrieved December 10, 2017 from http://english.alarabiya.net/en/media/digital/2017/01/24/What-is-the-Shamoon-virus-that-has-returned-to-hack-Saudi-networks-.html.

Al Wardi, Y. M., Jeevarathinam, S., & Al Sabei, S. H. (2016). Eastern bodies in western cockpits: An anthropometric study in the Oman military aviation. *Cogent Engineering* 3(1), available online at www.tandfonline.com/doi/full/10.1080/23311916.2016.1269384.

AMEinfo (2017). *The GCC can benefit from a sharing economy. Here's how.* Retrieved November 1, 2017 from http://ameinfo.com/media/digital/gcc-can-benefit-sharing-economy-heres/.

AMEInfo (2017). *UAE consumers have high appetite for AI: study.* Retrieved November 20, 2017 from http://ameinfo.com/technology/innovation/uae-consumers-high-appetite-ai-study/.

Arabian Gazette (2014). GCC To Invest USD 121.3bn In Land And Transport Infrastructure. Retrieved August 27, 2017 from https://arabiangazette.com/gcc-to-invest-usd-121-3bn-in-land-and-transport-infrastructure/.

Baldwin, D. (2017). Watch: Dubai tests world's first self-flying taxi. *Gulf News*. Retrieved November 15, 2017 from http://gulfnews.com/news/uae/transport/watch-dubai-tests-world-s-first-self-flying-taxi-1.2095961.

Bou Samra, J. (2017). The GCC cyber security market is booming. Cyber Defense Magazine (CDM). Retrieved October 15, 2017 from www.cyberdefensemagazine.com/the-gcc-cyber-security-market-is-booming/.

Budnick, P. & Osborne, K. (2012, Feb). Evolution of an Ergonomics Process Success Story. *The Ergonomics Report*™.

Buller, A. (2017). Virtual reality takes off in the Gulf Cooperation Council countries. *Computer Weekly.com*. Retrieved November 30, 2017 from www.computerweekly.com/news/450413539/Virtual-reality-takes-off-in-the-Gulf-Cooperation-Council-countries.

Carayon, P. (2017). *Handbook of Human Factors and Ergonomics in Health Care and Patient Safety*, Second Edition. Boca Raton, FL: CRC Press.

Chakravorti, B., BhallaRavi, A., & Shankar, C. (2017). 60 Countries' Digital Competitiveness, Indexed. *Harvard Business Review*. Retrieved August 24, 2017 from https://hbr.org/2017/07/60-countries-digital-competitiveness-indexed.

Charalambous. G., Fletcher, S. R., & Webb, P. (2017). The development of a Human Factors Readiness Level tool for implementing industrial human-robot collaboration. *International Journal of Advanced Manufacturing Technology* 91 (5–8), 2465–2475.

Diwakar, A. (2017). GCC Businesses Face a Major Cybersecurity Deficit: Experts. *Albawaba Business*. Retrieved November 20, 2017 from www.albawaba.com/business/gcc-businesses-are-facing-major-cybersecurity-deficit-986096.

Dul, J., Bruder, R., Buckle, P., Carayon, P., Falzon, P., Marras, W. S., Wilson, J. R., & van der Doelen, B. (2012). A strategy for human factors/ergonomics: Developing the discipline and profession. *Ergonomics* 55(4), 377–395.

Eid, S. (2017). GCC consumers need to be better protected. *Arabian Business*. Retrieved October 2, 2017 www.arabianbusiness.com/gcc-consumers-need-be-better-protected-665552.html.

FDA (2016). Applying Human Factors and Usability Engineering to Medical Devices: Guidance for Industry and Food and Drug Administration Staff. US Department of Health and Human Services Food and Drug Administration Center for Devices and Radiological Health Office of Device Evaluation. Retrieved November 4, 2017 from www.fda.gov/downloads/MedicalDevices/.../UCM259760.pdf.

Fitts, R. H., Riley, D. A., Costill, D. L., & Trappe, S. W. (2016). Effect of Prolonged Space Flight on Human Skeletal Muscle (Biopsy). NASA. Retrieved December 10, 2017 from www.nasa.gov/mission_pages/station/research/experiments/245.html#overview.

Fisk, A. D., Rogers, W. A., Charness, N., Czaja, S. J., & Sharit, J. (2009). *Designing for older adults: Principles and creative human factors approaches* (2nd ed.). Boca Raton, FL: CRC Press.

Flemming, S. A., Hilliard, A., & Jamieson, G. A. (2008). The Need for Human Factors in the Sustainability Domain. Human Factors and Ergonomics Society Annual Meeting Proceedings, 52, 748–752.

Greenwall, N. K. (2015, June). National Child Restraint Use Special Study. Traffic Safety Facts Research Note. Report No. DOT HS 812 157. Washington, DC: National Highway Traffic Safety Administration.

Giguashvili, N. (2017). IoT to Play a Central Role in Digital Transformation of GCC Hospitals. *IDC Community*. Retrieved November 13, 2017 from https://idccommunity.com/health/healthcare-transformation/iot_to_play_a_central_role_in_digital_transformation_of_gcc_hospital.

Hartson, R. & Pyla, P.S. (2012). *The UX Book: Process and guidelines for ensuring a quality user experience*. Amsterdam: Morgan Kaufmann.

Hedge, A. & Dorsey, J. (2013) Green Buildings need Good Ergonomics. *Ergonomics* 56(3), 492–506.

Helmreich, R. L. & Merritt, A. C. (1998). *Culture at work: National, organizational, and professional influences*. Aldershot: Ashgate.

Hendrick, H. W. (1996). *Good Ergonomics is Good Economics*. Santa Monica, CA: Human Factors and Ergonomics Society.

Hendrick, H. W. 1995. Future directions in macroergonomics. *Ergonomics* 38(8), 1617–1624.

Hutson, M. (2017). People don't trust driverless cars. Researchers are trying to change that. *Science Magazine*. Retrieved December 20, 2017 from www.sciencemag.org/news/2017/12/people-don-t-trust-driverless-cars-researchers-are-trying-change.

Ingham, D. (2017). GCC's healthcare prognosis is looking good. *Arabian Business*. Retrieved August 25, 2017 from www.arabianbusiness.com/gcc-s-healthcare-prognosis-is-looking-good-661119.html.

Issac, J. (2017a). Why GCC rail is a game-changer. *Khaleej Times*. Retrieved October 9, 2017 from www.khaleejtimes.com/news/transport/why-gcc-rail-is-a-game-changer.

Isaac, J. (2017b). GCC healthcare spending to hit Dh253.4b by 2020. *Khaleej Times*. Retrieved October 5, 2017 from www.khaleejtimes.com/business/local/gcc-healthcare-spending-to-hit-dh2534b-by-2020.

Kerr, S. (2016). Five goals of Saudi Arabia's ambitious transformation plans. *Financial Times*. Retrieved October 10, 2017 from www.ft.com/content/cbb86ed2-2e38-11e6-a18d-a96ab29e3c95.

Khaleej Times (2017). *Robots begin to replace humans in customer care services in Dubai*. Retreived September 20, 2017 from www.khaleejtimes.com/nation/dubai/robots-begin-to-replace-humans-in-customer-care-services-in-dubai-.

Khatib, H. (2017). Gulf space race: UAE and now KSA set sights high. *AMEinfo*. Retrieved November 7, 2017 from http://ameinfo.com/travel/saudi-gets-jump-uae-gcc-space-race/.

Kulbytė, T. (2017). 32 Customer Experience Statistics You Need to Know for 2018. Retrieved December 23, 2017 from www.superoffice.com/blog/customer-experience-statistics/.

Kuwait News Agency (KUNA) (2017). EQUATE sponsored forum to boost Arab Gulf Industrial sector pioneering role. Retrieved December 10, 2017 from www.kuna.net.kw/ArticleDetails.aspx?id=2662524&language=en.

Lalchandani, J. (2017). Digital transformation in GCC's health care industry. *Gulf News*. Retrieved November 5, 2017 from http://gulfnews.com/business/sectors/technology/digital-transformation-in-gcc-s-health-care-industry-1.2053444.

Lee, M., Carswell C. M., Seidelman W., & Sublette M. (2013). Green expectations: The story of a customizable lighting control panel designed to reduce energy use. Proceedings of the 2013 International Annual Meeting of the Human Factors and Ergonomics Society, San Diego, California, Sep 30–Oct 4: 57: 1353–1357.

Lemon, J. (2017). A car crash happens every minute in Saudi Arabia. *Step Feed*. Retrieved November 12, 2017 from https://stepfeed.com/a-car-crash-happens-every-minute-in-saudi-arabia-3428.

Lemzy, A. (2017). Omni-Channel Customer Service Best Practices. Retrieved October 10, 2017 from www.superoffice.com/blog/omni-channel-customer-service/.

Longanecker, C. (2016). Customer Experience Is the Future of Design. *UX Magazine*. Retrieved October 3, 2017 from https://uxmag.com/articles/customer-experience-is-the-future-of-design.

Martin, P. K. (2015). NASA's Efforts to Manage Health and Human Performance Risks for Space Exploration. Report No. IG-16-003. October 29, 2015. NASA Office of Inspector General, Office of Audits. Retrieved December 20, 2017 from https://oig.nasa.gov/audits/reports/FY16/IG-16-003.pdf.

Meulen, R. & Rivera, J. (2015). Gartner Says Organizations Are Changing Their Customer Experience Priorities. *Gartner*. Retrieved August 20, 2017 from www.gartner.com/newsroom/id/3072017.

McKenzie-Mohr, D. (2011). *Fostering sustainable behavior: An introduction to community-based social marketing* (3rd Edition). Gabriola Island, BC: New Society.

Moray, N. 1995. Ergonomics and the global problems of the 21st century. *Ergonomics* 38(8), 1691–1708.

Moray, N. (1994). Ergonomics and the Global Problems of the 21st Century, Keynote Address Presented at the 12th Triennial Congress of the International Ergonomics Association, Toronto, August, 1994.

Nickerson, R. S. (1992). What does human factors research have to do with environmental management? In Proceedings of the Human Factors Society 36th Annual Meeting (pp. 636–639). Santa Monica, CA: Human Factors and Ergonomics Society.

Noort, M. C., Reader, T. W., Shorrock, S., & Kirwan, B. (2015). The relationship between national culture and safety culture: Implications for international safety culture assessments. *Journal of Occupational and Organizational Psychology* 89, 515–538.

Parasuraman, R. & Riley, V. (1997). Humans and automation: Use, misuse, disuse, abuse. *Human Factors* 39, 230–253.

Pew, R. W. & Mavor, A. S. (eds) (2007). Human-System Integration in the System Development Process: A New Look. Committee on Human-System Design Support for Changing Technology, Committee on Human Factors, National Research Council.

Rasmussen, J. (2000). Human factors in a dynamic information society: Where are we heading? *Ergonomics* 43(7), 869–879.

Rung, G., Neron-Bancel, A., & Aoude, G. (2016). *The Social Impact Imperative: The Role of Private and Non Profit Sectors in the GCC*. Retrieved November 15, 2017 from http://www.oliverwyman.com/content/dam/oliver-wyman/global/en/2016/oct/Social-Impact-Imperative-Report.pdf.

Samman, S. N. (2010). Designing for Cultural Diversity. Invited conference workshop presented at the Human Factors and Ergonomics Society 54th Annual Meeting, San Francisco, CA, 27 September–1 October 2010.

Samman, S. N. (2016). Designing for Cultural Diversity. Invited conference workshop presented at the Chartered Institute of Ergonomics & Human Factors, DeVere Staverton Park, UK, April 19–22, 2016.

Shahbandari, S. (2017). New child car sear rules in UAE. *Thomas Reuters Zawya*. Retrieved September 2, 2017 from www.zawya.com/mena/en/story/New_child_car_seat_rules_in_UAE-GN_26072017_270743/.

Shostack, G. L. (1982). How to Design a Service. *European Journal of Marketing* 16(1). pp. 49–62.

Sethumadhavan, A. (2014a). Human factors and sustainability. *Ergonomics in Design* 22(2), 33.

Sethumadhavan, A. (2014b). Principles for Designing Sustainable Systems. *Ergonomics in Design* 22(4), 34.

Sethumadhavan, A. (2017). Self-Driving Cars: Enabling Safer Human–Automation Interaction. *Ergonomics in Design* 25(2), 25.

Soloman, M. (2014). 10 Trending Changes In Customers and Customer Service Expectations. *Forbes*. Retrieved September 15, 2017 from www.forbes.com/sites/micahsolomon/2014/08/08/10-trending-changes-in-customers-and-customer-service-expectations/#8837b817c38d.

Stevens, N. J. & Salmon, P. M. (2015). New Knowledge for Built Environments: Exploring Urban Design from Socio-technical System Perspectives. International Conference on Engineering Psychology and Cognitive Ergonomics, 8, 200–211.

Strauch, B. (2010). Can Cultural Differences Lead to Accidents? Team Cultural Differences and Sociotechnical System Operations. *Human Factors* 52(2), 246–263.

Talhouk, K. (2017). Manufacturing in the GCC can power economic growth. *The National*. Retrieved October 30, 2017 from www.thenational.ae/business/manufacturing-in-the-gcc-can-power-economic-growth-1.668032.

Thatcher, A., Waterson, P., Todd A., & Moray, N. (2017). State of Science: Ergonomics and global issues. *Ergonomics* 6, 1–17.

The Manufacturer (2016). *The role of human factors in the future of manufacturing.* Retrieved November 10, 2017 from www.themanufacturer.com/articles/the-role-of-human-factors-in-the-future-of-manufacturing/.

Trinh, K. & Jamieson, G. A. (2014). Feedback Design Heuristics for Energy Conservation. *Ergonomics in Design: The Quarterly of Human Factors Applications* 22(2), 13–21.

Vincenzi, D. A., Wise, J. A., Mouloua, M., & Hancock, P. A. (eds) (2009). *Human Factors in Simulation and Training.* New York: CRC Press.

Verhoef, P. C., Lemon, K. N., Parasuraman, A., Roggeveen, A., Tsiros, M., & Shlesinger, L. A. (2009). Customer experience creation: Determinants, dynamics and management strategies. *Journal of Retailing* 85(1), 31–41.

Wilson, J. R. (2000). Fundamentals of ergonomics in theory and practice. *Applied Ergonomics* 31(6), 557–567.

Zackowitz, I., Lenorovitz, D., Borzendowski, S., Rice, V., Calkins, L., Feeley, C., Milewski, J., & Knight, J. (2016). Mock Trail: A Demonstration of Human Factors Professionals Testifying on a Children's Transportation Safety Product. Proceedings of the Human Factors and Ergonomics Society Annual Meeting.

Zawya (2017). GCC loses 2.5% of national income in road accidents. Retrieved October 10, 2017 from www.zawya.com/mena/en/story/GCC_loses_25_of_national_income_in_road_accidents ZAWYA20170509044708/.

Case Study 12
MBRSC's Role in Building UAE's Space Program

Mariam Al Shamsi

THE SPACE MISSION

The space sector in the UAE represents a strong direction from the government to build a prosperous future and improve quality of life. It has become an instrumental proof of the leadership's desire to provide the best future for the country. By funding a relatively costly sector with a focus on science and technology research, the UAE is differentiating itself from the majority of countries in the region, becoming a hub for advanced research and a positive outlook for the future.

The Mohammed bin Rashid Space Centre (MBRSC) is a leading establishment in the UAE in the field of space research and development. As the leading center in the Middle East in the field of space, the center aims at making the UAE an influential member of the space sector globally. Since establishing the Emirates Institute for Advanced Science and Technology (EIAST) in 2006, and then the MBRSC in 2015 (to which EIAST is affiliated), the center focused on building human capabilities and knowledge of space systems development and utilization. This includes technical and scientific capacity in designing, manufacturing, integrating, and testing space systems.

MBRSC is developing and leading ambitious projects and programs under the umbrella of the UAE National Space Program. With an emphasis on a sustainable space program, MBRSC is working on several initiatives to create an awareness about space and develop programs in collaboration with other international partners on the basis of joint knowledge development and sharing. The four major programs include the Satellite Development Program, the Emirates Mars Mission–Hope Mission, the Mars 2117 program, and the UAE Astronauts Program. Throughout all of the programs MBRSC is introducing new efficient and effective methodologies in managing and developing space missions, which it hopes to be benchmarked globally.

THE SATELLITE DEVELOPMENT PROGRAM

The Knowledge Transfer Program undertaken by MBRSC was initiated in 2006 with a three-phase approach, during which the center was to build advanced space systems and a new generation of experts and engineers focusing on space technologies. In the first phase, a group of 12 engineers were involved in a unique knowledge transfer program with a practical live project running, DubaiSat-1. The engineers were involved in all activities from mission conceptualization up to the launch and actual operation of the mission in space. In the second phase, a totally new mission was designed based on initiatives and innovative solutions from the engineers. The mission goal was to raise the capability of the satellite and ground system, and to increase the involvement of the engineers in all phases of the project. This was clearly successful with the launch and operation of the DubaiSat-2 satellite. In the third phase, the full development of the system was transferred to the staff at MBRSC and the actual development took place in the UAE. KhalifaSat project represents the first satellite to be developed in the UAE and run as a project fully by MBRSC. This success was witnessed in the launch of one of the National Space Programs, which includes the previous and coming satellite projects.

THE EMIRATES MARS MISSION–HOPE MISSION

In 2014, the Emirates Mars Mission (EMM)–Hope Mission was announced, and MBRSC was the entity in charge of design, implementation, and supervision of all its phases. Hope is set to arrive at Mars in 2021, which will coincide with the fiftieth anniversary of the founding of the UAE, a truly significant date. The Hope probe will advance human knowledge about the atmosphere and climate on Mars, by simultaneously observing and investigating the lower and upper atmospheres of Mars. Views of most of the planet's surface at most times of day will enable unprecedented studies of the physical processes that drive the global atmospheric circulation, temperature structure, and the distribution and interaction of ice clouds, water vapor, and dust. In addition, EMM will reveal the connection between these conditions in the lower atmosphere and the escape of hydrogen and oxygen from the upper atmosphere, a process that may have been responsible for Mars' transition, over billions of years, from a thick atmosphere capable of sustaining liquid water on the surface, to the cold, thin, arid atmosphere we see today. This deep space mission represents the diversification strategy of the country for the space sector. In addition to enriching human knowledge, the strategic objectives of this mission are beyond reaching Mars; it is building human capabilities and Emirati scientists in STEM fields by providing research opportunities in the field of science. The yearly Research Experience for Undergraduates (REU) program is one of the EMM initiatives out of the many others that stimulate the students in the UAE to focus on planetary and interstellar sciences through the enhancement of research and development. MBRSC is following a similar strategy in the Emirates Mars Mission, as in all its missions, focusing on knowledge transfer. This program is collaborating with academic partnerships and research centers in the USA and other international institutions.

THE MARS 2117 PROGRAM

The UAE has long-term plans in the space sector. Mars 2117 is a program aiming to establish the first habitable human settlement on Mars by 2117, which shows the UAE's commitment to the space sector globally. MBRSC has been tasked with preparing a 100-year plan for its implementation, and Mars Science City is the first step.

Its main objective is leading the global scientific race to take people to Mars. The strategic objective of the City is to be the hub for research and development in finding solutions to human living challenges on Earth such as food, electricity, and water security. Finding solutions on Earth to these challenges will assist in implementing them when building the first city on Mars in the last phase of Mars 2117.

THE UAE ASTRONAUTS PROGRAMME

Another strategic initiative is the UAE Astronauts Programme. Its target is to build a sustainable astronaut programme consisting of a team of Emiratis fully prepared to carry out important scientific research in space. After 2021, the program is preparing to launch an astronaut in order to conduct vital experiments in the International Space Station (ISS). These experiments will address future challenges to further humanity's knowledge and to support future missions by conducting research that would prepare us for human exploration of our solar system beyond low Earth orbit (LEO) and further contribute to the international scientific community's understanding of human space flight. The potential scientific experiment will be proposed by the UAE scientific community and will be in one of the following science research areas; food security and sustainability, sustainable sources of water, human life support, energy and power management, advancing material science, and planetary contamination management.

THE SPACE PROJECTS

Building an advanced space program requires the knowledge, skills, and most importantly the right processes and culture. Our knowledge transfer program was not short of challenges. Starting from finding the right partners, who were willing to share the knowledge and work jointly, and the difficulty in demonstrating the effectiveness of space services to the government and private entities, all these became a wide range of hurdles that had to be overcome to ensure the sustainability of the space sector. Nevertheless, the clarity of the direction from the start, and the trust and confidence in a young and enthusiastic team, proved to be essential for the success.